■ The Editor would like to thank many people from the rail industry who have helped with information for this publication. A special 'thank you' also goes to Keith Ewins, Antony Christie, Spencer Conquest and Nathan Williamson for providing assistance and guidance. ■

ISBN: 978 1 80282 812 2

Editor: Colin J. Marsden

Senior editor, specials: Roger Mortimer
Email: roger.mortimer@keypublishing.com
Design: Colin J. Marsden
Advertising Sales Manager: Brodie Baxter
Email: brodie.baxter@keypublishing.com
Tel: 01780 755131
Advertising Production: Debi McGowan
Email: debi.mcgowan@keypublishing.com

Subscription/Mail Order
Key Publishing Ltd, PO Box 300, Stamford, Lincs, PE9 1NA
Tel: 01780 480404
Subscriptions email: subs@keypublishing.com
Mail Order email: orders@keypublishing.com
Website: www.keypublishing.com/shop

Publishing
Group CEO and Publisher: Adrian Cox

Published by
Key Publishing Ltd, PO Box 100, Stamford, Lincs, PE9 1XQ
Tel: 01780 755131 Website: www.keypublishing.com

Printing
Precision Colour Printing Ltd, Haldane, Halesfield 1, Telford, Shropshire. TF7 4QQ

Distribution
Seymour Distribution Ltd, 2 Poultry Avenue, London, EC1A 9PU
Enquiries Line: 02074 294000.

We are unable to guarantee the bona fides of any of our advertisers. Readers are strongly recommended to take their own precautions before parting with any information or item of value, including, but not limited to money, manuscripts, photographs, or personal information in response to any advertisements within this publication.

© Key Publishing Ltd 2023

All rights reserved. No part of this magazine may be reproduced or transmitted in any form by any means, electronic or mechanical, including photocopying, recording or by any information storage and retrieval system, without prior permission in writing from the copyright owner. Multiple copying of the contents of the magazine without prior written approval is not permitted.

Welcome to Rolling Stock Review 2023-2024

The constant changes to the UK traction and rolling stock fleets has continued over the last year and the quest for technical data and illustrations of fleet differences continues. Since the last edition of *RSR* covering the period 2022-2023 over 4,000 changes and amendments have been recorded.

It was always thought that as the number of loco and multiple unit classes reduced, the following and quest for information on the newer fleets would decline, but this has not been the case and today, we are seeing an increase in the number of younger enthusiasts seeking in-depth technical information, especially with new fleets such as the 197s, 777s and the very latest hi-tech Class 93 tri-mode locos.

The Editor is pleased to see an increase in the number of younger people showing interest in railways, following the subject and taking pictures.

In the modern traction period covering the BR era, technical data was provided by BR. However, after privatisation in the 1990s, the supply line changed, with official information provided by the owning lease companies, with support from the operators. Today, much of this information is deemed builder/customer sensitive and we have to depend on information supplied by the leading train builders, such as Alstom, CAF, Stadler, Siemens and Hitachi. In the main these companies are very supportive of the enthusiast movement, providing fact sheets and data and also supply complex information on request.

Currently, several of the newer builds are based on product 'platform' designs, such as the 'Civity' line supplied by CAF and the 'Aventra' product by Alstom (previously Bombardier), this is likely to continue, but with most operators now having modern trains, new train orders are likely set to slow down. *RSR* will however cover these developments as they happen.

From the 2023-2024 issue of *RSR* we will be including details of preserved rolling stock authorised for use over Network Rail lines.

For numeric information, names, formations, individual allocations and liveries, our sister publication *Rail123* includes this information.

The publication *MLIPlus*, includes regular updates on new rolling stock developments.

Colin J. Marsden,
Editor

Above: *During the past year, the 20 Class 769 tri-mode units leased by Porterbrook to Great Western were all returned to their owner for possible re-use, as GWR could not continue with the project. All sets are stored at Long Marston, where Nos. 769940, 769925 and 769936 were recorded in June 2023.* **CJM**

Top: *Development of the HydroFlex Class 799 unit continues with set No. 799201 operating a number of hydrogen test runs in 2023. On 21 June 2023 it was used to transport guests from Honeybourne to Long Marston for the Rail Live 2023 event. The set is seen from its fuel-cell vehicle.* **CJM**

Cover (main): *The first of the Stadler-built Class 93s for Rail Operations group, No. 93001 was delivered to the UK from Spain in summer 2023 and was under commissioning as RSR closed for press. The loco is seen on 26 July 2023 at Worksop.* **Richard Tuplin**

Cover (insets): *Left - To mark The Queens Platinum Jubilee in 2022, DB repainted No. 67007 in purple livery with event decals and branding. Right - The Stadler-built Class 777 fleet were introduced during 2023 in the Liverpool area, No. 777001 is shown at Kirkdale on 20 July 2023.* **Spencer Conquest / CJM**

Rolling Stock Review 2023-2024

The year 2022-2023 has seen a number of major changes in terms of locomotives and rolling stock. The year has also seen a major increase in the number of passengers returning to rail following the Covid-19 Pandemic and its gradual recovery.

In 2022-2023 Class 313s working for GTR Southern were finally withdrawn with Class 377s taking over their work, the final stored Class 317s moved to the scrap yard and Greater Anglia removed from service the Class 321 and 322 sets following squadron introduction of Class 720s. The first of 12 Class 720s for c2c were also commissioned and carried their first fare paying passengers.

New trains continued to be delivered, Stadler successfully introduced Class 777s on Merseyrail and Class 231s for Transport for Wales, The first of the bi-mode Class 756s were delivered and commissioned and should enter passenger service at the year end.

CAF introduced their 'Civity' products in the form of Class 196s for West Midlands Railway and Class 197s for Transport for Wales and were well accepted by the public. The first handful of CAF 'TramTrain' Class 398s sets were delivered to Transport for Wales at the new Taffs Well depot and dynamic testing commenced in mid summer.

The production of IET stock has continued at the Hitachi Newton Aycliffe plant with the first members of Class 805 and 807 for Avanti West Coast and 810 for East Midlands Railway emerging.

The saga of the Alstom (Bombardier) Class 701s for South Western Railway has continued, with testing of sets as built continuing, but with no date set for introduction, the fleet has been progressively moved to safe store sites around the country. The years late introduction of these sets has seen the SWR Class 455 and 458 fleets remain active.

Final testing of Class 730s for West Midlands Railway was progressing in mid-2023 and should be introduced by the end of the year.

The Class 769 bi- and tri-mode Porterbrook 'Flex' project has finally floundered, with the unreliable sets working for Transport for Wales being taken off lease and returned to Porterbrook at Long Marston, while the much larger Great Western Railway fleet were all returned to Porterbrook and are currently stored at Long Marston, due to ongoing technical issues and a changing passenger market.

At the end of 2022 Vivarail went into receivership, but Great Western Railway stepped in to purchase the intellectual rights of the fast-charge project, as well as a number of un-modified 'D' stock vehicles, plus sets 230001, 230003-005.

In terms of locos, the Class 69 conversion project by Progress Rail continued, very much slowed down due to technical issues, but in summer 2023 was back on stream.

The first of the new tri-mode locos for Rail Operations Group in the shape of Class 93s built by Stadler was delivered in June. The first of the GB Railfreight Class 99s was under early construction in Spain and due for delivery in September 2024.

In July 2023, two new tenders for stock were announced, 70 MUs for Chiltern and up to 450 trains for Northern. ■

Information updated to 20 August 2023

Left: *The Rail Adventure owned Class 43 HST power cars are now frequently used as 'locomotives' to move stock or engineering equipment, often on delivery runs from builders to operators. Its still quite hard to come to terms with HST power cars hauling freight and departmental trains. RailAdventure Nos. 43465 and 43484 are seen passing Scunthorpe on 24 February 2023 powering train 6Q75, the 10.00 Doncaster Belmont to Frodingham depot formed of new Matisa tamping machines Nos. DR75506 and DR75505. The tampers travelled with a Class 555 Tyne & Wear Metro set No. 555003 from Switzerland to the UK via the Channel Tunnel over the previous few days.* **Steve Thompson**

Trains On Order/Delivery 2022-2025

Class	Builder	Operator	Owner	Type	Number of sets/vehicles	Status
18	Clayton	Spot Hire	Beacon rail	Bi-Mode	15	Testing/introduction
69	Progress Rail	GB Railfreight	Progress Rail	Co-Co	16	In use/under conversion
93	Stadler	Rail Operations Group	Rail Operations Group	Bo-Bo	10 (+ 20 options)	Construction / delivery
99	Stadler	GB Railfreight	Beacon Rail	Co-Co	30 (+ options)	Construction
196/0	CAF	West Midlands Railway	Corelink Rail	DMU	12 x 2-car (24)	In use / delivery
196/1	CAF	West Midlands Railway	Corelink Rail	DMU	14 x 4-car (56)	In use / delivery
197/0	CAF	Transport for Wales	SMBC Leasing	DMU	51 x 2-car (102)	In use / delivery
197/1	CAF	Transport for Wales	SMBC Leasing	DMU	26 x 3-car (78)	In use / delivery
230	Vivarail	Transport for Wales	Vivarail	BMU	5 x 3-car (15)	In use
231	Stadler	Transport for Wales	SMBC Leasing	DMU	11 x 4-car (44)	In use
398	Stadler	Transport for Wales	SMBC Leasing	BMU±	36 x 3-car (108)	Construction/testing/delivery
458	Alstom	South Western Railway	Porterbrook	EMU	28 x 4-car (112)	Rebuild of Class 458 5-car stock
701/0	Bombardier	South Western Railway	Rock Rail	EMU	60 x 10-car (600)	Delivery/commissioning (delayed)
701/5	Bombardier	South Western Railway	Rock Rail	EMU	30 x 5-car (150)	Delivery/commissioning (delayed)
720	Bombardier	Greater Anglia	Angel Trains	EMU	133 x 5-car (665)	In use
720/6	Bombardier	c2c	Porterbrook	EMU	12 x 5-car (60)	In use
730/0	Bombardier	West Midlands Railway	Corelink Rail	EMU	48 x 3-car (144)	Construction/delivery/introduction
730/1	Bombardier	West Midlands Railway	Corelink Rail	EMU	36 x 5-car (180)	Construction/delivery/introduction
756/0	Stadler	Transport for Wales	SMBC Leasing	BMU■	7 x 3-car (21)	Construction/delivery/testing
756/1	Stadler	Transport for Wales	SMBC Leasing	BMU■	17 x 4-car (68)	Construction/delivery/testing
777	Stadler	MerseyRail	Liverpool City	EMU	53 x 4-car (212)	In use / delivery
805	Hitachi	Avanti West Coast	Rock Rail	BMU	13 x 5-car (65)	Construction/testing
807	Hitachi	Avanti West Coast	Rock Rail	EMU	10 x 7-car (70)	Construction/testing
810	Hitachi	Abellio East Midlands	Rock Rail	BMU	33 x 5-car (165)	Construction/testing
Awaited	Alstom/Hitachi	Avanti West Coast/HS2	Awaited	EMU	54 x 8-car (432)	Design (Delivery from 2027-2028)

± Bi-mode battery EMU   ■ Tri-mode battery EMU / DMU

Rolling Stock Review 2023-2024

Off Lease, Withdrawn and Cascaded Rolling Stock 2021-2025 (actual and proposed)

Class	Owner	2023 operator	Number of vehicles (sets)	Originally built	Off lease	Cascade or fate
Locomotives/Power cars						
43	Porterbrook/ Angel Trains/ First Group	GWR/XC	192 (as built)	1976-1982	2023-2024	GWR phased out 2023-2024 XC sets phased out 2023 PCs remain with ScotRail, Colas, Network Rail and Rail Adventure
91	Eversholt	LNE	31	1988-1992	2026	12 with LNER until around 2026-27
Diesel units						
150	Angel/Porterbrook	TfW	72 (36)	1986-1987	2023-2024	Store, available, some to WMR
153	Porterbrook	WMR	8 (8)	1988	2020	Disposal/TfW/Network Rail
	Angel/Porterbrook	EMR	13 (13)	1988	2022	Disposal/TfW/Network Rail
158	Angel Trains	TfW	48 (24)	1990-1992	2024	Store, available for re-lease
170	Porterbrook	ASR		1999-2002	2019-2020	Some sets to EMR
170	Porterbrook	WMR	6 MS	2000	2021	MS cars to CrossCountry
170	Porterbrook	TfW	32 (12)	2001	2023	To EMR
171	Porterbrook	GTR	4 (12)	2003-2005	2022-2023	To EMR (as 170s)
175	Angel Trains	TfW	70 (27)	1999-2000	2023-2024	Store, available for re-lease
180	Angel Trains	EMR	25 (5)	2001	2023	Available for re-lease
221	Beacon Rail	AWC	100 (20)	2001-2002	2022-2023	Possible to XC, 2 on short-hire to GCR
222	Eversholt	EMR	143 (27)	2004-2005	2024	Store, available for re-lease
230	WMR	WMR	6 (3)	2018	2023	Transferred to Great Western Rly
769	Porterbrook	TfW	36 (9)	2018	2023	Store
Electric units						
313/2	Eversholt	GTR	57 (19)	1976-1977	2023	Disposal
315	Eversholt	TfL	244 (61)	1980-1981	2019-2023	Disposal
317	Angel Trains	GAR	204 (51)	1981-1986	2021-2023	Disposal
319	Porterbrook	WMR/NOR	28 (7) + 52 (13)	1988	2022-2024	Disposal, or 'Flex' projects
321	Eversholt	GAR/NOR	428 (107)	1988-1991	2020-2022	Disposal or possible rebuild
322	Eversholt	NOR	25 (5)	1990	2022	To GAR, then disposed
323	Porterbrook	WMR	129 (43)	1992-1993	2023-2024	17 to Northern, 10 to store/scrap
332	Heathrow	HEX	61 (14)	1997-2002	2021	Disposal
350/2	Porterbrook	WMR	148 (37)	2008-2009	2024	Store, available for re-lease
360	Angel Trains	GAR	84 (21)	2002-2003	2021	To EMR
	Heathrow	HEX	25 (5)	2002-2006	2021	Disposal / GCRE Wales
365	Eversholt	GTR	84 (21)	1994-1995	2019-2022	Disposal
373	Eurostar	EUS	52 (½ trains)	1992-1996	2016-2021	Disposal
379	Akiem Rail	GAR	120 (30)	2011	2021-2022	Stored, available for re-lease
442	Angel	SWR	95 (19)	1988	2021	Disposal
455	Porterbrook	SWR	364 (91)	1982-1984	2021-2025	Store, disposal
455	Eversholt	GTR	184 (46)	1983-1983	2022	Disposal
456	Porterbrook	SWR	48 (24)	1990-1991	2021-2022	Disposal
458	Porterbrook	SWR	180 (36)	1990-1991	2022-2024	28 sets for rebuild as 4-car 68 vehicles for scrap
465	Eversholt / Angel	SET	Number awaited	1991-1993	2021-2024	Store
466	Angel	SET	Number awaited	1992-1994	2021-2024	Store
483	SWR	SWR	12 (6)	1989-1990	2021	Disposal
507	Angel	MER	96 (32)	1978-1980	2021-2024	Disposal
508	Angel	MER	81 (27)	1978-1980	2021-2024	Disposal
707	Angel Trains	SWR	150 (30)	2016-2017	2021-2024	To Southeastern
769	Portebrook	(GWR)	68 (190)	2020	2023	Returned never commissioned

ASR Abellio Scottish Railways
AWC Avanti West Coast
EMR East Midlands Railway
EUS Eurostar
GAR Greater Anglia Railway
GCR Grand Central Railway
GTR Govia Thameslink Railway
GWR Great Western Railway
HEX Heathrow Express
LNE London North Eastern Railway
NOR Northern
SET Southeastern Trains
SWR South Western Railway
TfL Transport for London
TfW Transport for Wales
WMR West Midlands Railway
XC CrossCountry

Right: *The Class 756 Stadler 'Flirt' sets are currently on delivery to Transport for Wales, these are a bi-mode version of the Class 231 fleet. Set No. 756003 is shown on display at Rail Live at Long Marston on 21 June 2023.* **CJM**

Left: *Class 37 equipment positions. 1: Number 1 end, 2: Number 2 end, 3: Nose section, housing traction motor blower and vacuum exhauster (if fitted), 4: Main electrical compartment, 5: Power unit and generator/alternator compartment, 6: Cooler group radiator with fan assembly on roof, 7: Nose section housing traction motor blower and air compressor, 8: Fire system activation handle, 9: Brake connection, 10: Radiator water filler port, 11: Fuel and lubricating oil filler port, 12: Main fuel tanks, 13: Fuel tank gauge, 14: Sandbox filler port, 15: Brake cylinder, 16: Hinged roof section providing access to power unit, 17: Nose section hinged access doors.*
Antony Christie

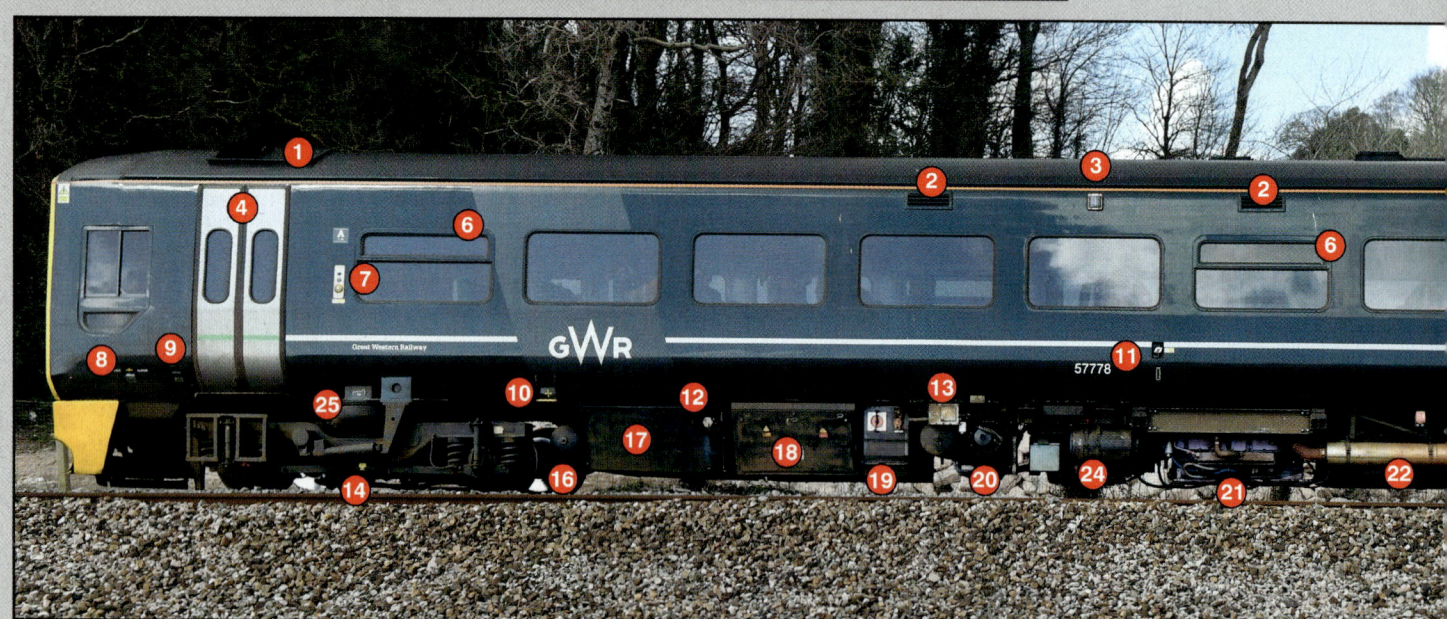

Above: *Class 158 equipment positions. 1: Air conditioning intake, 2: Passenger compartment air intake, 3: Hazard light (orange), 4: Bi-parting sliding plug doors, 5: Exhaust stack, 6: Staff operated hinged opening hopper, 7: Door open push button, 8: Local door release valve, 9: Emergency door release pulley, 10: Lifting point, 11: Cooling water filler and gauge, 12: Fuel inlet valve, 13: Auxiliary heater control panel, 14: BREL T4 bogie (un powered), 15: BREL P4 bogie (powered), 16: Surge reservoir, 17: Main fuel tank, 18: Battery box, 19: Battery isolating switch, 20: Auxiliary heater unit, 21: Engine (mounted centrally under train on power unit raft), 22: Exhaust system, 23: Voith hydrostatic transmission unit (mounted centrally under vehicle), 24: Air filter, 25: Secondary air suspension. Car No. 57778 from set No. 158959 is illustrated.* **CJM**

The main equipment positions on all locomotives and multiple units is roughly the same and the illustrations below serve as a guide to the locations of major equipment items, every fleet of train is slightly different and items of different shape and size will be found. Space in *Rolling Stock Review* does not allow us to include such detail for every class, but this information can be found in the of *Modern Locomotives Illustrated* magazine covering the class or type.

Above: *Class 70 equipment positions. 1: No. 1 end, 2: No. 2 end, 3: Sand filler port, 4: Sand delivery jet, 5: Cab door, 6: Cab footstep light, 7: Lifting point, 8: Removable inspection covers to electrical equipment, 9: Opening side inspection doors with engine and alternator behind, 10: Exhaust port, 11: Fire suppression system, 12: Emergency fuel cut off button, 13: Fuel filler port, 14: Fuel gauge, 15: Fuel tank, 16: Revised air intake, 17: Roof cooler fans, 18: Opening side inspection doors for cooler group. Loco illustrated No. 70803 when new.* **CJM**

Above: *Class 90 equipment positions side A. 1: No. 1 end, 2: No. 2 end, 3: Brecknell Willis high speed pantograph, 4: Sand filler port, 5: Air conditioning unit No. 1, 6: Power pack No. 2, 7: Control cubicle No. 2, 8: Communication cubicle, 9: Main transformer, 10: Rheostatic brake equipment, 11: Control cubicle No. 4, 12: Power pack No. 4, 13: Surge suppression equipment, 14: Air compressor, 15: Fire system.* **Mark V. Pike**

Multiple Unit Vehicle Type Codes

BDMSO	Battery Driving Motor Standard Open
DM	Driving Motor
DMBO	Driving Motor Brake Open
DMBS	Driving Motor Brake Standard
DMCL	Driving Motor Composite Lavatory
DMCO	Driving Motor Composite Open
DMF	Driving Motor First
DMFLO	Driving Motor First Luggage Open
DMRFO	Driving Motor Restaurant First Open
DMS	Driving Motor Standard
DMSL	Driving Motor Standard Lavatory
DMSO	Driving Motor Standard Open
DTCO	Driving Trailer Composite Open
DTPMV	Driving Trailer Parcels Mail Van
DTSO	Driving Trailer Standard Open
MBC	Motor Brake Composite
MBSO	Motor Brake Standard Open
MC	Motor Composite
MFL	Motor First Lavatory
MPMV	Motor Parcels Mail Van
MS	Motor Standard
MSL	Motor Standard Lavatory
MSLRB	Motor Standard Lavatory Restaurant Buffet
MSO	Motor Standard Open
MSRMB	Motor Standard Restaurant Micro Buffet
PTSO	Pantograph Trailer Standard Open
RB	Restaurant Buffet
TBFO	Trailer Brake First Open
TCO	Trailer Composite Open
TFO	Trailer First Open
TPMV	Trailer Parcels Mail Van
TSO	Trailer Standard Open
TSRMB	Trailer Standard Restaurant Micro Buffet

Left and Right: *In recent years the rail industry has fitted an increasing number of forward facing cameras to locos and multiple unit stock, to record the path ahead. The film is then available to inquiries or the coroner's court. The camera on a Class 318 and HST are marked in red. Both:* **CJM**

Rolling Stock Review : 2023-2024

Rolling Stock Review 2023-2024

Front End Detail Differences

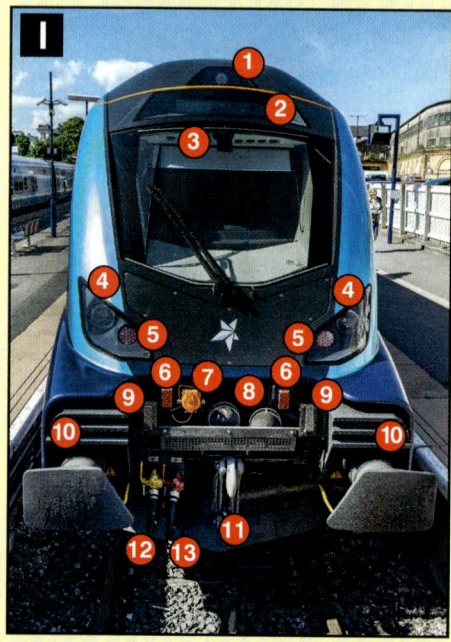

Rolling Stock Review 2023-2024

On modern rolling stock, there is a wide number of different front end designs in terms of equipment carried and how it is positioned. Over the years, we have often been told of trains being of a 'standard design', this is far from the case. Throughout *Rolling Stock Review*, we have included a number of detailed front ends, but for ease of comparison, nine varied designs are shown here. At the end of the day, the trains front end has three main functions, to provide a safe area for the driver, providing a good view of the track ahead, a method of attachment to another train and provide a warning of either the trains front or rear end using lights.

A. A traditional main line diesel, Class 47. 1: Warning horn grille, 2: Radio aerial, 3: Marker lights (white), 4: Tail lights (red), 5: Headlight, 6: Electric train supply jumper cable, 7: Main reservoir pipe, 8: Vacuum brake pipe, 9: Air brake pipe, 10: Coupling hook and shackle, 11: Electric train supply jumper socket.

B. Class 70. 1: Warning horns behind grilles, 2: High level marker light, 3: Combined marker (white) and tail light (red), 4: Windscreen washer filler port, 5: Headlight, 6: AAR type multiple control jumper socket (cable stowed in cab cross-walk). 7: Air brake pipe, 8: Main reservoir pipe, 9: Coupling hook and shackle.

C. Class 66. 1: Warning horns behind grille, 2: High level marker light, 3: AAR type multiple control jumper socket (cable stowed in engine compartment), 4: Marker light (white), 5: Headlight, 6: Tail light (red), 7: Combination coupler release mechanism, 8: Combination swing head coupling, 9: Coupling hook (shackle stowed in engine compartment), 10: Air brake pipe, 11: Main reservoir pipe.

D. Class 68 (also applicable to Class 88). 1: Warning horn, 2: High level marker light, 3: Multiple working jumper cable sockets (cable stowed in engine compartment), 4: Joint head, marker and tail light, 5: Electric train supply jumper cable, 6: Electric train supply jumper socket, 7: Main reservoir pipe, 8: Air brake pipe, 9: Coupling hook and shackle.

E. Class 710 (Bombardier/Alstom Aventra EMU). 1: Cant rail height marker light, 2: Destination display, 3: Combined white marker and red tail light (LED type), 4: Headlight, 5: Warning horns behing grille, 6: Emergency lamp bracket, 7: Dellner coupling, 8: Electrical connection box with roller cover.

F. Class 377, also applicable to Class 375 and 385. 1: High level marker light, 2: Gangway doors, 3: Headlight, 4: Combined white marker / red tail light, 5: Lamp bracket, 6: Track light, 7: Dellner coupling, 8: Electrical connection box, 9: Warning horns, located within coupling pocket.

G. Modern 'Turbostar' DMU, Class 172. 1: High level marker light, 2: Destination indicator, 3: Headlight, 4: Combined marker light (white) and tail light (red), 5: Forward facing camera, 6: Fold away lamp bracket, 7: BSI coupling, 8: Electrical connection box, 9: Warning horns.

H. Siemens Desiro City, Class 700. 1: High level destination indicator, 2: Route display, 3: Combined marker light (white) and tail light (red), 4: Headlight, 5: Lamp bracket, 6: Electrical connection box (roller front), 7: Dellner coupling, 8: Obstacle deflector plate (snowplough).

I. CAF TransPennine Driving Trailer. 1: High level marker light, 2: Destination display, 3: Forward facing camera, 4: Headlight, 5: Marker/tail light, 6: Control jumper socket, 7: Jumper socket, 8: Warning horns, 9: Lamp bracket, 10: Anti-climber blocks, 11: Coupling, 12: Main reservoir pipe, 13: Brake pipe. All images: **CJM**

Main Traction Depots

Depot	Code	Operator	Fleet (Allocation)
Allerton	AN	NOR (CAF)	319, 323, 331, 769
Ardwick	AK	TPT (SIE)	185
Ashford	AD	SET (HIT)	395
Aylesbury	AL	CRW	165, 168
Barrow Hill	BH	HNR	08, 09, 20, 37, 47
Bedford	BF	EMR	360
Birkenhead	BD	MER (STD)	230, 507, 508
Bounds Green	BN	LNE / HIT	801, 802/3
Bournemouth	BM	SWR	458
Bletchley	BY	WMR	730/1
Canton	CV/CF	TFW/COL	70, 150, 153, 231, 756
Carnforth	CS	WCR	08, 33, 37, 47, 57, 86
Central Rivers	CZ	AXC/AWC	220, 221
Chester	CH	CAF	175, 197
Coquelles	CO	Eurotunnel	Shuttle
Corkerhill	CK	SR	153, 156, 158
Craigentinny	EC	LNE/GBR/TPT (HIT)	73, 385, 802/2, 803
Crewe Basford	CB	FLR	47, 86, 90
Crewe Diesel	CL	LSL	08, 20, 37, 40, 47, 55, 73, 86, 87, 89, 90
Crewe Electric	CE	DBC	67, 90, 92, 325
Crewe Gresty	CR	DRS	68
Crown Point	NC	GAR	745, 755
Doncaster	DN	HIT	800, 801
Derby	DY	EMR	08, 170, 222, 810
Eastcroft	NM	EMR	158, 170
East Ham	EM	C2C	357, 720/6
Exeter	EX	GWR	150, 158
Haymarket	HA	SR	43, 170
Heaton	HT	NOR/GCR	156, 158, 180
Hornsey	HE	GTR	387, 717
Ilford	IL	GAR	720
Inverness	IS	SR	158
Kingmoor	KM	DRS	37, 57, 66, 88
Kirkdale	KK	MER (STD)	777
Laira	LA	GWR	08, 43
Leicester	LR	UKR/ROG	20, 37, 47, 56, 57, 58, (99)
Machynlleth	MN	CAF	158, 197
Manchester Int	MA	AWC, TPT	390, 397
Merehead	MD	MEN/FLR	08, 59
Midland Road	LD	FLR	66, 70
Neville Hill	NL	NOR/LNE	08, 91, 150, 155, 158, 170, 331, 333
New Cross	NG	LOG	378
Newton Heath	NH	NOR	150, 156, 195
Northampton	NN	WMR	319, 350
Northam	NT	SWR	444, 450
North Pole	NP	GWR (HIT)	800, 802
Nunnery	NU	SST	399
Old Oak Common	OC	CRO	345
Penzance	PZ	GWR	08, 57
Ramsgate	RM	SET	375, 377
Reading	RG	GWR	165, 230, 387
Roberts Road	RR	GBR (EMD)	66, 69
RTC Derby	ZA	NRL	08, 37 (97), 43, 73
Ryde	RY	SWR	484
Salisbury	SA	SWR	158, 159
Selhurst	SU	GTR	171, 377
Shields	GW	SR	318, 320, 334, 380
Slade Green	SG	SET	376, 465, 466, 707
Soho	SI	WMR	08, 323, 730/0
Southampton	SZ	FLR	08
Stewarts Lane	SL	GTR	387
Stourbridge Jn	SJ	WMR	139
St Philips Marsh	PM	GWR	08, 158, 165, 166
Taffs Well	(TW)	TFW	398
Temple Mills	TI	EUS	08, 373, 374
Three Bridges	TB	GTR (SEM)	700
Toton	TO	DBC/COL	60, 66
Tyseley	TS	AXC/WMR	08, 170, 172, 196
Wembley	WB	Alstom/GBR	08, 92
Willesden	WN	LOG	09, 710
Wimbledon	WD	SWR	455, 458, 701

In addition to the main depots above, a large number of stabling sidings exist with most operators, where locos and units can be found stabled or receiving light maintenance.

Rolling Stock Review 2023-2024

Know Your Couplings

Today, all modern multiple unit type trains, either electric or diesel powered, use automatic couplings to attach to other sets, thus saving the need to employ shunting staff and reduces the risk of staff having to work on the track around potentially moving trains. A large number of auto coupler types can be found in the UK.

Some of the earliest auto couplers were various types of the US-design buck-eye, which were used to couple loco-hauled and indeed HST vehicles together. In terms of multiple unit stock, buck-eye couplings were fitted to 1951-1966 design Southern Region (SR) EMUs and DEMU stock, and were also fitted to many Eastern, London Midland and Scottish Region MUs of the period. These buck-eye couplings however, only provided the physical attachment, with air and multiple control cables requiring to be connected manually. The full auto coupling emerged with the 1972-design PEP stock, giving physical, pneumatic and electrical connections. From this beginning, huge refinement has been seen over the years, with many different manufacturers now supplying auto couplings.

After using a fully automatic coupling on the PEP and subsequent 1972 stock, when the Class 455s were ordered for Southern Region use, a semi-automatic Tightlock coupling was fitted, with body mounted jumper cables and main reservoir pipes. These are now being phased out with withdrawal of Class 455 and 456 fleets.

More recently we have seen the use of London Underground style 'Wedgelock' couplers on Class 230 stock. The main equipment items for these are shown '**A**'. 1: Anti-climber plate, 2: Obstacle deflector plate, 3: Air horn, 4: Stabilizer for coupling alignment, 5: Electrical connections behind auto-open doors, 6: Pneumatic connections, 7: Tongue and Throat physical coupling.

Many of the second generation DMU sets incorporate a Bergische Stahl Industrie (BSI) coupling. This is a fully automatic coupling for physical, pneumatic and electrical connections. A BSI fitted to a Class 165 is shown on the right '**B**'. The main items are,
1: Emergency air pipe connection, 2: Main coupling with a 'gathering horn' on the right which guides the physical coupling together, 3: Electrical connection box, which when two couplings are pushed together the roller cover opens to allow the coupling pins to press together. 4: Main reservoir air connection, when the coupling snaps together these press together and form a leak-free union.

The Tightlock coupling can also be fitted for a fully automatic connection, in this case, the coupling would have a pneumatic/electrical connection box with a 'drum' cover, when the coupling is pushed together and the 'gathering horns' guide them together in the correct position, the 'drum' cover rolls open the air and electrical connections are pushed together. The coupling shown right '**C**' is from a Class 465. Items are, 1: Emergency air connection plug, 2: Manual/emergency release handle, 3: Semi-rotary electric/pneumatic cover, 4: Physical Tightlock.

The Dellner coupling has become very popular with modern MU builds. Illustration '**D**' shows a Class 360 set incorporating a Dellner with a under-slung electrical connection box.
1: Emergency air connector, 2: Dellner coupling face, 3: Pneumatic (main reservoir) connection, 4: Roller cover to electrical connections, 5: Emergency air supply line.

Illustration '**E**' shows a modern Dellner coupling on a Class 380 with an above coupling electrical connection.
1: Electrical connection box with roller cover, 2: Coupling face plate, 3: Main reservoir connection. All: **CJM**

Operator Codes

AFG	Arlington Fleet Group
ALS	Alstom
SR	ScotRail
ATC	Arriva Train Care
AWC	Avanti West Coast
AXC	Arriva CrossCountry
BEL	Belmond
BOD	Boden Rail Engineering
C2C	c2c Railway
COL	Colas Rail Freight
CRO	CrossRail
CRW	Chiltern Railways
DAS	Data Acquisition Services
DBC	DB-Cargo
DCR	Devon Cornwall Railway
DRS	Direct Rail Services
ELS	Elizabeth Line
EMR	East Midlands Railway
EPX	Europhoenix
EUS	Eurostar
FEC	First East Coast (Lumo)
FHT	First Hull Trains
FLR	Freightliner
GAR	Greater Anglia Railways
GBR	GB Railfreight
GTL	Grand Central Railway
GTR	Govia Thameslink Railway
GWR	Great Western Railway
HAN	Hanson
HEX	Heathrow Express
HNR	Harry Needle Railroad
HRA	Rail Adventure
LHG	L H Group Services
LNE	London North Eastern Railway
LNW	London North Western Railway
LOG	London Overground
LSL	Loco Services Ltd
MEN	Mendip Rail
MER	MerseyRail
MET	Meteor Power
NOR	Northern Rail
NRL	Network Rail
NRM	National Railway Museum
OLS	Off Lease Stock
PRI	Private
RIV	Riviera Trains
ROG	Rail Operations Group
RSS	Railway Support Services
SCS	Scottish Caledonian Sleeper
SET	Southeastern Trains
SST	Sheffield Super Tram
SWR	South Western Railway
TFL	Transport for London
TFW	Transport for Wales
TPT	TransPennine Trains
URL	UK Rail Leasing
VTN	Vintage Trains
WAB	Wabtec
WCR	West Coast Railway Co
WMR	West Midlands Railway

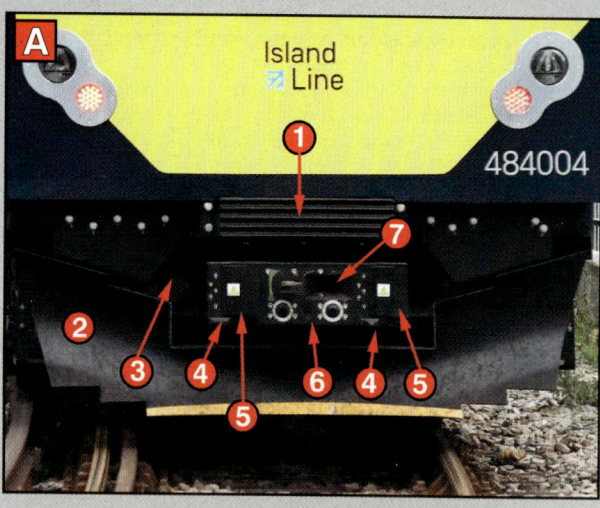

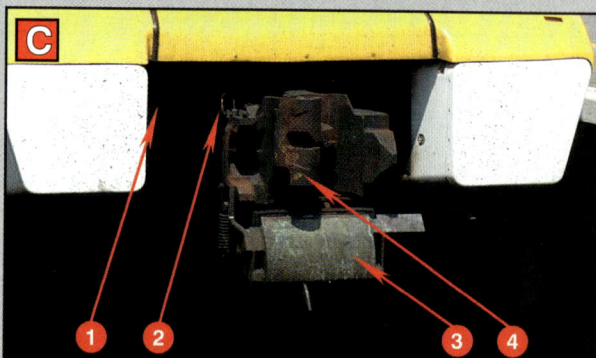
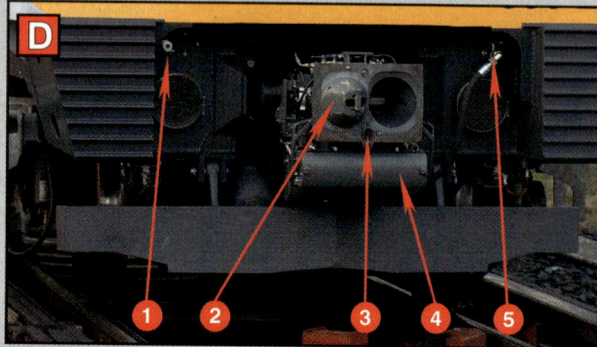
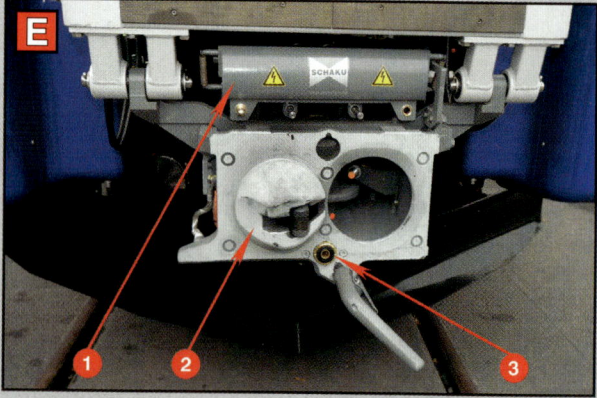

For history and pictures of this class, see *Modern Locomotives Illustrated* Issue 202

Class 08 and 09

2023 Fleet

Class:	08
Present number range:	08375-08994
Previous number series:	13000-13366, D3000-D4192
Former class codes:	DEJ4, then D3/2, 3/1
Built by:	BR workshops Derby, Crewe, Darlington, Doncaster, Horwich
Years introduced:	1953-1959
Wheel arrangement:	0-6-0
Maximum speed:	15mph (25km/h)
Length:	29ft 3in (8.91m)
Height:	12ft 8⅝in (3.88m)
Width:	8ft 6in (2.59m)
Weight:	49.6-50.4 tonnes
Wheelbase:	11ft 6in (3.50m)
Wheel diameter (new):	4ft 6in (1.37m)
Min curve negotiable:	3 chains (60.35m)
Engine type:	English Electric 6KT
Engine output:	400hp (298kW)
Power at rail:	260hp (194kW)
Tractive effort:	35,000lb (156kN)
Cylinder bore:	10in (250mm)
Cylinder stroke:	12in (300mm)
Transmission:	Electric
Main generator type:	EE801-8E or E801-14E
Aux generator type:	90V locos - EE736-2D or EE736-4E / 110V locos - EE906-3D
Traction motor type:	EE506-6A or EE506-7C
No. of traction motors:	2
Gear ratio:	Overall - 23.9:1, First train - 82:15, Second train - 70:16
Brake type:	Originally Vacuum, later Dual or Air
Brake force:	19 tonnes
Route availability:	5
Heating type:	Not fitted
Multiple coupling type:	Not fitted
Fuel tank capacity:	668gal (3,036lit)
Lub oil capacity:	45gal (204lit)
Cooling water capacity:	140gal (636lit)
Sanding equipment:	Pneumatic
Sub-class variations:	Standard loco, a descendant of LMS design

08375	RSS	08568	RSS	08650	FLR	08764	ALS	08877	HNR
08389	HNR	08571	EMS	08652	RSS	08765	HNR	08879	HNR
08401	EMS	08573	RMS	08653	HNR	08774	DAW	08885	RMS
08405	RSS	08575	FLR	08663	RSS	08780	LSL	08887	ALS
08411	RSS	08578	HNR	08669	EMS	08782	HNR	08891	FLR
08418	WCR	08580	RSS	08670	RSS	08783	EMR	08892	HNR
08423	BAR	08585	FLR	08676	HNR	08785	RSS	08899	RSS
08428	HNR	08588	RMS	08678	WCR	08786	HNR	08903	SEM
08441	RSS	08593	RSS	08682	HNR	08787	HAN	08904	HNR
08442	ATC	08596	EMS	08683	RSS	08788	RMS	08905	HNR
08445	EMS	08598	RSS	08685	HNR	08790	ALS	08908	EMR
08447	RUS	08600	AVD	08690	EMR	08798	HNR	08912	AVD
08451	ALS	08602	HNR	08691	FLR	08802	HNR	08913	EMR
08454	ALS	08605	RSS	08696	ALS	08805	WMR	08918	HNR
08460	RSS	08611	ALS	08700	RMS	08807	DAW	08921	RSS
08472	EMS	08613	BAR	08701	HNR	08809	RMS	08922	RSS
08480	RSS	08615	EMS	08703	RSS	08810	ATC	08924	HNR
08483	LSL	08616	WMR	08704	RIV	08818	HNR	08925	GBR
08484	RSS	08617	ALS	08709	RSS	08822	GWR	08927	RSS
08485	WCR	08622	BAR	08711	HNR	08823	EMS	08933	AGI
08499	COL	08623	HNR	08714	HNR	08824	HNR	08934	GBR
08500	HNR	08624	FLR	08721	ALS	08825	GBR	08936	RMS
08507	RSS	08629	GBR	08724	EMS	08834	HNR	08939	RSS
08502	HNR	08630	HNR	08730	EMS	08836	GWR	08943	HNR
08511	RSS	08631	LSL	08735	ATC	08846	RSS	08947	LHG
08516	ATC	08632	RSS	08737	LSL	08847	PHR	08948	EUS
08523	EMD	08635	UOB	08738	RSS	08853	EMS	08950	EMR
08525	EMR	08641	GWR	08743	SEM	08865	HNR	08954	HNR
08527	HNR	08643	EMS	08752	RSS	08868	HNR	08956	TFL
08530	FLR	08644	GWR	08754	RMS	08870	ERS		
08531	FLR	08645	GWR	08756	RMS	08871	RMS		
08536	RSS	08648	RMS	08757	RSS	08872	HNR		
08567	AFG	08649	MET	08762	ERS	08874	RMS		

Left: *Smaller companies and industrial users are the main operators of the Class 08/ 09 fleets today, with a few still used by TOC and FOCs. Owned and operated by West Midlands Trains for depot pilot work at Soho depot, Birmingham, No. 08805 carries Railfreight grey livery. This is a dual-brake example and sports a standard headlight.* **Antony Christie**

Below: *Loco Services Ltd (LSL), based at Crewe operate four Class 08s for loco and stock movements. This broad side view shows No. 08631 from its nose or radiator end, displaying standard 1970-80s BR rail blue livery. This example also retains dual vacuum/air brakes. The hinged doors on the bodyside give access to the cooler, engine and generator compartments. Batteries and the air compressor are box mounted on the frame.* **Spencer Conquest**

Class 08 Key Facts

- Originally the largest fleet of diesel locos in the UK, including Class 08, 09, 10 and some unclassified locos, with a fleet size of 1,193 locos.
- Few now remain in main stream service. Most are 'spot-hire' locos.
- A large number are in industrial service or preserved.
- Private owners now provide locos on short time or spot-hire contracts.
- Some locos have with knuckle or buck-eye couplings.
- Several are fitted with remote control equipment, operated from ground level 'belt packs'.

Rolling Stock Review : 2023-2024

Class 08 and 09

Class:	09/0	09/1 and 09/2
Present number range:	09002-09023	09106, 09201, 09204
Previous number range:	D3665-D3671, D3719-D3721, D4099-D4114	Rebuilt from Class 08
Former class codes:	DEJ4, then 3/1	Class 08
Built by:	BR Darlington/Horwich	Rebuilt: RFS Kilnhurst
Years introduced:	1959-1962	1992-1993
Wheel arrangement:	0-6-0	0-6-0
Maximum speed:	27½mph (34km/h)	27½mph (34km/h)
Length:	29ft 3in (8.91m)	29ft 3in (8.91m)
Height:	12ft 8⅝in (3.88m)	12ft 8⅝in (3.88m)
Width:	8ft 6in (2.59m)	8ft 6in (2.59m)
Weight:	50 tonnes	50 tonnes
Wheelbase:	11ft 6in (3.50m)	11ft 6in (3.50m)
Wheel diameter (new):	4ft 6in (1.37m)	4ft 6in (1.37m)
Min curve negotiable:	3 chains (60.35m)	3 chains (60.35m)
Engine type:	English Electric 6KT	English Electric 6KT
Engine output:	400hp (298kW)	400hp (298kW)
Power at rail:	269hp (201kW)	269hp (201kW)
Tractive effort:	25,000lb (111kN)	25,000lb (111kN)
Cylinder bore:	10in (250mm)	10in (250mm)
Cylinder stroke:	12in (300mm)	12in (300mm)
Transmssion:	Electric	Electric
Main generator type:	EE801-13E or EE801-14E	EE801-13E or EE801-14E
Aux generator type:	EE906-3D	EE906-3D
Traction motor type:	EE506-10C	EE506-10C
No. of traction motors:	2	2
Gear ratio:	Overall - 23.9:1 First train - 82:15 Second train - 70:16	Overall - 23.9:1 First train - 82:15 Second train - 70:16
Brake type:	Dual	Dual
Brake force:	19 tonnes	19 tonnes
Route availability:	5	5
Heating type:	Not fitted	Not fitted
Multiple coupling type:	Not fitted	Not fitted
Fuel tank capacity:	668gal (3,037lit)	668gal (3,037lit)
Lub oil capacity:	45gal (204lit)	45gal (204lit)
Cooling water capacity:	140gal (636lit)	140gal (636lit)
Sanding equipment:	Pneumatic	Pneumatic
Sub-class variations:	Higher-speed version of standard Class 08 originally allocated to the former BR Southern Region	Modified from standard Class 08

2023 Fleet

09002 HNR	09106 HNR
09006 HNR	09201 HNR
09007 LOG	09204 ATC
09009 GBR	
09014 HNR	
09022 POB	
09023 EMR	

Left: *Class 08/09 front end layout. 1: Radiator, 2: Radiator water filler, 3: Headlight, 4: Original style White marker/red tail lights, 5: Main reservoir pipe (yellow), 6: Air brake pipe (red), 7: Coupling shackle. Some locos are fitted with high level dual-cock air pipes, originally installed for attachment to MU stock. A number still retain vacuum train brakes.* **CJM**

Above: *A dual position driving cab is fitted, applicable to both Class 08 and 09 locos. The main driving position is on the right (near) side. The power controller and master switch are towards the middle and the brake controller to the right. Power and brake controllers on both sides of the cab are connected and operate as one.* **CJM**

Left: *The requirement for shunting locos is ever decreasing, with train engines now booked for most train formation work. This example, No. 08629 shunting at Eastleigh is operated by GB Railfreight and was previously used at Railcare Wolverton Works where this blue, white and green livery was applied. This example does not have high-level pipes or a headlight, with just two small joint marker (white) and tail light (red) lamps on the body end.* **Antony Christie**

Left: *A major supplier of standard 0-6-0 diesel-electric shunting locos to industry and operators is the Harry Needle Railroad Company (HNRC) offering a variety of good quality, often refurbished locos of both the standard Class 08 design or the higher speed Class 09. No. 09002 is seen at Barrow Hill from the cab end, and shows a different marker head light style, using modern LED marker and tail lights, with a standard headlight in a shroud. This loco is air brake fitted only and has lost its high-level duplicate air pipes.* **Antony Christie**

Class 09 Key Facts

- Higher speed (27mph, 34km/h) version of standard BR 0-6-0 diesel-electric shunting loco, introduced for use on BR Southern Region.
- Some Class 09s were fitted with high level air pipes to allow coupling to 1951-1966 design SR EMU and DEMUs.
- Some are fitted with remote control equipment, usually identified by yellow beacons on side and roof.
- Extra Class 09/1 and 09/2s were converted from Class 08s in the 1990s.

For history and pictures of this class, see *Modern Locomotives Illustrated* Issue 203

Class 20

Sub-class:	20/0	20/3	20/9
Present number range:	20007-20227	20311, 20314	20901-20906
Former number range:	D8000-D8199, D8300-D8327 series	From 20/0 fleet	From 20/0 fleet
Former class codes:	D10/3, then 10/3	20/0	20/0
Built by:	English Electric, Vulcan Foundry or Robert Stephenson & Hawthorn	Rebuilt: Brush Traction or RFS Doncaster	Rebuilt: Hunslet Barclay
Years introduced:	1957-1968	As 20/3: 1995-1998	As 20/9: 1989
Maximum speed:	75mph (121km/h)	75mph (121km/h)	60mph (97km/h)
Wheel arrangement:	Bo-Bo	Bo-Bo	Bo-Bo
Length:	46ft 9¼in (14.26m)	46ft 9¼in (14.26m)	46ft 9¼in (14.26m)
Height:	12ft 7⅝in (3.85m)	12ft 7⅝in (3.85m)	12ft 7⅝in (3.85m)
Width:	8ft 9in (2.67m)	8ft 9in (2.67m)	8ft 9in (2.67m)
Weight:	73 tonnes	73 tonnes	73 tonnes
Wheelbase:	32ft 6in (9.90m)	32ft 6in (9.90m)	32ft 6in (9.90m)
Bogie wheelbase:	8ft 6in (2.59m)	8ft 6in (2.59m)	8ft 6in (2.59m)
Bogie pivot centres:	24ft 0in (7.31m)	24ft 0in (7.31m)	24ft 0in (7.31m)
Wheel diameter (new):	3ft 7in (1.09m)	3ft 7in (1.09m)	3ft 7in (1.09m)
Min curve negotiable:	3.5 chains (70.40m)	3.5 chains (70.40m)	3.5 chains (70.40m)
Engine type:	English Electric 8SVT Mk2	English Electric 8SVT Mk2	English Electric 8SVT Mk2
Engine output:	1,000hp (746kW)	1,000hp (746kW)	1,000hp (746kW)
Power at rail:	770hp (574kW)	770hp (574kW)	770hp (574kW)
Tractive effort:	42,000lb (187kN)	42,000lb (187kN)	42,000lb (187kN)
Cylinder bore:	10in (220mm)	10in (220mm)	10in (220mm)
Cylinder stroke:	12in (350mm)	12in (350mm)	12in (350mm)
Transmission:	Electric	Electric	Electric
Main generator type:	EE819-3C	EE819-3C	EE819-3C
Aux generator type:	EE911-2B	EE911-2B	EE911-2B
ETS generator type:	Not fitted	Not fitted	Not fitted
Traction motor type:	EE526-5D or EE526-8D	EE526-8D	EE526-8D
No. of traction motors:	4	4	4
Gear ratio:	63:17	63:17	63:17
Brake type:	Vacuum, later dual	Air	Air
Brake force:	35 tonnes	31 tonnes	35 tonnes
Route availability:	5	5	5
Heating type:	Not fitted	Not fitted	Not fitted
Multiple coupling type:	Blue star	DRS system	Blue star, DRS system
Fuel tank capacity:	380gal (1,727lit)	1,080gal (4,909lit)	380-1,040gal (1,727-4,727lit)
Lub oil capacity:	100gal (455lit)	100gal (455lit)	100gal (455lit)
Cooling water capacity:	130gal (591lit)	130gal (591lit)	130gal (591lit)
Sanding equipment:	Pneumatic	Pneumatic	Pneumatic
Sub-class variations:	Standard as-built locos.	Refurbished locos for DRS. Modified cab equipment.	Modified from Class 20/0 originally for Hunslet Barclay.

Above: Class 20 cab end detail (showing DRS modified loco). 1: Horns behind grille, 2: Radio aerial, 3: Screen washer jets, 4: White marker light, 5: Headlight, 6: Red tail light, 7: DRS jumper cable socket, 8: Main reservoir pipe, 9: Air brake pipe, 10: Engine control air pipe, 11: Coupling shackle. Some locos still sport blue star multiple control jumper sockets, vacuum pipes and standard headlights.
Brian Garrett

2023 Fleet

20007	EEP
20056	HNR
20066	HNR
20096	LSL
20107	LSL
20118	LSL
20121	HNR
20132	LSL
20142	189
20168	HNR
20189	189
20205	C2L
20227	C2L
20311	HNR
20314	HNR
20901	HNR
20905	HNR
20906	HNR

Above Left, Above Right & Left: The Class 20s are of single cab full width design, having driving controls facing either direction. The ends either carry a disc or headcode box reporting system, depending on the year of build. Head lights are carried on both ends of main line certified locos. Some locos have standard light clusters. Above left No. 20142, with a four-position headcode box is seen in London Transport maroon livery. This is a 'spot-hire' loco owned by 20189 Ltd. Above right are two of the Loco Services Ltd locos, Nos. D8086 (20096) and D8107 (20107) displaying 1960s green with a small yellow end. This is now an air brake only loco. On the left is No. 20189, owned by 20189 Locomotives, it carried BR rail blue livery, is dual braked and used for 'spot-hire' work on the main line.
Cliff Beeton / Spencer Conquest (2)

Key Facts

- Over 200 'standard' design single cab Type 1s were built in 1957-1968.
- 18 currently certified for main line use.
- Some locos fitted with 'trip-cock' apparatus for use on LUL lines.
- Many Class 20s are preserved and can be found on light railways, others operate as 'spot-hire' locos.

Rolling Stock Review : 2023-2024

Class 31

For history and pictures of this class, see *Modern Locomotives Illustrated* Issue 189

Sub Class:	31/1
TOPS Number range:	31106 & 31128
1957 BR number range:	D5524 & D5546
Former class codes:	D14/2, then 14/2
Built by:	Brush Ltd, Loughborough
Years introduced:	1959
Wheel arrangement:	A1A-A1A
Weight:	107-111 tonnes
Height:	12ft 7in (3.84m)
Length:	56ft 9in (17.30m)
Width:	8ft 9in (2.67m)
Wheelbase:	42ft 10in (13.05m)
Bogie wheelbase:	14ft 0in (4.27m)
Bogie pivot centres:	28ft 10in (8.79m)
Wheel diameter:	Powered - 3ft 7in (1.09m) Pony - 3ft 3½in (1m)
Min curve negotiable:	4.5 chains (90.5m)
Engine type:	English Electric 12SVT
Engine output:	1,470hp (1,096kW)
Power at rail:	1,170hp (872kW)
Tractive effort:	35,900lb (160kN)
Cylinder bore:	10in (254mm)
Cylinder stroke:	12in (305mm)
Maximum speed:	90mph (145km/h)
Brake type:	Dual
Brake force:	49 tonnes
Route availability:	5
Heating type:	Steam - Spanner Mk 1
Multiple coupling type:	Blue star
Main generator type:	Brush TG160-48
Aux generator type:	Brush TG69-42
Traction motor type:	Brush TM73-68
No of traction motors:	4
Gear ratio:	60:19
Fuel tank capacity:	530gal (2,409lit)
Cooling water capacity:	156gal (709lit)
Lub oil capacity:	110gal (500lit)
Boiler water capacity:	600gal (2,727lit)
Boiler fuel capacity:	100gal (454lit)
Sanding equipment:	Pneumatic

Key Facts

- One of the original Modernisation Plan classes of the 1950s.
- In 2023 just two locos remain certified for use on Network Rail tracks.
- Originally five sub-classes existed, Class 31/0, 31/1, 31/4, 31/5 and 31/6.
- A sizeable number of Class 31s are preserved.

Above: When the first Brush Type 2s emerged from the Loughborough Works of Brush Traction in late 1957, they were fitted with Mirrlees engines, due to failures these were replaced with English Electric engines in the 1960s. The first Brush Type 2 (Class 31) locos were built with disc train reporting which continued on some locos up to D5561, thereafter four-character headcode displays were fitted. Originally front gangway doors were also provided to allow inter loco access. Originally vacuum train braking was fitted, this was progressively changed to dual brake equipment. No. 31106, one of the original disc fitted locos is seen at Derby from its No. 2 end. This loco is currently out of traffic at the East Lancs Railway. **Antony Christie**

2023 Fleet

31106	PRI
31128	NEM

Below: Nemesis Rail, based in Burton-on-Trent own No. 31128, a headcode box fitted example which is currently named Charybdis. The loco is available for 'spot-hire' and is fully certified for main line operation. It carries BR 1960s rail blue livery, with its TOPS number. The loco is dual brake fitted and in 2023 was fitted with three section miniature snowploughs at both ends. The loco is viewed from its No. 1 or cooler end in May 2022. **Antony Christie**

Rolling Stock Review : 2023-2024

For history and pictures of this class, see *Modern Locomotives Illustrated* Issue 184

Class 33

Right: Class 33 front end equipment. Style applicable to members of Class 33/0 and 33/2, locos from Class 33/1 (of which none are currently working on the main line, are fitted with buck-eye couplings and high level connections). 1: Horns behind grille, 2: Route indicator, 3: Lamp bracket, 4: White marker light with red hinged shade, 5: Blue star control jumper socket, 6: Engine control air pipe, 7: Electric train heat jumper socket, 8: Main reservoir pipe, 9: Air brake pipe, 10: Coupling hook and shackle, 11: Vacuum pipe, 12: AWS pick-up magnet, 13: Electric train heat jumper cable (missing). CJM

2023 Fleet

33/0
33012 PRI
33025 WCR
33029 WCR

33/2
33207 WCR

Sub-class:	33/0	33/2
Present number range:	33012, 33025, 33029	33207
Previous number range:	D6500-D6585	D6586-D6597
Former class codes:	D15/1, later 15/6	D15/2, later 15/6A
Southern Region code:	KA	KA-4C
Built by:	Birmingham RC&W	Birmingham RC&W
Years introduced:	1960-1962	1962
Wheel arrangement:	Bo-Bo	Bo-Bo
Maximum speed:	85mph (137km/h)	85mph (137km/h)
Length:	50ft 9in (15.47m)	50ft 9in (15.47m)
Height:	12ft 8in (3.86m)	12ft 8in (3.86m)
Width:	9ft 3in (2.81m)	8ft 8in (2.64m)
Weight:	77 tonnes	77 tonnes
Wheelbase:	39ft 0in (11.89m)	39ft 0in (11.89m)
Bogie wheelbase:	10ft 0in (3.04m)	10ft 0in (3.04m)
Bogie pivot centres:	29ft 0in (8.84m)	29ft 0in (8.84m)
Wheel diameter (new):	3ft 7in (1.09m)	3ft 7in (1.09m)
Min curve negotiable:	4 chains (80.46m)	4 chains (80.46m)
Engine type:	Sulzer 8LDA28A	Sulzer 8LDA28A
Engine output:	1,550hp (1,156kW)	1,550hp (1,156kW)
Power at rail:	1,215hp (909kW)	1,215hp (909kW)
Tractive effort:	45,000lb (200kN)	45,000lb (200kN)
Cylinder bore:	11.02in (270mm)	11.02in (270mm)
Cylinder stroke:	14.17in (360mm)	14.17in (360mm)
Transmission:	Electric	Electric
Main generator type:	Crompton Parkinson CG391-B1	Crompton Parkinson CG391-B1
Aux generator type:	Crompton Parkinson CAG193-A1	Crompton Parkinson CAG193-A1
ETS generator type:	Crompton Parkinson CAG392-A1	Crompton Parkinson CAG392-A1
Traction motor type:	Crompton Parkinson C171-C2	Crompton Parkinson C171-C2
No. of traction motors:	4	4
Gear ratio:	62:17	62:17
Brake type:	Dual	Dual
Brake force:	35 tonnes	35 tonnes
Route availability:	6	6
Heating type:	Electric - index 48	Electric - index 48
Multiple coupling type:	Blue star	Blue star
Fuel tank capacity:	750gal (3,410lit)	750gal (3,410lit)
Lub oil capacity:	108gal (491lit)	108gal (491lit)
Cooling water capacity:	230gal (1,046lit)	230gal (1,046lit)
Sanding equipment:	Pneumatic	Pneumatic
Sub-class variations:	Basic locomotive	Hastings line profile

Above: A fleet of 98 BRC&W Type 3s, later Class 33s, were built, four are currently authorised for main line operation, three with West Coast Railway (WCR) and one owned privately, based at the Swanage Railway. Class 33/0 No. 33025, owned by WCR, is seen at Reading from its No. 2 end, carrying standard WCR maroon livery. The loco carries a three section snowplough set. CJM

Below: No. 33012 is privately owned, based on the Swanage Railway in Dorset. It is refurbished and main line certified. In 2023 it was painted in 1960s BR green with a white body line and roof, a small yellow warning end, red buffer beams and three section plough set, it is shown on the Swanage Railway. Antony Christie

Key Facts

- Introduced in 1960-1962 to replace steam traction on BR Southern Region.
- Three sub-classes operaed, 33/0 standard locos, 33/1 push-pull locos and 33/2 for narrow bodied 'Hastings' locos.
- Four locos main line certified, operated by Carnforth-based West Coast Railway Co and the Swanage Railway.
- A total of 25 locos from all sub-classes are preserved and can be found working on many light railways, carrying a number of historic liveries.
- No push-pull modified Class 33/1s are main line certified, but some are preserved.

Right: The final 12 Class 33s were built to the narrower 'Hastings' body profile of 8ft 8in (2.64m), allowing operation over the restricted clearances of the Tonbridge to Hastings route. These locos were classified as 33/2. One is main line certified, No. 33207 operated by West Coast Railway. The loco carries the name Jim Martin. No. 33207 is seen operating on the preserved Chinnor and Princes Risborough Railway. Antony Christie

Rolling Stock Review : 2023-2024

Class 37

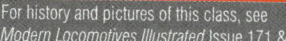

For history and pictures of this class, see *Modern Locomotives Illustrated* Issue 171 & 238

Sub-class:	37/0 & 97/3	37/4	37/5	37/6	37/7
Present number range:	37025-37259, 97301-97304¤	37401-37425	37510-37521, 37667-37688	37601-37612	37703-37884
Previous number range:	D6600-D6999	From Class 37/0 fleet	From Class 37/0 fleet	From Class 37/5 fleet	From Class 37/0 fleet
Former class codes:	D17/1, then 17/3	37/0	37/0	37/5	37/0
Built by:	English Electric, Vulcan Foundry or Robert Stephenson & Hawthorn	-	-	-	-
Refurbished by:	-	BREL Crewe	BREL Crewe	BRML Doncaster	BREL Crewe
Years introduced:	1960-1965, 97/3 - 2008	1985-1986	1986-1987	1994-1996	1986-1987
Wheel arrangement:	Co-Co	Co-Co	Co-Co	Co-Co	Co-Co
Maximum speed:	80mph (129km/h)	80mph (129km/h)	80mph (129km/h)	90mph (145km/h)	80mph (129km/h)
Length:	61ft 6in (18.74m)	61ft 6in (18.74m)	61ft 6in (18.74m)	61ft 6in (18.74m)	61ft 6in (18.74m)
Height:	Between 12ft 9⅛in and 13ft 0¾in (3.89m-3.96m) #	13ft 0¾in (3.96m)	13ft 0¾in (3.96m)	13ft 0¾in (3.96m)	13ft 0¾in (3.96m)
Width:	8ft 11⅝in (2.73m)	8ft 11⅝in (2.73m)	8ft 11⅝in (2.73m)	8ft 11⅝in (2.73m)	8ft 11⅝in (2.73m)
Weight:	Between 102-108 tonnes	107 tonnes	Between 106-110 tonnes	108 tonnes	120 tonnes
Wheelbase:	50ft 8in (15.44m)	50ft 8in (15.44m)	50ft 8in (15.44m)	50ft 8in (15.44m)	50ft 8in (15.44m)
Bogie wheelbase:	13ft 6in (4.11m)	13ft 6in (4.11m)	13ft 6in (4.11m)	13ft 6in (4.11m)	13ft 6in (4.11m)
Bogie pivot centres:	37ft 2in (11.32m)	37ft 2in (11.32m)	37ft 2in (11.32m)	37ft 2in (11.32m)	37ft 2in (11.32m)
Wheel diameter (new):	3ft 7in (1.09m)	3ft 7in (1.09m)	3ft 7in (1.09m)	3ft 7in (1.09m)	3ft 7in (1.09m)
Min curve negotiable:	4 chains (80.46m)	4 chains (80.46m)	4 chains (80.46m)	4 chains (80.46m)	4 chains (80.46m)
Engine type:	English Electric 12CSVT	English Electric 12CSVT	English Electric 12CSVT	English Electric 12CSVT	English Electric 12CSVT
Engine output:	1,750hp (1,304kW)	1,750hp (1,304kW)	1,750hp (1,304kW)	1,750hp (1,304kW)	1,750hp (1,304kW)
Power at rail:	1,250hp (932kW)	1,254hp (935kW)	1,250hp (932kW)	1,250hp (932kW)	1,250hp (932kW)
Tractive effort:	55,500lb (245kN)	57,440lb (256kN)	55,590lb (247kN)	55,500lb (245kN)	62,000lb (276kN)
Cylinder bore:	10in (250mm)	10in (250mm)	10in (250mm)	10in (250mm)	10in (250mm)
Cylinder stroke:	12in (300mm)	12in (300mm)	12in (300mm)	12in (300mm)	12in (300mm)
Transmission:	Electric	Electric	Electric	Electric	Electric
Main generator type:	EE822-10G, EE822-13G or EE822-16J	-	-	-	-
Main alternator type:	-	Brush BA1005A	Brush BA1005A	Brush BA1005A	Brush BA1005A@
Aux generator type:	EE911/5C	-	-	-	-
Aux alternator type:	-	Brush BA606A	Brush BA606A	Brush BA606A	Brush BA606A
ETS alternator type:	-	Brush BAH701	-	-	-
Traction motor type:	EE538-1A	EE538-5A	EE538-5A	EE538-5A	EE538-5A
No. of traction motors:	6	6	6	6	6
Gear ratio:	53:18	59:16	59:16	59:16	59:16
Brake type:	Dual or Air	Dual or Air	Dual or Air	Air	Dual or Air
Brake force:	50 tonnes	50 tonnes	50 tonnes	50 tonnes	50 tonnes
Route availability:	5	5	5	5	7
Heating type:	Not fitted	Electric - index 30	Not fitted	Through wired	Not fitted
Multiple coupling type:	Blue star	Blue star	Blue star	Blue star, plus DRS	Blue star
Fuel tank capacity:	890gal (4,046lit)*	1,690gal (7,682lit)	1,690gal (7,682lit)	1,690gal (7,682lit)	1,690gal (7,682lit)
Lub oil capacity:	120gal (545lit)	120gal (545lit)	120gal (545lit)	120gal (545lit)	120gal (545lit)
Cooling water capacity:	160gal (727lit)	160gal (727lit)	160gal (727lit)	160gal (727lit)	160gal (727lit)
Sanding equipment:	Pneumatic	Pneumatic	Pneumatic	Pneumatic	Pneumatic
Sub-class variations:	Basic locomotive	Refurbished loco fitted with electric train supply and CP7 bogies	Refurbished standard freight locos Phase 1 - 375xx Phase 2 - 376xx	Refurbished locos, for Nightstar and Eurostar tractor use. Fitted with UIC jumpers. Later sold to DRS and other operators, fitted with cab top-marker lights	Refurbished heavy-weight freight locos

Notes: ¤ ERTMS fitted locos
Depending on body style
* 37059/37069/37254 fitted 1,690gal (7,682lit) tanks
@ 37800 has GEC G564AZ

Left: *Class 37 front end equipment, positions in roughly the same places on all sub classes. This example still retains the nose end doors. 1: Air warning horns behind grilles, 2: Front end doors, 3: Former route indicator boxes, fitted with white marker light, 4: Headlight, 5: Clips to secure front doors open. 6: Red tail light, 7: Lamp bracket, 8: Main reservoir pipe (yellow), 9: Position of vacuum pipe, 10: Blue star multiple control jumper, 11: Air brake pipe (red), 12: Coupling hook and shackle, 13: Blue star multiple control jumper socket.* **Antony Christie**

Above Right: *The English Electric Type 3, later Class 37 were introduced from 1960 and became one of the most reliable BR designs. In 2023, over 60 were main line registered. The basic Class 37/0 fleet is represented by the two body designs, solid headcode and split box. When Colas Rail Freight was short of power some years ago, a number were purchased/leased from the preservation sector, including No. 37099. The loco usually operates Network Rail test trains, and is seen with its No. 1 end leading at Didcot.* **Spencer Conquest**

Key Facts

- One of the most successful diesel designs.
- After mid-life refurbishment and modification, eight sub-classes existed.
- Currently, the main operators of '37's are WCRC, DRS, Colas, LSL and ROG.
- A total of 32 locos are preserved.
- Some locos are fitted with drop-head Dellner or Tightlock couplings to enable attachment to modern EMU and DMUs.

Rolling Stock Review : 2023-2024

Class 37

2023 Fleet

Sub-class:	37/9
TOPS number range:	37901-37906
1957 BR number range	From main fleet
Former class codes:	37/0
Refurbished by:	BREL Crewe
Years introduced:	1986-1987
Wheel arrangement:	Co-Co
Weight:	120 tonnes
Height:	13ft 0¾in (3.96m)
Length:	61ft 6in (18.74m)
Width:	8ft 11⅝in (2.73m)
Wheelbase:	50ft 8in (15.44m)
Bogie wheelbase:	13ft 6in (4.11m)
Bogie pivot centres:	37ft 2in (11.32m)
Wheel diameter:	3ft 7in (1.09m)
Min curve negotiable:	4 chains (80.4m)
Engine type:	Mirrlees MB275T §
Engine output:	1,800hp (1,340kW)
Power at rail:	1,300hp (940kW)
Tractive effort:	62,680lb (279kN)
Cylinder bore:	10½in (275mm)
Cylinder stroke:	12¼in (305mm)
Maximum speed:	80mph (129km/h)
Brake type:	Dual
Brake force:	50 tonnes
Route availability:	7
Heating type:	Not fitted
Multiple coupling type:	Blue Star
Main alternator type:	Brush BA1005A
Traction motor type:	EE538-5A
No of traction motors:	6
Gear ratio:	59:16
Fuel tank capacity:	1,690gal (7,682lit)
Cooling water capacity:	160gal (727lit)
Lub oil capacity:	120gal (545lit)
Sanding equipment:	Pneumatic
Sub-class variations:	Refurbished heavy weight freight locos, fitted with experimental engines 37901-904 - Mirrlees MB275T 37905-906 - Ruston RK270T

Above & Below: In the mid-1980s a batch of 31 standard Class 37/0s were modified at BREL Crewe Works to provide Electric Train Supply (ETS), to the increasing number of electrically heated coaches used on secondary services. The locos were re-classified 37/4 and are recognisable by the addition of electric heating cables and socket on the nose end. In 2023, a fleet of 13 Class 37/4s remain main line certified, operating with DRS, HNRC and Colas. Above is No. 37425 operated by DRS painted in 1990s BR Regional Railways blue stripe livery, seen from its No 2 end at York. Below, is Harry Needle Railroad operated No. 37405, used for spot hire contract operations and in 2023 displays HNRC orange livery, the loco is seen from its No. 1 end. **CJM / Cliff Beeton**

Left: No. 37418 is privately owned and in 2023 was main line certified operating with Derby-based Loram, being frequently used to power inspection saloon 975025 Caroline. This loco displays BR large logo livery with Highland stag logos on the cab sides. It is seen in September 2022 passing Dawlish Warren, with its No. 2 end leading. **CJM**

Right: As part of the Class 37 refurbishment project, 64 standard Class 37/0s were converted for the freight sector, re-classified as 37/5, numbered 37501-521/667-699. Eleven locos remain in service today. No. 37688 is part of the Crewe-based Loco Services Ltd fleet, and currently displays Railfreight Construction livery. The loco is a frequent operator on the main line powering charter passenger services on contract work. It is shown from its No. 2 end passing Dawlish Warren on 19 October 2022 leading a York to Paignton charter. The loco retains dual air/vacuum brake equipment. **CJM**

Class 37/0
37025	COL
37038	DRS
37057	COL
37059	DRS
37069	DRS
37099	COL
37116	COL
37175	COL
37207(1)	EPX
37218	DRS
37219	COL
37240	BOD
37254	COL
37259	DRS

Class 37/4
37401	DRS
37402	DRS
37403	DRS
37405	HNR
37407	DRS
37409	DRS
37418	PRI
37419	DRS
37421	COL
37422	DRS
37423	DRS
37424	DRS
37425	DRS

Class 37/5
37510	ROG
37516	WCR
37517(S)	WCR
37518	WCR
37521	LSL
37667	LSL
37668	WCR
37669	WCR
37676	WCR
37685	WCR
37688	LSL

Class 37/6
37601	EPX
37602(S)	DRS
37603(S)	HNR
37604(S)	HNR
37606(S)	PRI
37607	COL
37608	ROG
37609(S)	HNR
37610	COL
37611	EPX
37612	COL

Class 37/7
37703(S)	HNR
37706	WCR
37712(S)	WCR
37716	DRS
37800	ROG
37884	ROG

Class 37/9
37901	EPX
37905	EPX
37906	URL

Class 97/3
97301	(S)
97302	NRL
97303	NRL
97304	NRL

(S) Stored

(1) VLR Centre, Dudley Hybrid Electric Power development loco

Rolling Stock Review : 2023-2024 17

Class 37 and 97/3

Left: Carnforth-based West Coast Railway operates a fleet of seven Class 37/5s, which are used to power charter passenger services, as long as a generator or another loco is provided to give a train supply, or operate contract duties such as the annual RHTT services. The WCRC fleet are finished in company deep maroon livery with gold branding. No. 37668 is illustrated in October 2022 when it was on hire to Colas for use on an RHTT service. **CJM**

Left: Between 1994-1996, 12 Class 37/6 locos were introduced, rebuilt from Class 37/5s at BRML Doncaster, to power Channel Tunnel 'Nightstar' services within the UK. After the project was abandoned the fleet was progressively sold to Direct Rail Services. Here the locos were modified and fitted with DRS style jumper equipment. In recent years, as other more modern power has taken over DRS duties, the 37/6s have been sold on and now operate with HNRC, Europhoenix, Rail Operations Group and Colas. No. 37601, owned by Europhoenix and leased to Rail Operations Group is seen at Crewe from its No. 2 end. The loco sports modified jumper equipment and a two section snow plough. **Spencer Conquest**

Left: A batch of 44 Class 37/7 'heavy weight' freight locos were refurbished from Class 37/0s between 1986-1987 to power heavy freight services, these carried extra ballast weights to improve adhesion. The locos were numbered in the 37/7 (phase 1 locos) and 37/8 (phase 2 locos) series. Today, six remain, with four operational, one works with West Coast Railway, two with Europhoenix, working with Rail Operations Group and one with DRS. No. 37800, is shown in full ROG livery from its No. 1 end. This loco sports a drop head Tightlock coupling for attachment to multiple unit stock for depot to depot and depot to scrap yard movements. The 37/7s are air brake only. **Spencer Conquest**

Left: In 2008, four standard Class 37s, were modified as Class 97/3s for use by Network Rail and fitted with Cambrian Coast route, European Rail Traffic Management System (ERTMS) cab signalling equipment. As well as allowing the powering of engineering trains over the then recently re-signalled route, the locos also operate any non multiple unit passenger or freight services over the equipped line. The four are otherwise standard Class 37/0s. All carry Network Rail yellow livery and are based at the Railway Technical Centre, Derby. They can also be found out-based at Shrewsbury depot. No. 97303 is seen at Cardiff powering a Network Rail test train. These locos retain dual brake equipment. **Spencer Conquest**

Class 43 - HST

The Class 43 HST power car fleet is quickly coming to an end, with the Great Western and CrossCountry sets being phased out of service by the end of 2023/24. The Scottish sets will continue until new train orders are made and the trains delivered. Network Rail/ Colas operated power cars will continue for the foreseeable future.

Above: After finishing long distance work on Great Western, 52 Class 43 power cars were overhauled and transferred to ScotRail. In 2023, 50 remain in traffic, based at Edinburgh Haymarket, they operate the Inter7City (I7C) services on the Edinburgh/ Glasgow to Inverness/Aberdeen corridor. The fleet carries pictogram-based I7C livery, No. 43003, the oldest operational member, is shown at Haymarket. **CJM**

Above: The Great Western fleet is rapidly reducing in 2023, with the plan to withdraw all HST 'Castle' sets by the end of 2024. The fleet is allocated to Plymouth Laira depot, operating with the four-vehicle 'Castle' sets. All sport green GW green livery and all carry cast 'Class 255 Castle' nameplates. Above, No. 43154 Compton Castle is seen at Newport, working a Penzance to Cardiff service. The inset image shows the nameplate Cardiff Castle as applied to No. 43187. Some slight livery variations exist within the class, mainly is respect of the position of running numbers, a small number carry promotional branding. **CJM / Antony Christie**

Class	43
Number range:	43003-43484
Built by:	BREL Crewe
Refurbished:	Brush Traction, Loughborough
Years introduced:	1976-1982
Refurbished:	2006-2009
Wheel arrangement:	Bo-Bo
Maximum speed:	125mph (201km/h)
Length:	58ft 5in (17.80m)
Height:	12ft 10in (3.91m)
Width:	8ft 11in (2.72m)
Weight:	70.25 tonnes
Wheelbase:	42ft 4in (12.90m)
Bogie wheelbase:	8ft 7in (2.62m)
Bogie pivot centres:	33ft 9in (10.29m)
Wheel diameter (new):	3ft 4in (1.02m)
Min curve negotiable:	4 chains (80.46m)
Engine type:	MTU 16V4000 R41R*
Engine output:	2,250hp (1,678kW) at 1,500rpm
Power at rail:	1,770hp (1,320kW)
Tractive effort:	17,980lb (80kN)
Cont tractive effort:	10,340lb (46kN)
Cylinder bore:	6½in (165mm)
Cylinder stroke:	7½in (190mm)
Transmission:	Electric
Main alternator type:	Brush BA1001B
Traction motor type:	Brush TMH68-46
No. of traction motors:	4
Gear ratio:	59:23
Brake type:	Air
Brake force:	35 tonnes
Route availability:	5
Bogie type:	BP16
Heating type:	Electric - three phase 415V
Multiple coupling type:	Not supported
Fuel tank capacity:	1,030gal (4,682lit)
Lub oil capacity:	75gal (341lit)
Cooling water capacity:	163gal (741lit)
Sanding equipment:	Not fitted
Luggage capacity:	2.5 tonnes
Special fittings:	43013/014/423-484 fitted with buffers and hook
Notes:	* Some fitted with Paxman 12VP185 engine

Right: HST power car front end equipment. 1: Combined white marker and red tail light, 2: Headlight, 3: Warning horns behind grille, 4: Emergency coupling lug behind cover. 5: High-level marker light. Some vehicle retain three section light clusters with individual marker/tail lights. **Antony Christie**

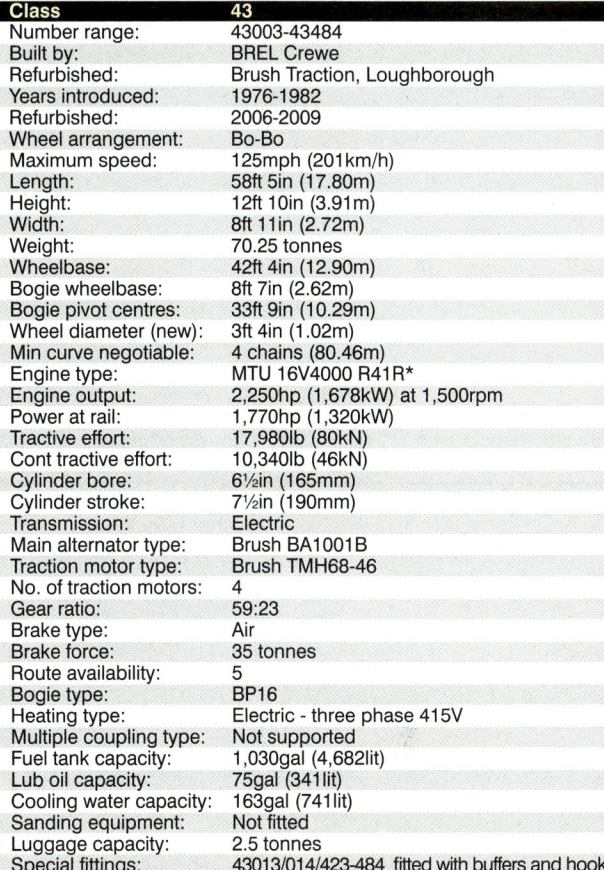

Above: In mid-2023, Arriva CrossCountry operated two or three HST rakes, based at Plymouth Laira, the sets operate on the Plymouth-Leeds-Newcastle-Edinburgh corridor. All vehicles carry standard XC, except No. 43184 which carries InterCity 'Executive colours. Standard-liveried No. 43378 is shown. CrossCountry plan to withdrawn their entire HST fleet from September 2023, power cars and coaches are likely to be scrapped. **CJM**

2023 Fleet

43003	ASR	43058	LSL	43144	ASR	43188	GWR
43004	GWR	43059	LSL	43145	ASR	43189	GWR
43007	AXC	43062	NRL	43146	ASR	43192	GWR
43008	AXC	43066	DAS	43147	ASR	43194	GWR
43009	GWR	43076	DAS	43148	ASR	43198	GWR
43010	GWR	43083	LSL	43149	ASR	43239	AXC
43012	ASR	43088	OLS	43150	ASR	43251	COL
43013	NRL	43092	GWR	43151	ASR	43257	COL
43014	NRL	43093	GWR	43152	ASR	43272	COL
43015	ASR	43094	GWR	43153	GWR	43274	COL
43016	GWR	43097	GWR	43154	GWR	43277	COL
43021	ASR	43098	GWR	43155	GWR	43285	AXC
43022	EXP	43122	GWR	43156	GWR	43290	NRL
43026	ASR	43124	ASR	43158	EXP	43296	HRA
43027	GWR	43125	ASR	43160	OLS	43299	NRL
43028	ASR	43126	ASR	43162	GWR	43301	AXC
43029	GWR	43127	ASR	43163	ASR	43303	AXC
43031	ASR	43128	ASR	43164	ASR	43304	AXC
43032	ASR	43129	ASR	43168	ASR	43308	HRA
43033	ASR	43130	ASR	43169	ASR	43321	AXC
43034	ASR	43131	ASR	43170	EXP	43357	AXC
43035	ASR	43132	ASR	43172	GWR	43366	AXC
43036	ASR	43133	ASR	43175	ASR	43378	AXC
43037	ASR	43134	ASR	43176	ASR	43423	HRA
43040	GWR	43135	ASR	43177	ASR	43465	HRA
43042	GWR	43136	ASR	43179	ASR	43467	HRA
43046	LSL	43137	ASR	43181	ASR	43468	HRA
43047	LSL	43138	ASR	43182	ASR	43480	HRA
43049	LSL	43139	ASR	43183	ASR	43484	HRA
43052	DAS	43141	ASR	43184	AXC		
43054	DAS	43142	ASR	43186	GWR		
43055	LSL	43143	ASR	43187	GWR		

Key Facts

- Production power-car fleet derived from 1972 Class 41 (252) prototype.
- In 2023 operated by GWR, ASR, CrossCountry, Network Rail, Colas, Loco Services Limited and Rail Adventure.
- Eight power carry standard draw gear, and now operated by Rail Adventure and Network Rail.
- All power cars originally fitted with Paxman 'Valenta' 12RP200L engines.

Class 43 - HST

Above & Left: During 2020, Locomotive Services Ltd (LSL), based in Crewe, rolled-out an overhauled HST set as the 'Midland Pullman', fitted with a luxury all first class Pullman interior. Externally the train was painted in a version of the 1960s Midland 'Blue' Pullman in Nankin Blue. The set, based at Crewe, is used on high-end charter services. Four power cars, fitted with Paxman engines (43046, 43047, 43055 and 43059) have modernised front ends to incorporate a roof level marker light, obviating the need for a yellow end. Led by power car No. 43046, the Midland Pullman set is seen (above) between Torre and Newton Abbot in July 2022. LSL also operate several other Class 43s, including No. 43049, which in early 2022 was overhauled and repainted in BR InterCity Swallow livery by Leeds Neville Hill depot, It was also named Leeds Neville Hill. This power car is sometimes used as a 'stand-in' for 'Midland Pullman' operations. Both: **CJM**

Left: In July 2022, CrossCountry HST power car No. 43384 was out-shopped by its home depot of Laira in InterCity 125 'Executive' colours, to mark 40 years of HST deployment on the NorthEast-SouthWest cross-country route. The power car was officially renumbered back to its original identity of 43184 and also carries its original Class 253 set No. 253051, as applied when first delivered in February 1982. No. 43184 only sees limited operation on the passenger duties and often spends weeks on end at Laira depot, or as in Spring 2023 used for drivers route refreshers for diversion workings. In early 2023 it was named Laira Diesel Depot in recognition of the long association that the Plymouth depot has had with the 125 fleet. In this view it is seen with fellow named power car No. 43366 HST 40 near Cullompton running light from Laira to Neville Hill on 8 April 2023. To mark the end of XC HST operations in September 2023, two further repaints into heritage liveries took place to Nos. 43207 (43007) and 43208 (43008). **CJM**

Class 43 - HST

Right: In April 2021 a partnership was set up between UK based Hanson & Hall and German operator RailAdventure GmbH, purchasing eight Class 43 power cars. Six carry standard draw gear at the leading end were overhauled by Arlington Fleet Services, Eastleigh, being coupled 'back-to-back' to work as three double-locomotives. Originally it was announced that the latest translator technology would be fitted at their inner ends, to enable coupling to and hauling multiple different types of stock, by mid 2023 this has not been done. The loco pairs are used to haul stock movements in the UK and are available to operate to/from European test tracks. Of the six purchased, Nos. 43296 and 43308 are considered as a source of spare parts. The operational pairs carry international numbers. No. 92 70 00 43465-8 with No. 92 70 00 43468-2 are seen at Long Marston in June 2022. In 2022-2023 the power twins have been used for the delivery of Class 555 and 777 stock as well as some departmental freight train operations. **CJM**

Right: With a significant number of spare Class 43s after their replacement on GWR, LNER and EMR services, new operators emerged. Colas Rail, the traction and crew provider for a number of Network Rail test trains, took over some to operate services. Most kept their previous LNER or EMR liveries with Colas branding, but in 202 No. 43277 emerged from at overhaul at the South Devon Railway, Buckfastleigh, sporting a stunning orange and yellow Colas livery, with the orange blending through a black radiator panel into Network Rail yellow at the inner end. No. 43277 is seen in company with Colas branded LNER-liveried No. 43251 at Attenborough on 2 March 2023.
Steve Donald

New Measurement Train

Network Rail operates a New Measurement Train (NMT) formation, formed of ex-Mk3 hauled and HST coaches, it works over the entire main line system on a programmed basis. The train is powered by Class 43s. Usually three power cars are assigned to this role, Nos. 43013, 43014 and 43062 but in 2020-2022 two additional vehicles were taken on from the displaced LNER fleet to help out with while major upgrade work was done to the three usual vehicles including the installation of ERTMS equipment. Nos. 43290 and 43299 were the additional power cars. These retain their LNER red swirl colours with Network Rail branding added. Right upper, a very tatty looking 43290 is recorded. Right below is the full NMT set powered by dedicated power car No. 43062, which sports a high level marker light and nose mounted camera equipment. The three dedicated power cars sport Network Rail yellow livery. Both: **CJM**

Class 47

Sub-class:	47/0 & 47/3	47/4	47/7	47/4 (47/8)
Present number range:	47237-47270, 47355-47375	47488-47614	47701-47787	47798, 47802-47854
Previous number range:	D1521-D1998	From Class 47/0 fleet	From Class 47/4 fleet	From Class 47/4 main fleet
Former class code:	27/2	27/2	47/4	47/4
Built by:	Brush, Loughborough and BR Crewe Works	Brush, Loughborough	Modified by BR depots	Rebuilt BREL Crewe or BR depots
Years introduced:	1965-1966	As 47/4 1974, 1980	1993-1995	1989-1995
Wheel arrangement:	Co-Co	Co-Co	Co-Co	Co-Co
Maximum speed:	95mph (153km/h) some 75mph (121km/h)	95mph (153km/h)	95mph (153km/h)	95mph (153km/h)
Length:	63ft 6in (19.35m)	63ft 6in (19.35m)	63ft 6in (19.35m)	63ft 6in (19.35m)
Height:	12ft 10^3/$_8$in (3.92m)	12ft 10^3/$_8$in (3.92m)	12ft 10^3/$_8$in (3.92m)	12ft 10^3/$_8$in (3.92m)
Width:	9ft 2in (2.79m)	9ft 2in (2.79m)	9ft 2in (2.79m)	9ft 2in (2.79m)
Weight:	111-121 tonnes	120-125 tonnes	119-121 tonnes	124 tonnes
Wheelbase:	51ft 6in (15.69m)	51ft 6in (15.69m)	51ft 6in (15.69m)	51ft 6in (15.69m)
Bogie wheelbase:	14ft 6in (4.42m)	14ft 6in (4.42m)	14ft 6in (4.42m)	14ft 6in (4.42m)
Bogie pivot centres:	37ft 0in (11.28m)	37ft 0in (11.28m)	37ft 0in (11.28m)	37ft 0in (11.28m)
Wheel diameter (new):	3ft 9in (1.14m)	3ft 9in (1.14m)	3ft 9in (1.14m)	3ft 9in (1.14m)
Min curve negotiable:	4 chains (80.46m)	4 chains (80.46m)	4 chains (80.46m)	4 chains (80.46m)
Engine type:	Sulzer 12LDA28C	Sulzer 12LDA28C	Sulzer 12LDA28C	Sulzer 12LDA28C
Engine output:	2,580hp (1,924kW)	2,580hp (1,924kW)	2,580hp (1,924kW)	2,580hp (1,924kW)
Power at rail:	2,080hp (1,551kW)	2,080hp (1,551kW)	2,080hp (1,551kW)	2,080hp (1,551kW)
Tractive effort:	60,000lb (267kN)	60,000lb (267kN)	60,000lb (267kN)	60,000lb (267kN)
Cylinder bore:	11in (270mm)	11in (270mm)	11in (270mm)	11in (270mm)
Cylinder stroke:	14in (350mm)	14in (350mm)	14in (350mm)	14in (350mm)
Transmission:	Electric	Electric	Electric	Electric
Main generator type:	Brush TG160-60 Mk4, or TG172-50 Mk1	Brush TG160-60 Mk4, or TG172-50 Mk1	Brush TM172-50 Mk1A	Brush TG160-60 Mk4, or TG172-50 Mk1
Aux generator type:	Brush TG69-20 or Brush TG69-28Mk2	Brush TG69-20 or Brush TG69-28Mk2	Brush TG69-20 or Brush TG69-28Mk2	Brush TG69-20 or Brush TG69-28Mk2
ETS Alternator type:	-	Brush BL100-30	Brush BL100-30	Brush BL100-30
Traction motor type:	Brush TM64-68	Brush TM64-68	Brush TM64-68	Brush TM64-68
No. of traction motors:	6	6	6	6
Gear ratio:	66:17	66:17	66:17	66:17
Brake type:	Dual or Air	Dual	Dual or Air	Dual or Air
Brake force:	61 tonnes	61 tonnes	61 tonnes	61 tonnes
Route availability:	6	7	6	6
Heating type:	Not fitted	Electric - index 66	Electric - index 66	Electric - index 66
Multiple coupling type:	Not fitted	Not fitted	Not fitted - TDM wired	Not fitted
Fuel tank capacity:	720-1,295gal (3,273-5,887lit)	720-1,295gal (3,273-5,887lit)	1,295gal (5,887lit)	1,295gal (5,887lit)
Lub oil capacity:	190gal (864lit)	190gal (864lit)	190gal (864lit)	190gal (864lit)
Cooling water capacity:	300gal (1,364lit)	300gal (1,364lit)	300gal (1,364lit)	300gal (1,364lit)
Sanding equipment:	Not fitted	Not fitted	Not fitted	Not fitted
Sub-class variations:	47/0 - Original locomotives 47/3 - Original no heat locos	Constructed with either dual (steam and electric) or electric train heating	Modified 47/4s with RCH Time Division Multiplex push-pull equipment	Locomotives refurbished for passenger services, fitted with ETS, refurbished from 47/4 fleet
Notes:	Engine output originally 2,750hp			

2023 Fleet

Class 47/0		Class 47/4		Class 47/7		47746	WCR	47787(S)	WCR	47812	WCR	47832	WCR
47237	WCR	47501	LSL	47703	HNR	47749	GBR	**Class 47/4**		47813	WCR	47848	WCR
47245	WCR	47593	LSL	47712	LSL	47760	WCR	47802	WCR	47815	WCR	47851	WCR
47270§	WCR	47614	LSL	47714	HNR	47768	WCR	47804	WCR	47818(S)	AFG	47854	WCR
				47715	HNR	47772	WCR	47805	LSL	47826	WCR		
Class 47/3				47727	GBR	47773	VTN	47810	LSL	47828§	LSL	§ Privately owned	
47375	EXP			47739	GBR	47786	WCR			47830	FLR		

Above: *The basic Class 47/0 'standard' design of Brush Type 4 loco now consists of just three main line certified examples, operated by Carnforth-based, West Coast Railway. No. 47237 is shown from its No. 1 end, carrying standard WCR maroon livery. This loco also sports green spot and DRS type control jumper sockets on the front end. This loco retains its dual vacuum/air brake capability.* **Antony Christie**

Key Facts

- Originally the largest fleet of BR main line locos, with 512 built.
- Backbone of passenger/freight services from the mid-1960s to 2000s.
- Three main types built, 47/0 - basic loco, steam heat, 47/3 no heat and 47/4 having dual or electric train heating. Several extra sub-classes introduced.
- Built with vacuum brakes, later converted to dual air/vacuum braking, many now air only.
- GBRf locos 47739 and 47749 fitted with drop-head Dellner couplings for direct coupling to modern multiple unit stock.
- Considerable number of locos preserved, some are certified for main line operation.

Class 47

Right: In 2023, Freightliner still retains one operational Class 47 on its roster, No. 47830 Beeching's Legacy. Painted in two-tone green, the loco is usually kept at Crewe Basford Hall, and normally used for driver's route training. It is occasionally used to operate short Freightliner trains between depots. It is seen light loco at Crewe on 3 February 2022. **Spencer Conquest**

Below: During 2022, Loco Services Ltd, returned Class 47/7, No. 47712 (one of the original Edinburgh-Glasgow push-pull locos) to full main line operation and restore its push-pull capability with a Mk2 DBSO and Mk3 stock. Painted in ScotRail grey and blue livery and named Lady Diana Spencer, the loco is available or charter train use. It is seen from its No. 1 end at Crewe. **Spencer Conquest**

Left & Right: GB Railfreight Rail Services division operate three Class 47/7s, for rolling stock movements. Nos. 47739 and 47749 are fitted with drop-head Dellner couplings and jumper sockets allowing coupling direct to modern MUs, without the need of a translator vehicle. No. 47739 in its unique blue livery is seen right, while left, the drop-head coupling is shown in the raised position. Both: **CJM**

Right Middle & Right Below: The most numerous sub-class of '47' fleet still in operation are the electric train supply (ETS) or electric train heat (ETH) fitted Class 47/4s, covering locos numbered in the 474xx, 475xx, 477xx and 478xx series. These locos come from a diverse background and thus have a number of modifications and different front end equipment styles, especially in terms of jumper cables for remote and multiple working. Several locos carry two or thee section mini-snowploughs and some have revised buffer beam surrounds with or without step plates. Liveries are also diverse, ranging from 1960s two-tone BR loco green with small yellow panels through various BR livery styles to those of the private operators. In the right middle image No. 47804 operated by West Coast Railway displays WCRC maroon livery with gold/yellow branding. The locos No. 1 end is on the left. Right below is Loco Services Ltd rail-blue liveried No. 47614, which is sometimes referred to as No. 47853, this is another number it once carried. This loco is air brake only and carries a 'green spot' jumper socket on the front end and a three section snowplough. The loco is seen from its No. 1 end. Both: **CJM**

Rolling Stock Review : 2023-2024

Class 50

For history and pictures of this class, see *Modern Locomotives Illustrated* Issue 193

Class:	50
Number range:	50007-50050
Previous number range:	D400-D449
Former class codes:	27/3
Built by:	English Electric, Vulcan Foundry
Years introduced:	1967-1968
Wheel arrangement:	Co-Co
Maximum speed:	100mph (161km/h)
Length:	68ft 6in (20.88m)
Height:	12ft 10¾in (3.92m)
Width:	9ft 1¼in (2.77m)
Weight:	117 tonnes
Wheelbase:	56ft 2in (17.12m)
Bogie wheelbase:	13ft 6in (4.11m)
Bogie pivot centres:	42ft 8in (13.00m)
Wheel diameter (new):	3ft 7in (1.09m)
Min curve negotiable:	4 chains (80.4m)
Engine type:	English Electric 16CSVT
Engine output:	2,700hp (2,013kW)
Power at rail:	2,070hp (1,543kW)
Tractive effort:	48,500lb (216kN)
Cylinder bore:	10in (254mm)
Cylinder stroke:	12in (305mm)
Transmission:	Electric
Main generator type:	EE840-4B
Aux generator type:	EE911-5C
ETS generator type:	EE915-1B
Traction motor type:	EE538-5A
No. of traction motors:	6
Gear ratio:	53:18
Brake type:	Dual
Brake force:	59 tonnes
Route availability:	6
Heating type:	Electric - index 61
Multiple coupling type:	Orange square
Fuel tank capacity:	1,055gal (4,800lit)
Lub oil capacity:	130gal (591lit)
Cooling water capacity:	280gal (1,274lit)
Sanding equipment:	Electro-pneumatic, later isolated

2023 Fleet

50007	GBR
50008	HRA
50044	50A
50049	GBR
50050	BOD

Right Top: *Five Class 50s are certified for main line use in 2023, with two, Nos. 50007 and 50049 owned by the Class 50 Alliance and operated in partnership with GB Railfreight. They are used for spot hire and charter workings and based at the Severn Valley Railway. The pair carry GBRf livery. No. 50007* Hercules *is seen passing Daventry, on spot hire work for Alstom/Porterbrook.* **CJM**

Right Middle: *No. 50008* Thunderer *is owned by Hanson & Hall Rail, the loco carries Hanson & Hall grey livery, and is used to provided a spot hire service to the rail industry, Often being involved in stock movements. No. 50008 is seen at on the outskirts of Derby on 15 January 2023.* **Antony Christie**

Right Lower: *Owned by the Class 50 Fund and based on the Severn Valley Railway, No. 50044* Exeter *has been cosmetically 'un-refurbished' It is mainline certified and can operate charter services as required. To conform with main line standards a high-intensity headlight has been fitted into one of the original tail light spaces. The loco carries BR rail blue livery.* **Antony Christie**

Key Facts

- Production design based on English Electric DP2 prototype.
- The 50 locos were originally used over the northern section of the West Coast Main Line, later transferred to the Western Region.
- Replaced Class 52 'Western' locos on Western Region and later worked on the Waterloo-Salisbury-Exeter route.
- In 2023 five are privately owned with a 100mph (161km/h) capability for 'spot hire'. Two, Nos. 50007/049 are operated by GB Railfreight.

Above: *No. D400 (50050), currently owned by Nottingham-based Boden Rail is fully main line certified and is used for spot hire work if required. In 2023 the loco was up for sale as an operational loco, it currently carries BR rail blue livery, with the Fearless name. The loco is shown from its No. 2 end at Plymouth powering a charter train on 24 August 2022.* **Antony Christie**

Class 56

2023 Fleet

Class 56/0			
56049 COL	56091 DCR	56113 COL	56303 GBR
56051 COL	56094 COL	56115 EXP	
56078 COL	56096 COL	56117 EXP	
56081 GBR	56098 GBR		
56087 COL	56101 EXP	**Class 56/3**	
56090 COL	56103 DCR	56301 UKR	
	56105 COL	56302 COL	

Sub-class:	56/0	56/3
Number range:	56049-56117	56301-56303
Previous number range:	56001-56135	From Class 56/0 fleet
Built by:	56001-56030 - Electroputere at Craiova in Romania 56031-56135 - BREL Doncaster/Crewe	
Refurbished by:	-	Brush Traction/FM Rail
Years introduced:	1976-1984	1976-1984
Years refurbished:	-	2006-2010
Wheel arrangement:	Co-Co	Co-Co
Maximum speed:	80mph (129km/h)	80mph (129km/h)
Length:	63ft 6in (19.35m)	63ft 6in (19.35m)
Height:	13ft 0in (3.96m)	13ft 0in (3.96m)
Width:	9ft 2in (2.79m)	9ft 2in (2.79m)
Weight:	126 tonnes	126 tonnes
Wheelbase:	47ft 10in (14.58m)	47ft 10in (14.58m)
Bogie wheelbase:	13ft 5⅞in (4.11m)	13ft 5⅞in (4.11m)
Bogie pivot centres:	37ft 8in (11.48m)	37ft 8in (11.48m)
Wheel diameter (new):	3ft 9in (1.14m)	3ft 9in (1.14m)
Min curve negotiable:	4 chains (80.46m)	4 chains (80.46m)
Engine type:	Ruston Paxman 16RK3CT	Ruston Paxman 16RK3CT
Engine output:	3,250hp (2,424kW)	3,250hp (2,424kW)
Power at rail:	2,400hp (1,790kW)	2,400hp (1,790kW)
Tractive effort:	61,800lb (275kN)	61,800lb (275kN)
Cylinder bore:	10in (250mm)	10in (250mm)
Cylinder stroke:	12in (300mm)	12in (300mm)
Transmission:	Electric	Electric
Main alternator type:	Brush BA1101A	Brush BA1101A
Aux alternator type:	Brush BAA602A	Brush BAA602A
Traction motor type:	Brush TMH73-62	Brush TMH73-62
No. of traction motors:	6	6
Gear ratio:	63:16	63:16
Brake type:	Air	Air
Brake force:	60 tonnes	60 tonnes
Route availability:	7	7
Heating type:	Not fitted	Not fitted
Multiple coupling type:	Red diamond	Red diamond
Fuel tank capacity:	1,150gal (5,228lit)	1,150gal (5,228lit)
Lub oil capacity:	120gal (546lit)	120gal (546lit)
Cooling water capacity:	308gal (1,400lit)	308gal (1,400lit)
Sanding equipment:	Pneumatic	Pneumatic
Sub-class variations:	As built	Refurbished

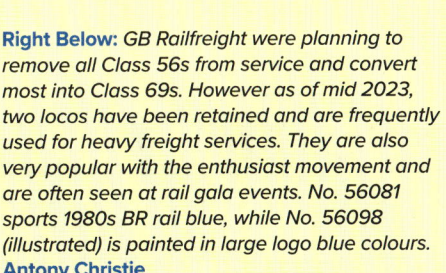

Above: Just 19 Class 56s of the original fleet of 135 remain in front line service, operating with Colas Rail Freight, GB Railfreight, DC Rail and UK Rail Leasing. The operational fleet consists only of BREL built examples. In DC Rail Freight grey with with white bodyside branding No. 56091 is shown from its No. 2 end. **Stuart Hood**

Right Middle: In 2023, 10 Class 56 were operated by Colas Rail Freight, working on their heavier freight flows, mainly in the north Midlands and North West. They sport Colas orange and yellow livery, off-set by a black upper band. The locos are on paper allocated to Nottingham Eastcroft depot. No. 56078 is recorded at Stafford in mid-2022.
Spencer Conquest

Right Below: GB Railfreight were planning to remove all Class 56s from service and convert most into Class 69s. However as of mid 2023, two locos have been retained and are frequently used for heavy freight services. They are also very popular with the enthusiast movement and are often seen at rail gala events. No. 56081 sports 1980s BR rail blue, while No. 56098 (illustrated) is painted in large logo blue colours.
Antony Christie

Key Facts

- Originally 135 locos were built between 1976-1984 for Railfreight operations.
- Based on the Class 47 design, with revised power unit and alternator.
- The first 30 locos were built in Romania by Electoputere.
- After privatisation in 1996, the fleet fell from favour, replaced by Class 66s.
- In 2023 the fleet is operated by Colas Rail Freight, GB Railfreight, DC Rail Freight and UK Rail Leasing.
- A total of 16 Class 56s are currently being rebuilt as Class 69s for GB Railfreight by Progress Rail, Longport.

Class 57

For history and pictures of this class, see *Modern Locomotives Illustrated* Issue 182

Sub-class:	57/0	57/3	57/6	57/6
Number range:	57001-57012	57301-57316	57601	57602-57605
Rebuilt by:	Brush Traction	Brush Traction	Brush Traction	Brush Traction
Originally built by:	Brush	Brush	Brush	Brush
Years introduced – as Class 47:	1962-1964	1962-1964	1962-1964	1962-1964
Years introduced – as Class 57:	1998-1999	2002-2005	2001	2002-2003
Wheel arrangement:	Co-Co	Co-Co	Co-Co	Co-Co
Maximum speed:	75mph (121km/h)	95mph (153km/h)	95mph (153km/h)	95mph (153km/h)
Length:	63ft 6in (19.35m)	63ft 6in (19.35m)	63ft 6in (19.35m)	63ft 6in (19.35m)
Height:	12ft 10⅜in (3.92m)	12ft 10⅜in (3.92m)	12ft 10⅜in (3.92m)	12ft 10⅜in (3.92m)
Width:	9ft 2in (2.79m)	9ft 2in (2.79m)	9ft 2in (2.79m)	9ft 2in (2.79m)
Weight:	120.6 tonnes	117 tonnes	121 tonnes	117 tonnes
Wheelbase:	51ft 6in (15.70m)	51ft 6in (15.70m)	51ft 6in (15.70m)	51ft 6in (15.70m)
Bogie wheelbase:	14ft 6in (4.41m)	14ft 6in (4.41m)	14ft 6in (4.41m)	14ft 6in (4.41m)
Bogie pivot centres:	37ft 0in (11.28m)	37ft 0in (11.28m)	37ft 0in (11.28m)	37ft 0in (11.28m)
Wheel diameter (new):	3ft 9in (1.14m)	3ft 9in (1.14m)	3ft 9in (1.14m)	3ft 9in (1.14m)
Min curve negotiable:	4 chains (80.46m)	4 chains (80.46m)	4 chains (80.46m)	4 chains (80.46m)
Engine type:	General Motors 645-12E3	General Motors 645-12F3B	General Motors 645-12E3	General Motors 645-F3B-12
Engine output:	2,500hp (1,864kW)	2,750hp (2,051kW)	2,500hp (1,864kW)	2,750hp (2,051kW)
Power at rail:	2,025hp (1,510kW)	2,200hp (1,641kW)	2,025hp (1,510kW)	2,200hp (1,641kW)
Tractive effort:	55,000lb (244.6kN)	55,000lb (244.6kN)	55,000lb (244.6kN)	55,000lb (244.6kN)
Cylinder bore:	9¹⁶⁄₁₆in (230mm)	9¹⁶⁄₁₆in (230mm)	9¹⁶⁄₁₆in (230mm)	9¹⁶⁄₁₆in (230mm)
Cylinder stroke:	10in (250mm)	10in (250mm)	10in (250mm)	10in (250mm)
Transmission:	Electric	Electric	Electric	Electric
Main alternator type:	Brush BA1101A	Brush BA1101A	Brush BA1101E	Brush BA1101G
Aux alternator type:	Brush BAA602A	Brush BAA602A	Brush BAA602A	Brush BAA602A
ETS alternator type:	-	Brush BAA	Brush	Brush
Traction motor type:	Brush TM68-46	Brush TM68-46	Brush TM68-46	Brush TM68-46
No. of traction motors:	6	6	6	6
Gear ratio:	66:17	66:17	66:17	66:17
Brake type:	Air	Air	Air	Air
Brake force:	80 tonnes	60 tonnes	60 tonnes	60 tonnes
Route availability:	6	6	6	6
Heating type:	Not fitted	Electric - index - 100	Electric - index - 95	Electric - index - 100
Multiple control type:	DRS system	DRS system	Not fitted	Not fitted
Fuel tank capacity:	1,221gal (5,551lit)	1,295gal (5,887lit)	720gal (3,273lit)	1,295gal (5,887lit)
Lub oil capacity:	190gal (864lit)	190gal (864lit)	190gal (864lit)	190gal (864lit)
Cooling water capacity:	298gal (1,355lit)	298gal (1,355lit)	298gal (1,355lit)	298gal (1,355lit)
Sanding equipment:	Pneumatic	Pneumatic	Pneumatic	Pneumatic
Sub-class variations:	Porterbrook sponsored rebuild of Class 47 with rebuilt GM power units supplied by VMV	Porterbrook/Virgin funded '47' rebuilds for 'Thunderbird' work. Operated today by DRS, ROG & WCRC	Revised specification with prototype electric train supply	Rebuilt for use on Great Western sleeper services

Right: *The Class 57/0 front end layout. 1: Warning horns behind grille, 2: Out of use ariel connection, 3: Lamp bracket, 4: Marker light, 5: Tail light, 6: Headlight, 7: Jumper socket, 8: Main reservoir pipe (yellow), 9: Brake pipe (red), 10: Coupling hook and shackle.* **Antony Christie**

Below: *The Class 57 emerged following a major rebuild project, which saw 33 Class 47s rebuilt by Brush Traction, in partnership with Porterbrook Leasing. The '57s' incorporated upgraded General Motors 645 power units. The original project covered 12 Class 57/0s for Freightliner, these later transferred to DRS and WCR and in 2022 are on the books of WCR, DRS and LSL. The design, incorporated a 645-12E3 engine with a Brush alternator, delivering 2,500hp (1,864kW). No. 57012 operated by West Coast Railway is sen passing Crewe. All locos are air brake only.* **Spencer Conquest**

Class 57

2023 Fleet

Class 57/0		Class 57/3			
57001	WCR	57301	DRS	57312	DRS
57002	DRS	57302	LSL	57313	WCR
57003	LSL	57303	DRS	57314	WCR
57005	WCR	57304	DRS	57315	WCR
57006	WCR	57305	DRS	57316	WCR
57007	LSL	57306	DRS	**Class 57/6**	
57008	WCR	57307	DRS	57601	WCR
57009	WCR	57308	DRS	57602	GWR
57010	WCR	57309	DRS	57603	GWR
57011	WCR	57310	DRS	57604	GWR
57012	WCR	57311	LSL	57605	GWR

Right Top & Right Middle Upper: The second sub-class of Class 57s were 16 Class 57/3s, originally converted as rescue and tractor power for Virgin Trains. They were later sold, with 12 going to Direct Rail Services and four to West Coast Railway. By 2023 10 locos are still operated by DRS, with some available for sale. Two are owned by LSL, and four are with WCR. Of the DRS fleet, some are maintained as rescue locos for Avanti West Coast. In the right top image, DRS No. 57304 *Lady Penelope* shows its drop-head Dellner coupling. The middle view shows DRS No. 57305 in all-blue livery on hire to Great Western to assist covering Class 57/6 rosters. Its Dellner coupling is retained but covered. It is seen from its No. 2 end. The WCR locos have had their drop-head Dellner couplings removed, but the recess space has been retained. LSL No. 57311 has had the drop-head removed and the space plated over. Both: **Antony Christie**

Right Middle Lower: Porterbrook Leasing in conjunction with Brush, did a speculative conversion of a Class 57 fitted with electric train supply in 2001. This trial, led to the building of the 16 Class 57/3s, see above, and four for GWR. No. 57601 which operated on GWR services for a while, was later purchased by West Coast Railway and is now part of their fleet carrying 'Pullman' colours and named *Windsor Castle*. The locomotive is seen from its No. 2 end at Reading. **Spencer Conquest**

Right Bottom: To power the 'Night Riviera' between London and Penzance, Great Western via Porterbrook, ordered four Class 57/6s Nos. 57602-57605. These are currently based at Penzance Long Rock depot, with Reading depot providing stabling at the London end. The '57/6s' are fitted with a General Motors 645-F3B-12 engine of 2,750hp (2,051kW). They have an electric train supply of index of 100 and provide hotel power for the Mk3 stock. No. 57605 *Totnes Castle* is shown, with an inset of the new style nameplate *Restormel Castle* applied to No. 57602. Both: **Antony Christie**

Key Facts

- Designed by a partnership of Brush Traction, Porterbrook Leasing and Freightliner to upgrade Class 47s with 'refreshed' EMD 645 series engines.

- No. 57601 was a prototype designed to supply head end power, worked for Great Western and now with WCRC.

- Two electric train supply fleets, Class 57/3 for Virgin Trains and 57/6 for Great Western were originally introduced.

- Virgin locos later sold to Direct Rail Services and in 2023 operate with DRS, LSL and WCR.

- DRS hold a contract to supply 57/3s to Avanti West Coast as 'Thunderbird' locos on an 'as required' basis.

Rolling Stock Review : 2023-2024

Class 59

For history and pictures of this class, see *Modern Locomotives Illustrated* Issue 173

Sub-class:	59/0	59/1	59/2
Number range:	59001-59005	59101-59104	59201-59206
GM model:	JT26CW-SS	JT26CW-SS	JT26CW-SS
Built by:	GM-EMD, La Grange, Illinois, USA	GM-DD, London, Ontario, Canada	GM-DD, London, Ontario, Canada
Years introduced:	1985-1989	1990	1994-1995
Wheel arrangement:	Co-Co	Co-Co	Co-Co
Maximum speed:	60mph (97km/h)	60mph (97km/h)	75mph (121km/h)
Length:	70ft 0½ in (21.34m)	70ft 0½ in (21.34m)	70ft 0½ in (21.34m)
Height:	12ft 10in (3.91m)	12ft 10in (3.91m)	12ft 10in (3.91m)
Width:	8ft 8¼ in (2.65m)	8ft 8¼ in (2.65m)	8ft 8¼ in (2.65m)
Weight:	121 tonnes	121 tonnes	121 tonnes
Wheelbase:	56ft 9in (17.29m)	56ft 9in (17.29m)	56ft 9in (17.29m)
Bogie wheelbase:	13ft 7in (4.14m)	13ft 7in (4.14m)	13ft 7in (4.14m)
Bogie pivot centres:	43ft 6in (13.25m)	43ft 6in (13.25m)	43ft 6in (13.25m)
Wheel diameter (new):	3ft 6in (1.06m)	3ft 6in (1.06m)	3ft 6in (1.06m)
Min curve negotiable:	4 chains (80.46m)	4 chains (80.46m)	4 chains (80.46m)
Engine type:	EMD 16-645E3C	EMD 16-645E3C	EMD 16-645E3C
Engine output:	3,000hp (2,237kW)	3,000hp (2,237kW)	3,000hp (2,237kW)
Power at rail:	2,533hp (1,889kW)	2,533hp (1,889kW)	2,533hp (1,889kW)
Tractive effort:	122,000lb (542kN)	122,000lb (542kN)	122,000lb (542kN)
Cylinder bore:	9 1/16 in (230mm)	9 1/16 in (230mm)	9 1/16 in (230mm)
Cylinder stroke:	10in (250mm)	10in (250mm)	10in (250mm)
Transmission:	Electric	Electric	Electric
Traction alternator:	EMD AR11	EMD AR11	EMD AR11
Companion alternator:	EMD D14A	EMD D14A	EMD D14A
Aux alternator:	EMD 3A8147	EMD 3A8147	EMD 3A8147
Traction motor type:	EMD D77B	EMD D77B	EMD D77B
No. of traction motors:	6	6	6
Gear ratio:	62:15	62:15	62:15
Brake type:	Air	Air	Air
Brake force:	69 tonnes	69 tonnes	69 tonnes
Route availability:	7	7	7
Heating type:	Not fitted	Not fitted	Not fitted
Multiple coupling type:	AAR (59, 66, 67, 69, 70, 73)	AAR (59, 66, 67, 69, 70, 73)	AAR (59, 66, 67, 69, 70, 73)
Fuel tank capacity:	1,000gal (4,546lit)	1,000gal (4,546lit)	1,000gal (4,546lit)
Lub oil capacity:	202gal (918lit)	202gal (918lit)	202gal (918lit)
Cooling water capacity:	212gal (964lit)	212gal (964lit)	212gal (964lit)
Sanding equipment:	Pneumatic	Pneumatic	Pneumatic
Sub-class variations:	Original fleet purchased by Foster Yeoman, now owned by Freightliner	Second batch purchased ARC Southern, now owned by Freightliner	Final derivative for National Power, now owned by Freightliner

Key Facts

- First US-built main line locos in England, ordered by Foster Yeoman to increase reliability of traction on Mendip aggregate traffic.

- ARC Southern and National Power later ordered near identical fleets. These were built in Canada, by 2023 all sub-classes are operated by Freightliner on Mendip flows.

- No. 59003 worked between May 1997 and June 2000 on a joint Foster Yeoman/DB contract in Germany, then sold to Heavy Haul Power International. Returned to UK in August 2014 after purchase by GB Railfreight.

- Nos. 59001 and 59201 carry brass US-style (non-operational) bells on one cab end.

Below: All four Class 59/0s carry Aggregate Industries livery. No. 59005 *Kenneth J Painter* approaches Reading. The stars, centrally positioned on the buffer beam above the coupling, represent the loco number, ie 5 stars = No. 59005. **Spencer Conquest**

Inset: Class 59/0 front end equipment. These five locos have slightly different equipment positions to members of Class 59/1 and 59/2. 1: Warning horns behind grille, 2: Non-operational US style bell, 3: Red tail light, 4: White marker light, 5: Twin headlights, 6: Lamp bracket, 7: Association of American Railroads (AAR) multiple control jumper socket, 8: Air brake pipe, 9: Coupling hook and shackle, 10: Main reservoir pipe. **CJM**

Class 59

2023 Fleet

59/0
59001 FLR
59002 FLR
59003 GBR
59004 FLR
59005 FLR

59/1
59101 FLR
59102 FLR
59103 FLR
59104 FLR

59/2
59201 FLR
59202 FLR
59203 FLR
59204 FLR
59205 FLR
59206 FLR

Above: The second importer of Class 59s was ARC-Southern (later part of the Hanson Group), who ordered four, classified Class 59/1. These had different front ends, with a BR Group Standard marker/headlight cluster. In 2023, the fleet operates as one common Mendip Rail pool with the 59/0s and 59/2s, owned by Freightliner. In Hanson livery, with cast number and nameplates. No. 59101 passes Kensington Olympia in August 2022. **CJM**

Right: National Power, purchased '59s' to work power station traffic in Yorkshire. Built in London, Canada, the fleet was later sold to EWS, later DB-Cargo and in September 2019 were sold to Freightliner, after they won the Mendip aggregate contract. No. 59206 in the latest Genesee & Wyoming/Freightliner livery, is shown from its silencer end near Swindon. **CJM**

Below: In May 1997, Foster Yeoman No. 59003 *Yeoman Highlander* was exported to Germany to operate a joint Foster Yeoman/DB contract in Eastern Europe. When this concluded, the loco was sold to Heavy Haul Power International (HHPI) and used in Eastern Europe. In 2014 it was sold to GB Railfreight, who returned it to the UK. In GBRf livery, it usually operates in and around the Westbury/Eastleigh area. The loco is seen at Dawlish. **CJM**

Rolling Stock Review : 2023-2024

Class 60

Class:	60
Number range:	60001-60100
Built by:	Brush Traction, Loughborough
Years introduced:	1989-1993
Wheel arrangement:	Co-Co
Maximum speed:	60mph (97km/h)
Length:	70ft 0in (21.33m)
Height:	12ft 11½in (3.95m)
Width:	8ft 8in (2.64 m)
Weight:	129 -131 tonnes
Wheelbase:	56ft 3⅛in (17.15m)
Bogie wheelbase:	13ft 6½in (4.13m)
Bogie pivot centres:	42ft 9¾in (13.04m)
Wheel diameter (new):	3ft 7in (1.09m)
Min curve negotiable:	4 chains (80.46m)
Engine type:	Mirrlees MB275T
Engine output:	3,100hp (2,311kW)
Power at rail:	2,415hp (1,800kW)
Tractive effort:	106,500lb (474kN)
Cylinder bore:	10¼in (275mm)
Cylinder stroke:	12¼in (305mm)
Main alternator type:	Brush BA1006
Aux alternator type:	Brush BAA702A
Traction motor type:	Brush TM2161A
No. of traction motors:	6
Gear ratio:	19:97
Brake type:	Air
Brake force:	74 tonnes (62 tonnes*)
Route availability:	8
Heating type:	Not fitted
Multiple control type:	Within class only
Fuel tank capacity:	990gal (4,500lit*)
Lub oil capacity:	220gal (1,000lit)
Cooling water capacity:	125gal (567lit)
Sanding equipment:	Pneumatic

* 60002-05/07/09/10/12/15/17/20/23/25/28/30/33/37/38/41/46/51/52/54-56/58/59/64/70/71/74/77/80/81/89-92/96-98, fitted with 1,150gal (5,228lit) fuel tanks, are thus heavier and have reduced brake force.

Above: *Class 60 front end equipment positions. 1: Warning horns, 2: Multiple control jumper socket (behind flap), 3: White marker light, 4: Headlight, 5: Red tail light, 6: Air brake pipe, 7: Main reservoir pipe, 8: Coupling hook and shackle, 9: Lamp bracket.* **CJM**

2023 Fleet

60001	DBC
60002	GBR
60007	DBC
60010	DBC
60011	DBC
60015	DBC
60017	DBC
60020	DBC
60021	GBR
60024	DBC
60026	GBR
60028	DCR
60029	DCR
60040	DBC
60044	DBC
60046	DCR
60047	GBR
60055	DCR
60056	DBC
60059	DBC
60062	DBC
60065	DBC
60066	DBC
60074	DBC
60076	GBR
60079	DBC
60085	GBR
60087	GBR
60092	DBC
60094	DBC
60095	GBR
60096	GBR

Left & Below: *A fleet of 100 Class 60s were introduced between 1989-1993, built by Brush Traction of Loughborough for BR Railfreight. Today only 32 remain active with DB-C, DCR and GB Railfreight. On the left DB-C No. 60044 is seen from its No. 1 end from the single grille side. Below, No. 60062 is shown from the two-grille side from the No. 2 end. This loco was given a promotional 'Steel on Steel' livery in 2022 to promote the UK steel industry. The Class 60s with their excellent haulage power are used on some of the heaviest trains, including oil and steel flows.*
CJM / Spencer Conquest

Key Facts

- Fleet of 100 state-of-the-art high output diesel-electric locos ordered by BR Railfreight as the answer to the US design Class 59.

- Privatisation and under DB control, a small number were refurbished as 'Super 60s' powering some of DBs heaviest freight flows.

- A batch of 10 were sold to Colas in 2014, refurbished as 'Super 60s' and later sold to GB Railfreight.

- In 2019, four locos were refurbished as 'Super 60s' and sold to DC Rail. In 2022 DC Rail purchased a large number of stored DB '60s' for possible refurbishment.

Class 60

Above & Right: *Freight operator DC Rail Freight, part of the Cappagh Group, originally purchased four Class 60s in 2019 for heavy freight flows. The locos were overhauled by DB, receiving 'Super 60' modifications. The fleet is allocated to Toton depot and maintained under contract by DB. In the image above, No. 60046 William Wilberforce shows DC Rail Freight grey with white branding, powering a rake of Cappagh-branded high-capacity box wagons at Cockwood Harbour. The image right below, shows No. 60028 painted in parent company Cappagh blue livery. Note this loco has received modified light clusters with a combined marker/tail light on either side.*
CJM / Spencer Conquest

Right Below and Right Bottom: *In 2014, DB sold 10 Class 60s to competitor freight business Colas Rail Freight. These were overhauled to 'Super 60' standards by DB Toton and repainted into Colas Rail Freight livery. The fleet powered the heaviest trains operated by the company. In June 2018, all 10 locos were again sold, this time to GB Railfreight, with the following year, a further four locos purchased direct from DB. Slowly, the fleet are being progressively repainted in various GB Railfreight schemes. Right, No 60026 Penyghent displays all over mid blue livery with GBRf bodyside branding. It is viewed from the cooler group end at Warrington Bank Quay. Right below, No. 60002 sports full GBRf blue and orange livery, it is viewed from the No. 2 or silencer end. This loco was named Graham Farish after the model company in 2020 and sports the companies branding on the non-driving cab side.*
Spencer Conquest (2) / CJM / Antony Christie

Rolling Stock Review : 2023-2024

Class 66

For history and pictures of this class, see Modern Locomotives Illustrated Issues 204 & 205

Sub-class	66/0	66/3 and 66/4	66/5	66/6
Number range:	66001-66249	66301-66316*, 66411-66434	66501-66599	66601-66625
Built by:	General Motors, EMDD, London, Canada	General Motors, EMDD, London, Canada	General Motors, EMDD, London, Canada	General Motors, EMDD, London, Canada
GM model:	JT-42-CWR	JT-42-CWR	JT-42-CWR	JT-42-CWR
Years introduced:	1998-2000	2003-2008	1999-2007	2000-2007
Wheel arrangement:	Co-Co	Co-Co	Co-Co	Co-Co
Design speed:	87.5mph (141km/h)	87.5mph (141km/h)	87.5mph (141km/h)	87.5mph (141km/h)
Maximum speed:	75mph (121km/h)	75mph (121km/h)	75mph (121km/h)	65mph (105km/h)
Length:	70ft 0½in (21.34m)	70ft 0½in (21.34m)	70ft 0½in (21.34m)	70ft 0½in (21.34m)
Height:	12ft 10in (3.91m)	12ft 10in (3.91m)	12ft 10in (3.91m)	12ft 10in (3.91m)
Width:	8ft 8¼in (2.65m)	8ft 8¼in (2.65m)	8ft 8¼in (2.65m)	8ft 8¼in (2.65m)
Weight:	126 tonnes	126 tonnes	126 tonnes	126 tonnes
Wheelbase:	56ft 9in (17.29m)	56ft 9in (17.29m)	56ft 9in (17.29m)	56ft 9in (17.29m)
Bogie wheelbase:	13ft 7in (4.14m)	13ft 7in (4.14m)	13ft 7in (4.14m)	13ft 7in (4.14m)
Bogie pivot centres:	43ft 6in (13.26m)	43ft 6in (13.26m)	43ft 6in (13.26m)	43ft 6in (13.26m)
Wheel diameter (new):	3ft 6in (1.06m)	3ft 6in (1.06m)	3ft 6in (1.06m)	3ft 6in (1.06m)
Min curve negotiable:	4 chains (80.46m)	4 chains (80.46m)	4 chains (80.46m)	4 chains (80.46m)
Engine type:	GM 12N-710G3B-EC	GM 12N-710G3B-U2	GM 12N-710G3B-EC§	GM 12N-710G3B-EC
Engine output:	3,300hp (2,460kW)	3,245hp (2,420kW)	3,300hp (2,460kW)	3,300hp (2,460kW)
Power at rail:	3,000hp (2,238kW)	3,000hp (2,238kW)	3,000hp (2,238kW)	3,000hp (2,238kW)
Tractive effort (max):	92,000lb (409kN)	92,000lb (409kN)	92,000lb (409kN)	105,080lb (467kN)
Tractive effort (cont):	58,390lb (260kN)	58,390lb (260kN)	58,390lb (260kN)	66,630lb (296kN)
Cylinder bore:	9^{1}/$_{16}$ (230mm)	9^{1}/$_{16}$ (230mm)	9^{1}/$_{16}$ (230mm)	9^{1}/$_{16}$ (230mm)
Cylinder stroke:	11in (279mm)	11in (279mm)	11in (279mm)	11in (279mm)
Traction alternator:	GM-EMD AR8	GM-EMD AR8	GM-EMD AR8	GM-EMD AR8
Companion alternator:	GM-EMD CA6	GM-EMD CA6	GM-EMD CA6	GM-EMD CA6
Traction motor type:	GM-EMD D43TR	GM-EMD D43TRC	GM-EMD D43TR	GM-EMD D43TR
No. of traction motors:	6	6	6	6
Gear ratio:	81:20	81:20	81:20	83:18
Brake type:	Air, Westinghouse PBL3	Air, Westinghouse PBL3	Air, Westinghouse PBL3	Air, Westinghouse PBL3
Brake force:	68 tonnes	68 tonnes	68 tonnes	68 tonnes
Bogie type:	HTCR Radial	HTCR Radial	HTCR Radial	HTCR Radial
Route availability:	7	7	7	7
Heating type:	Not fitted	Not fitted	Not fitted	Not fitted
Multiple coupling type:	AAR	AAR	AAR	AAR
Fuel tank capacity:	1,440gal (6,546lit)	1,133gal (5,150lit)	66501- 66584 1,440gal (6,546lit) 66585-66599 1,133gal (5,150lit)	66601-66622 1,440gal (6,546lit) 66623-66625 1,133gal (5,150lit)
Lub oil capacity:	202 gal (918lit)	202 gal (918lit)	202 gal (918lit)	202 gal (918lit)
Sanding equipment:	Pneumatic	Pneumatic	Pneumatic	Pneumatic
Special fittings:	EM2000 Q-Tron, GPS Combination coupler+, SSC	EM2000 Q-Tron, GPS low emission	EM2000 Q-Tron, GPS	EM2000 Q-Tron, GPS
Sub-class variations:	Standard DB-C	GBRf & Freightliner	Standard Freightliner	Freightliner modified gearing
Note:	+ 66001-002 unable to be fitted with Combination couplers	* Some locos differ in technical configuration	§ 66585-66599 have a GM 12N-710G3B-U2 engine and GM-EMD D43TRC traction motors	

Key Facts

- Structurally based on the Class 59, the Class 66s were first ordered by EWS after privatisation.

- Follow-on orders were placed by Freightliner, GBRf, Fastline and DRS.

- Later built locos, fitted with low emissions technology, with an extra bodyside (5th) door, as no through-way was possible to walk cab to cab.

- A number of different end light clusters have been used on different builds, or after upgrade work, most locos now have combined marker/tail light.

- Some DB and Freightliner locos have been exported to Mainland Europe (see table).

Above: *After the sale of most of the UK freight operations to US-based Wisconsin Central in 1996, new locos were ordered, consisting of 250 EMD JT-42-CWR (UK Class 66), a development of the Class 59. Originally painted in EWS maroon and gold, the fleet is now being repainted into DB red and grey. No. 66010 from its single grille side, shows the now standard DB red/grey colours, all but two (66001/002) of the DB fleet have swinghead knuckle couplers.* **CJM**

Left: *No. 66089 shows the two grill side of the loco and the original EWS maroon livery, still found on some locos. This loco retains the original light clusters.* **CJM**

Rolling Stock Review : 2023-2024

Class 66

66/7	66/8	66/9
66351-66360, 66701-66799	66846-66850	66951-66957
General Motors, EMDD, London, Canada & Caterpillar, Muncie, IN	General Motors, EMDD, London, Canada	General Motors, EMDD, London, Canada
JT-42-CWR	JT-42-CWR	JT-42-CWR
2001-2014 (2021)	Orig: 2004, 66/8: 2011	2004, 2008
Co-Co	Co-Co	Co-Co
87.5mph (141km/h)	87.5mph (141km/h)	87.5mph (141km/h)
75mph (121km/h)	75mph (121km/h)	75mph (121km/h)
70ft 0½in (21.34m)	70ft 0½in (21.34m)	70ft 0½in (21.34m)
12ft 10in (3.91m)	12ft 10in (3.91m)	12ft 10in (3.91m)
8ft 8¼in (2.65m)	8ft 8¼in (2.65m)	8ft 8¼in (2.65m)
126 tonnes	126 tonnes	126 tonnes
56ft 9in (17.29m)	56ft 9in (17.29m)	56ft 9in (17.29m)
13ft 7in (4.14m)	13ft 7in (4.14m)	13ft 7in (4.14m)
43ft 6in (13.26m)	43ft 6in (13.26m)	43ft 6in (13.26m)
3ft 6in (1.06m)	3ft 6in (1.06m)	3ft 6in (1.06m)
4 chains (80.46m)	4 chains (80.46m)	4 chains (80.46m)
GM 12N-710G3B-EC§	GM 12N-710G3B-EC	GM 12N-710G3B-U2
3,300hp (2,460kW)	3,300hp (2,460kW)	3,300hp (2,460kW)
3,000hp (2,238kW)	3,000hp (2,238kW)	3,000hp (2,238kW)
92,000lb (409kN)	92,000lb (409kN)	92,000lb (409kN)
58,390lb (260kN)	58,390lb (260kN)	58,390lb (260kN)
9¹¹⁄₁₆ (230mm)	9¹¹⁄₁₆ (230mm)	9¹¹⁄₁₆ (230mm)
11in (279mm)	11in (279mm)	11in (279mm)
GM-EMD AR8	GM-EMD AR8	GM-EMD AR8
GM-EMD CA6	GM-EMD CA6	GM-EMD CA6
GM-EMD D43TR	GM-EMD D43TR	GM-EMD D43TR
6	6	6
81:20	81:20	81:20
Air, Westinghouse PBL3	Air, Westinghouse PBL3	Air, Westinghouse PBL3
68 tonnes	68 tonnes	68 tonnes
HTCR Radial	HTCR Radial	HTCR Radial
7	7	7
Not fitted	Not fitted	Not fitted
AAR	AAR	AAR
66701-66717/ 66733-66746/ 66780-66789 - 1,440gal (6,546lit) 66718-66722 – 1,220gal (5,546lit) 66723-66732/66747-66749/ 66773-66799 1,133gal (5,150lit)	1,440gal (6,546lit)	1,133gal (5,150lit)
202 gal (918lit)	202 gal (918lit)	202 gal (918lit)
Pneumatic	Pneumatic	Pneumatic
EM2000 Q-Tron, GPS	EM2000 Q-Tron, GPS	EM2000 Q-Tron, GPS
66718-772 - low emission GBRf	Ex-Freightliner now with Colas Rail Freight	Freightliner Low-emission locos 66951-66952 low-emission demo locos

§ 66718-66746 have a GM 12N-710G3B-U2 engine and GM-EMD D43TRC traction motors. 66351-360, 66747-751/790-799 from Europe

Above: Class 66 front end layout. 1: Warning horns behind grille, 2: High level marker light, 3: Lamp bracket, 4: Association of American Railroad (AAR) multiple control jumper socket (cable kept inside loco, 5: Headlight, 6: Combined white marker and red tail light, 7: Air brake pipe, 8: Main reservoir pipe, 9: Coupling hook and shackle. The loco illustrated is DRS No. 66428. DB locos (except 66001/002), 10 of the GBRf fleet and five DRS locos carry 'swing-head' combination couplers. Many different light cluster designs can be found on the fleet, but the main items of equipment are in the same positions. **CJM**

Above: The second UK FOC to buy Class 66s was Freightliner, who took delivery of their first loco, a Class 66/5 in 1999. Eventually in excess of 100 were introduced with standard gearing, numbered in the 665xx and 66953-957 range. No. 66567, a standard design four-door loco, is seen from its silencer end, in original Freightliner green and yellow livery. **CJM**

Left Upper: DB, as with most freight operators, have a small number of locos decorated in promotional liveries. In 2022, No. 66004 emerged from the Toton paint-shop displaying this attractive 'Im a Climate Hero' green livery in promotion of it being powered by waste vegetable oil (HVO). The loco is seen from its radiator end at Didcot. **Spencer Conquest**

Left Lower: When built at the EMD plant in London, Ontario, Canada, the first two locos did not emerge from the main production line and were assembled slightly differently, in doing, they were not suitable for fitting of 'Swinghead' couplers in the subsequent installation program, thus Nos. 66001 and 66002 retain just a simple draw hook coupling. No. 66001 has subsequently been fitted with plumbing for fitting a LUL style 'trip-cock' for powering RHTT trains over the Chiltern/LUL tracks. The pair are also used, with an adapter coupling to attach to Dellner fitted stock on delivery from Europe, especially the Stadler 'Flirt' stock. No. 66001 is seen at Didcot attached to a Transport for Wales Class 756 in February 2023. **Spencer Conquest**

Class 66

2023 Fleet

66/0				
66001 DBC	66096 DBC	66189 DBC§	66421 DRS	
66002 DBC	66097 DBC	66190 DBC	66422 DRS	
66003 DBC	66098 DBC	66191 DBC*	66423 DRS	
66004 DBC	66099 DBC	66192 DBC	66424 DRS	
66005 DBC	66100 DBC	66193 DBC*	66425 DRS	
66006 DBC	66101 DBC	66194 DBC	66426 DRS	
66007 DBC	66102 DBC	66195 DBC	66427 DRS	
66009 DBC	66103 DBC	66196 DBC§	66428 DRS	
66010 DBC	66104 DBC	66197 DBC	66429 DRS	
66011 DBC	66105 DBC	66198 DBC	66430 DRS	
66012 DBC	66106 DBC	66199 DBC	66431 DRS	
66013 DBC	66107 DBC	66200 DBC	66432 DRS	
66014 DBC	66108 DRS	66201 DBC*	66433 DRS	
66015 DBC	66109 DBC	66202 DBC*	66434 DRS	
66017 DBC	66110 DBC	66203 DBC*		
66018 DBC	66111 DBC	66204 DBC*	**66/5**	
66019 DBC	66112 DBC	66205 DBC	66501 FLR	
66020 DBC	66113 DBC	66206 DBC	66502 FLR	
66021 DBC	66114 DBC	66207 DBC	66503 FLR	
66022 DBC*	66115 DBC	66208 DBC*	66504 FLR	
66023 DBC	66116 DBC	66209 DBC*	66505 FLR	
66024 DBC	66117 DBC	66210 DBC*	66506 FLR	
66025 DBC	66118 DBC	66211 DBC*	66507 FLR	
66026 DBC*	66119 DBC	66212 DBC*	66508 FLR	
66027 DBC	66120 DBC	66213 DBC*	66509 FLR	
66028 DBC	66121 DBC	66214 DBC*	66510 FLR	
66029 DBC	66122 DRS	66215 DBC*	66511 FLR	
66030 DBC	66123 DBC*	66216 DBC*	66512 FLR	
66031 DRS	66124 DBC	66217 DBC*	66513 FLR	
66032 DBC	66125 DBC	66218 DBC*	66514 FLR	
66033 DBC*	66126 DRS	66219 DBC*	66515 FLR	
66034 DBC	66127 DBC	66220 DBC§	66516 FLR	
66035 DBC	66128 DBC	66221 DBC	66517 FLR	
66036 DBC*	66129 DBC	66222 DBC*	66518 FLR	
66037 DBC	66130 DBC	66223 DBC*	66519 FLR	
66038 DBC*	66131 DBC	66224 DBC	66520 FLR	
66039 DBC	66133 DBC	66225 DBC*	66522 FLR	
66040 DBC	66134 DBC	66226 DBC*	66523 FLR	
66041 DBC	66135 DBC	66227 DBC§	66524 FLR	
66042 DBC*	66136 DBC	66228 DBC*	66525 FLR	
66043 DBC	66137 DBC	66229 DBC*	66526 FLR	
66044 DBC	66138 DBC	66230 DBC	66527 EXP	
66045 DBC*	66139 DBC	66231 DBC*	66528 FLR	
66047 DBC	66140 DBC	66232 DBC*	66529 FLR	
66049 DBC*	66142 DBC	66233 DBC*	66530 EXP	
66050 DBC	66143 DBC	66234 DBC*	66531 FLR	
66051 DBC	66144 DBC	66235 DBC*	66532 FLR	
66052 DBC*	66145 DBC	66236 DBC*	66533 FLR	
66053 DBC	66146 DBC*	66237 DBC*	66534 FLR	
66054 DBC	66147 DBC	66239 DBC*	66535 EXP	
66055 DBC	66148 DBC	66240 DBC*	66536 FLR	
66056 DBC	66149 DBC	66241 DBC*	66537 FLR	
66057 DBC	66150 DBC	66242 DBC*	66538 FLR	
66059 DBC	66151 DBC	66243 DBC	66539 FLR	
66060 DBC	66152 DBC	66244 DBC	66540 FLR	
66061 DBC	66153 DBC§	66245 DBC	66541 FLR	
66062 DBC*	66154 DBC	66246 DBC*	66542 FLR	
66063 DBC	66155 DBC	66247 DBC*	66543 FLR	
66064 DBC	66156 DBC	66248 DBC§	66544 FLR	
66065 DBC	66157 DBC§	66249 DBC§	66545 FLR	
66066 DBC§	66158 DBC		66546 FLR	
66067 DBC	66159 DBC§	**66/3**	66547 FLR	
66068 DBC	66160 DBC	66301 GBR	66548 FLR	
66069 DBC	66161 DBC	66302 GBR	66549 FLR	
66070 DBC	66162 DBC	66303 GBR	66550 FLR	
66071 DBC*	66163 DBC§	66304 GBR	66551 FLR	
66072 DBC*	66164 DBC	66305 GBR	66552 FLR	
66073 DBC	66165 DBC	66306 GBR	66553 FLR	
66074 DBC	66166 DBC*	66307 GBR	66554 FLR	
66075 DBC	66167 DBC	66308 GBR	66555 FLR	
66076 DBC	66168 DBC	66309 GBR	66556 FLR	
66077 DBC	66169 DBC	66310 GBR	66557 FLR	
66078 DBC	66170 DBC	66311 GBR	66558 FLR	
66079 DBC	66171 DBC	66312 GBR	66559 FLR	
66080 DBC	66172 DBC	66313 GBR	66560 FLR	
66082 DBC	66173 DBC§	66314 GBR	66561 FLR	
66083 DBC	66174 DBC	66315 GBR	66562 FLR	
66084 DBC	66175 DBC	66316 GBR	66563 FLR	
66085 DBC	66176 DBC		66564 FLR	
66086 DBC	66177 DBC	**66/4**	66565 FLR	
66087 DBC	66178 DBC§	66411 EXP	66566 FLR	
66088 DBC	66179 DBC	66412 EXP	66567 FLR	
66089 DBC	66180 DBC§	66413 FLR	66568 FLR	
66090 DBC	66181 DBC	66414 FLR	66569 FLR	
66091 DRS	66182 DBC	66415 FLR	66570 FLR	
66092 DBC	66183 DBC	66416 FLR	66571 FLR	
66093 DBC	66184 DBC	66417 EXP	66572 FLR	
66094 DBC	66186 DBC	66418 FLR	66582 EXP	
66095 DBC	66187 DBC	66419 FLR	66583 EXP	
	66188 DBC	66420 FLR	66584 EXP	

Above: To improve haulage power, Freightliner ordered a batch of 25 '66s' with 105,080lb tractive effort, having revised gearing of 83:18, this reduced the maximum speed to 65mph (105km/h). To keep these separate from the main fleet, they were classified as Class 66/6 and numbered in the 666xx series. No. 66623 is shown in the latest Genesee & Wyoming livery reflecting the parent owner of Freightliner. **CJM**

Above: In 2009, Freightliner adopted a 'Powerhaul' green and yellow livery, based on the Class 70s, this is shown on No.66420, a five door example inherited from DRS. This example sports the revised front light cluster arrangement with a joint LED marker/tail light and larger headlight. The loco is seen in early 2023 passing Daventry. **CJM**

Above: Direct Rail Services operates 14 Class 66s, now trading as Nuclear Transport Solutions (NTS). The fleet are all of the 'five-door' style and carry DRS blue livery. With the lastest style of light cluters, Nos. 66427 and 66428 are seen from their two grille, three door side. **CJM**

Above: Colas Rail Freight took over a fleet of five Class 66s from Freightliner in 2011, classified 66/8, as Nos. 66846-66850 and based at Hoo Junction, Kent. The fleet is used on Network Rail infrastructure duties and can frequently be found in the Westbury, Eastleigh and Hoo Junction areas. The locos are of the standard four-door design and were previously numbered 66573-66577. The fleet carries standard Colas Rail Freight colours, and sport a combined marker/tail light and large headlight on each side. No. 66846 is seen at Newport. **CJM**

Class 66

2023 Fleet

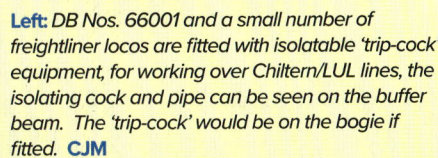

66585	FLR	**66/7**		66742	GBR	66784	GBR	
66586	EXP	66701	GBR	66743	GBR	66785	GBR	
66587	FLR	66702	GBR	66744	GBR	66786	GBR	
66588	FLR	66703	GBR	66745	GBR	66787	GBR	
66589	FLR	66704	GBR	66746	GBR	66788	GBR	
66590	FLR	66705	GBR	66747	GBR	66789	GBR	
66591	FLR	66706	GBR	66748	GBR	66790	GBR	
66592	FLR	66707	GBR	66749	GBR	66791	GBR	
66593	FLR	66708	GBR	66750	GBR	66792	GBR	
66594	FLR	66709	GBR	66751	GBR	66793	GBR	
66595	EXP	66710	GBR	66752	GBR	66794	GBR	
66596	FLR	66711	GBR	66753	GBR	66795	GBR	
66597	FLR	66712	GBR	66754	GBR	66796	GBR	
66598	FLR	66713	GBR	66755	GBR	66797	GBR	
66599	FLR	66714	GBR	66756	GBR	66798	GBR	
		66715	GBR	66757	GBR	66799	GBR	
66/6		66716	GBR	66758	GBR			
66601	FLR	66717	GBR	66759	GBR	**66/8**		
66602	FLR	66718	GBR	66760	GBR	66846	COL	
66603	FLR	66719	GBR	66761	GBR	66847	COL	
66604	FLR	66720	GBR	66762	GBR	66848	COL	
66605	FLR	66721	GBR	66763	GBR	66849	COL	
66606	FLR	66722	GBR	66764	GBR	66850	COL	
66607	FLR	66723	GBR	66765	GBR			
66608	EXP	66724	GBR	66766	GBR	**66/9**		
66609	EXP	66725	GBR	66767	GBR	66951	FLR	
66610	FLR	66726	GBR	66768	GBR	66952	FLR	
66611	EXP	66727	GBR	66769	GBR			
66612	EXP	66728	GBR	66770	GBR	**66/5**		
66613	FLR	66729	GBR	66771	GBR	66953	FLR	
66614	FLR	66730	GBR	66772	GBR	66954	FLR	
66615	FLR	66731	GBR	66773	GBR	66955	FLR	
66616	FLR	66732	GBR	66774	GBR	66956	FLR	
66617	FLR	66733	GBR	66775	GBR	66957	FLR	
66618	FLR	66734	GBR	66776	GBR			
66619	FLR	66735	GBR	66777	GBR	* in France		
66620	FLR	66736	GBR	66778	GBR	§ in Poland		
66621	FLR	66737	GBR	66779	GBR	± Numbers		
66622	FLR	66738	GBR	66780	GBR	allocated		
66623	FLR	66739	GBR	66781	GBR			
66624	EXP	66740	GBR	66782	GBR			
66625	EXP	66741	GBR	66783	GBR			

Left Upper, Left Middle & Left Below: GB Railfreight (GBRf) are a large user of Class 66s and in addition to locos purchased new via various funding houses, have also imported a number of former European Class 66s to the UK, upgrading them to UK standards. Some locos carry standard GBRf blue, orange and yellow, but a significant number are painted in special or promotional colours, supporting partnership businesses, the rail industry or for special events. In the left upper image standard liveried No. 66764 with rounded lines on the cab side is seen from its silencer end. Left middle, No. 66781 shows the square end livery design with the blue ending at the cab door. The left lower image shows one of the various light cluster differences found on the recent European import. No. 66750 shows a capped marker/headlight cluster, oblong buffers and dual main reservoir and brake pipes, with the lamp bracket on the drivers side. All: **CJM**

Left: GB Railfreight operate a number of locos painted in non standard liveries. No. 66783, a loco inherited from DB and originally numbered 66058 carried Biffa red and yellow livery. This loco retained its combination couplers from DB days and retains its original marker/headlight clusters. This was one of five Class 66s fitted with semi auto-release couplers and nose mounted coupling illuminating lights for use on the Lickey bank. The light has been retained in the GBRf days. **CJM**

Left: DB Nos. 66001 and a small number of freightliner locos are fitted with isolatable 'trip-cock' equipment, for working over Chiltern/LUL lines, the isolating cock and pipe can be seen on the buffer beam. The 'trip-cock' would be on the bogie if fitted. **CJM**

Right: If wheelsets become defective they can be mounted in 'wheel skates' to allow movement by rail. The equipment is seen fitted to No. 66100. **Spencer Conquest**

Class 67

Class:	67
Number range:	67001-67030
Built by:	Alstom/General Motors, Valencia, Spain
GM model:	JT-42-HWHS
Years introduced:	1999-2000
Wheel arrangement:	Bo-Bo
Design speed	125mph (201km/h)
Maximum speed:	110mph (177km/h)
Length:	64ft 7in (19.68m)
Height:	12ft 9in (3.88m)
Width:	8ft 9in (2.66m)
Weight:	90 tonnes
Wheelbase:	47ft 3 in (14.40m)
Bogie wheelbase:	9ft 2 in (2.79m)
Bogie pivot centres:	38ft 1in (11.63m)
Wheel diameter (new):	3ft 2in (965mm)
Min curve negotiable:	3.8 chains (75m)
Engine type:	GM 12N-710G3B-EC
Engine output:	3,200hp (2,386kW)
Power at rail:	2,493hp (1,860kW)
Maximum tractive effort:	31,750lb (141kN)
Continuous tractive effort:	20,200 (89.8kN) (with HEP active)
Cylinder bore:	9$\frac{1}{16}$ in (230mm)
Cylinder stroke:	11in (279mm)
Traction alternator:	GM-EMD AR9A
Companion alternator:	GM-EMD CA6HEX
Traction motor type:	GM-EMD D43FM
No. of traction motors:	4
Gear ratio:	59:28
Brake type:	Air, Westinghouse PBL3
Brake force:	78 tonnes
Bogie type:	Alstom high speed
Route availability:	8
Head end power (heating):	Electric - index 66 (in multiple each loco - 48)
Multiple coupling type:	AAR
Fuel tank capacity:	1,201gal (5,460lit)
Lub oil capacity:	202gal (918lit)
Cooling water capacity:	212gal (964lit)
Sanding equipment:	Pneumatic

Notes:
67004/007/009/011/030 fitted with Radio Electronic Token Block equipment.
67008/010/012/013/014/015/017/020/022/025/029 modified to operae push-pull with TfW Mk 4 stock.

Right: *Class 67 front end, as modified to work with Mk4 stock for Transport for Wales. 1: Warning horns behind grilles, 2: High level marker light, 3: RCH jumper cables, 4: AAR jumper socket (cable stowed in engine compartment), 5: Mk4 stock jumper socket, 6: Marker light, 7: Headlight, 8: Tail light, 9: Brake pipe (red), 10: Main Reservoir pipe (yellow), 11: Swing-head Combination coupler (in stowed position), 12: Electric train heat jumper socket, 13: Electric train heat jumper cable.* **Antony Christie**

Right & Below: *As part of total fleet modernisation, EWS ordered 30, 125mph (201 km/h) Bo-Bo Class 67s for passenger and mail traffic. Built by Alstom, under a GM contract in Spain, they have the same GM710 engine as the Class 66. The fleet, soon lost most of their work when the Royal Mail rail transport contract was lost. The fleet was originally painted in EWS maroon and gold, but most have now been repainted in DB red or promotional/business colours. In the above view, No. 67021 carries Pullman livery (also applied to No. 67024) for working with the Belmond Pullman, No. 67021 is shown from its silencer end with the Belmond Pullman at Basingstoke. Below No. 67013 displays DB red and grey livery and also bodyside branding 'First Choice for rail freight in the UK' It is viewed from the cooler group end, while working on TfW Mk4 duties. The extra Mk4 jumper socket is seen to the right of the AAR connection.* **Spencer Conquest / CJM**

Class 67

67001	DBC
67002	DBC
67003	DBC
67004	DBC
67005	DBC
67006	DBC
67007	DBC
67008	DBC
67009	DBC
67010	DBC
67011	DBC
67012	DBC
67013	DBC
67014	DBC
67015	DBC
67016	DBC
67017	DBC
67018	DBC
67019	DBC
67020	DBC
67021	DBC
67022	DBC
67023	COL
67024	DBC
67025	DBC
67026	DBC
67027	COL
67028	DBC
67029	DBC
67030	DBC

Above Left & Above Right: Mk4 passenger stock, displaced from LNER services by the introduction of IETs have cascaded to Transport for Wales and are now operating Cardiff-Holyhead/Manchester services. A batch of 11 Class 67s Nos. 67008/010/012/013/014/015/017/020/022/025/029 have been modified to operate with this stock in a 'push-pull' mode, with a Mk4 DVT at the remote end. Locos carry a variety of liveries, DB red, TfW grey/black and TfW black. Above left No. 67017 displays the grey scheme, while above right No. 67020 shows the black livery, this colour was applied to some '67s' and Mk4s for a proposed Grand Central service between London and Blackpool which was abandoned. **CJM / Spencer Conquest**

Below: Nos. 67005 and 67006 are dedicated to Royal Train use and carry 'Royal Claret' livery. The Royal set is only used by the King and Queen and The Prince of Wales and his family. The train normally operates in a 'push-pull' mode with a loco at either end of the 7-9 car formation. No. 67005 is sen at Crewe. **Spencer Conquest**

Above: In 2022 to mark The Queens Platinum Jubilee, DB repainted No. 67007 in the Jubilee events purple colours with bodyside decals and branding celebrating the event. The loco is deployed in the general pool. **Spencer Conquest**

Right: After a period working with Colas, Nos. 67023/027 owned by Beacon Rail moved to GB Railfreight in 2022, with No. 67027 outshopped in GBRf colours. In summer 2023 both were taken off-lease by GBRf and returned to Colas control. No. 67027 receiving Colas decals applied to the GB Railfreight blue. The loco is seen at Westbury on 9 August 2023. **Andy Curtis**

Below: The Class 67 driving cab is very different from that of the Class 66, using a 'wrap around' style. The standard design EMD power controller is on the right side, while the brake controllers are on the left. The cab from No. 67003 is shown. **CJM**

Key Facts

- High-speed locos designed for Royal Mail, parcels and passenger traffic.
- Nos. 67005 and 67006 dedicated to Royal Train use, painted in Royal Claret livery.
- Nos. 67023 and 67027 owned by Beacon Rail operated for a short time with GBRf, returning to Colas.
- Nos. 67021 and 67024 in Pullman livery, for use with Belmond British Pullman train.
- A fleet of 11 locos are modified for use with TfW Mk4 stock to enable push-pull operation.

Class 68

Class:	68
Number range:	68001-68034
Built by:	Vossloh Espania, Valencia, Spain
Classification:	UK-Light
Years introduced:	2014-2017
Wheel arrangement:	Bo-Bo
Maximum speed:	100mph (160km/h)
Length:	67ft 3in (20.50m)
Height:	12ft 6½in (3.82m)
Width:	8ft 10in (2.69m)
Weight:	85 tonnes
Wheelbase:	48ft 10in (14.90m)
Bogie wheelbase:	9ft 3in (2.80m)
Bogie pivot centres:	38ft 8in (11.83m)
Wheel diameter (new):	43¼in (1,100mm)
Min curve negotiable:	4 chains (80.46m)
Engine type:	Caterpiller C175-16
Engine output:	3,750hp (2,800kW) at 1,740rpm
Max tractive effort:	71,260lb (317kN)
Continuous tractive effort:	56,200lb (250kN)
Cylinder bore:	6.9in (175mm)
Cylinder stroke:	8.7in (220mm)
Transmission:	Electric
Traction alternator:	ABB WGX560
Traction motor type:	ABB 4FRA6063
No. of traction motors:	4
Brake type:	Air Disc, EP, dynamic regen
Brake force:	73 tonnes
Bogie type:	Vossloh Fabricated
Route availability:	7
Axle load:	21.4 tonnes per axle
Heating type:	Electric - 500kW, Index - 96
Multiple coupling:	Within class and Class 88
Fuel tank capacity:	1,320gal (6,000lit)
Lub oil capacity:	117gal (530lit)
Water capacity:	67gal (304lit)
Sanding equipment:	Pneumatic

Notes:
68008-68015 modified to operate push-pull with Mk3 stock
68019-68034 modified to operate push-pull with Mk5 stock

Above: The 34-strong Class 68 fleet, operated by Carlisle-based Direct Rail Services (DRS) were built by Vossloh (Stadler) in Spain, in 2023 there are the most modern diesels in use in the UK. All but two are owned by Beacon Rail, with the final two locos funded by DRS. The 3,750hp (2,800kW) locos are leased to Direct Rail Services with two batches sub-leased to TransPennine Trains and Chiltern Railways. The fleet is of the mixed traffic type. No. 68004 shows the standard DRS colours, with the locos No. 2 end nearest the camera. **Spencer Conquest**

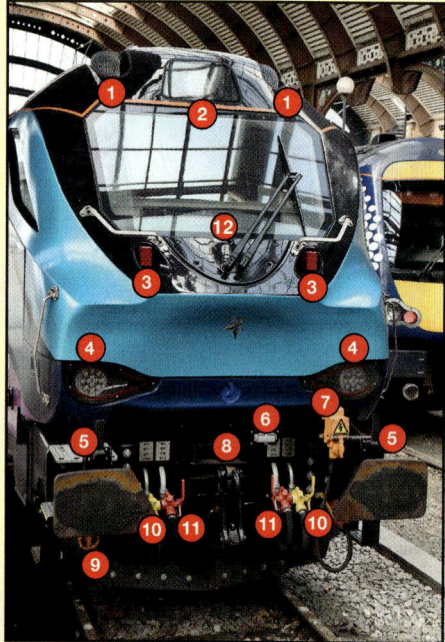

Right: Class 68 front end equipment, showing changes for working with First TransPennine Mk5a stock. 1: Warning horns, 2: High level marker light, 3: Multiple control jumper cable sockets (cables kept in engine bay), 4: Combined LED head, white marker and red tail light group, 5: Lamp bracket, 6: Jumper cable socket for operation with Mk5a stock, 7: Electric train heat (supply) jumper cable, 8: Coupling hook and shackle, 9: Electric train heat (supply) jumper socket, 10: Main reservoir pipe (yellow), 11: Air brake pipe (red), 12: Destination indicator. Note, locos modified to operate with Chiltern Mk3 sets and DVTs have a different shape jumper socket at point No. 6. **CJM**

Left: Multiple working coupling between two Class 68s or a 68 and 88. The two thin wires at the top provide all connections. **Howard Lewsey**

Left: Direct Rail Services supply TransPennine Trains with up to 14 Class 68s from the batch 68019-68032, to power Mk5a push-pull passenger stock on TransPennine services each day. The locos are modified to operate with CAF-built Mk5a stock, with the provision of extra jumpers and a destination indicator at the bottom of the cab front window. These carry TransPennine colours, without a yellow warning end. No. 68020 Reliance is seen passing Dore & Totley with a train bound for Manchester. Nos. 68033 and 68034 are also modified for TPT working, but retain standard DRS livery. When not required for TransPennine use, the locos can be rostered for freight or charter work. In 2023 a number of the Mk5 sets were long term out of use and thus the number of locos sub-leased was reduced. **CJM**

Class 68

2023 Fleet

68001	DRS
68002	DRS
68003	DRS
68004	DRS
68005	DRS
68006	DRS
68007	DRS
68008	DRS
68009	DRS
68010	CRW
68011	CRW
68012	CRW
68013	CRW
68014	CRW
68015	CRW
68016	DRS
68017	DRS
68018	DRS
68019	TPT
68020	TPT
68021	TPT
68022	TPT
68023	TPT
68024	TPT
68025	TPT
68026	TPT
68027	TPT
68028	TPT
68029	TPT
68030	TPT
68031	TPT
68032	TPT
68033	DRS
68034	DRS

Above: DRS Class 68s took over a loco provision contract with Chiltern Railways in December 2014, with six leased to the TOC. The locos, from the batch Nos. 68010-68015, are modified with Mk3 push-pull equipment, and carry an extra (round) jumper socket on the upper buffer beam, to the left of the train heating cable. The locos were repainted in Chiltern Railways two-tone grey livery. To cover any traction shortfall, DRS-liveried, Nos. 68008 and 68009 are also modified for Mk3 push-pull operation. Chiltern-liveried No. 68011 is seen with a Mk3 formation at Birmingham Moor Street. **CJM**

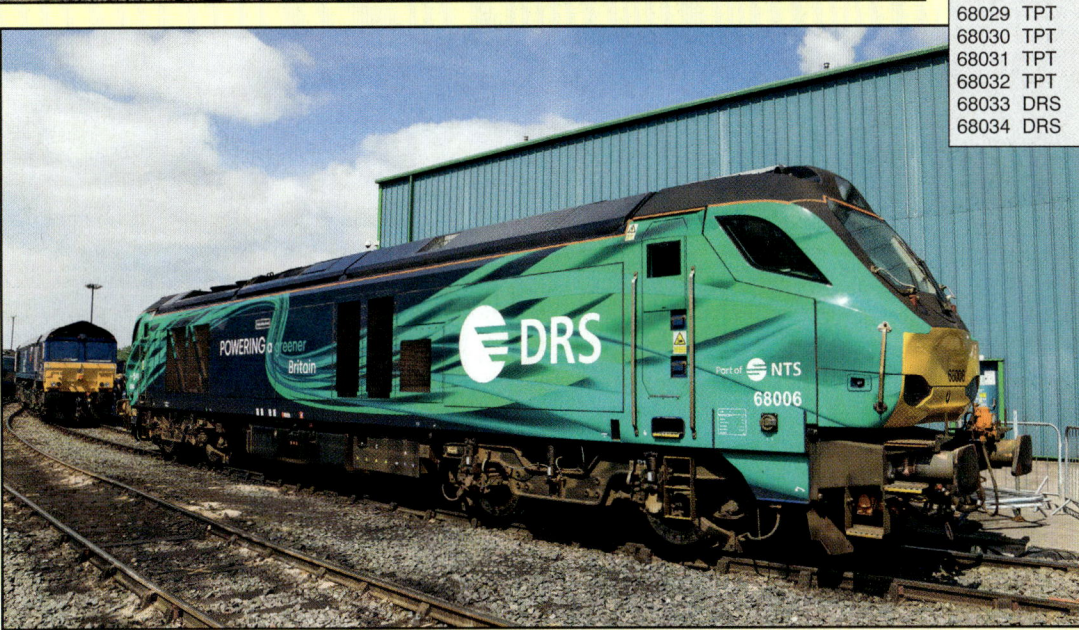

Above: As part of the COP 26 climate event held in Glasgow in November 2021, DRS re-purposed No. 68006 to run on Hydrotreated Vegetable Oil (HVO) fuel. It was named *Pride of the North* and given a revised blue and fleck green livery. It is not clear if this loco still burns HVO or has reverted to diesel fuel. The loco is seen stabled at Crewe Gresty Bridge. **Steve Donald**

Above: Class 68 cab. The cab style is based on the standard Vossloh (now Stadler) 'Eurolight' design, adapted for UK service to include TPWS and AWS. The same design is also used for the Class 88 bi-mode fleet. The main power controller is to the drivers right and the brake controllers to the left. A Train Management System (TMS) interface is provided and the locos can be upgraded for full ERTMS in the future. The cab of No. 68002 is shown. **CJM**

Key Facts

- Owned by Beacon Rail (32) and DRS (2), the 34-strong fleet are operated by Direct Rail Services.

- Six can be sub-leased to Chiltern Railways for use on Marylebone-Birmingham services and carry Chiltern livery.

- 14 locos can be sub-leased to TransPennine to operate with Mk5a stock on Transpennine - East Coast services. These carry full TPT colours.

- Facilities exist for locos to operate in multiple with other Class 68s or Class 88 electro-diesel locos.

Rolling Stock Review : 2023-2024

Class 69

Class:	69
Number range:	69001-69016
Previous number range:	56xxx
Rebuilt by:	Progress Rail, Stoke
Originally built:	1976-1984
As Class 69:	2020-2024
Wheel arrangement:	Co-Co
Maximum speed:	80mph (129km/h)
Length:	63ft 6in (19.35m)
Height:	13ft 0in (3.96m)
Width:	9ft 2in (2.79m)
Weight:	126 tonnes
Wheelbase:	47ft 10in (14.58m)
Bogie wheelbase:	13ft 5⅞in (4.11m)
Bogie pivot centres:	37ft 8in (11.48m)
Wheel diameter (new):	3ft 9in (1.14m)
Min curve negotiable:	4 chains (80.46m)
Engine type:	EMD/Caterpillar 12N 710G3B-T2
Engine output:	3,200hp (2,386kW)
Power at rail:	2,789hp (2,080kW)
Tractive effort:	62,900lb (280kN)
Cylinder bore:	9¹¹⁄₁₆ (230mm)
Cylinder stroke:	11in (279mm)
Transmission:	Electric
Main alternator type:	EMD AR10/CA6
Traction motor type:	Brush TMH73-62
No. of traction motors:	6
Gear ratio:	63:16
Brake type:	Air
Brake force:	60 tonnes
Route availability:	7
Heating type:	Not fitted
Multiple coupling type:	AAR
Fuel tank capacity:	1,150gal (5,228lit)
Lub oil capacity:	120gal (546lit)
Cooling water capacity:	308gal (1,400lit)
Sanding equipment:	Pneumatic

Below: Class 69 front end layout. 1: High level marker light, 2: AAR jumper socket, cable stowed in engine bay, 3: Headlight, 4: Combined marker/tail light, 5: Lamp bracket, 6: Air brake pipe (red), 7: Coupling hook/shackle, 8: Main reservoir pipe (yellow), 9: Air warning horns. **CJM**

Top, Above & Below: In early 2019, GB Railfreight announced that a fleet of 16 'new' locos would be rebuilt from Class 56s, classified 69. Rebuild work was undertaken by Progress Rail in Longport. The 'new' engine being a 12-cylinder EMD/Caterpillar 12N 710G3B-T2, a stage IIIa emission-compliant unit, an updated version of a Class 66 engine using electronics. The '56' underframe, bogies, traction motors and brake system were kept. New compressors, traction motor blowers and electronic cubicles were installed. A new design driving cab, based on a '66 is fitted. Front ends were modified, with a new marker light below the window height, and two standard light clusters with a headlight and marker/tail light above each buffer. AAR multiple jumper sockets were fitted. Air brakes are fitted, with one brake pipe and two main reservoir pipes either end. Loco deliveries have been slow and some technical issues encountered. At the top is No. 69001 with its No.1 end leading. The image above of No. 69002 shows the opposite side of the loco, with a new side grille panel inserted adjacent to the exhaust silencer group. Below, No. 69008 shows a modified front end, applied to Nos. 69008-69013, with a jumper socket to the left of the AAR connection, used to connect to modern multiple units. These six locos can also be fitted with drop-head Dellner or Tightlock couplings, hence the modified draw gear area. This loco showns the single grille side. On paper The '69s' are based at Tonbridge.
CJM / Antony Christie / Spencer Conquest

Class 69

Above: Although converted at EMD Longport, locos have been painted at Arlington Fleet Services, Eastleigh. No. 69004 emerged painted in BR Research red/blue. It is seen passing High Brooms on a Gypsum train on 16 May 2022. **CJM**

Right: Cast nameplate *Pathfinder Railtours* as applied to No. 69006. **CJM**

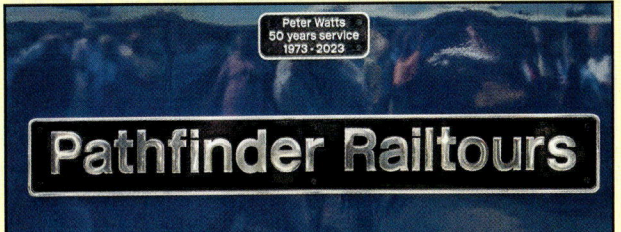

Below: When the Class 69s visited the paint shop at Arlington Fleet Services at Eastleigh after conversion several non-standard colour schemes were applied, including No. 69005 emerging in 1960s style BR green, in the style as applied to the Class 33s when new. The loco was also named *Eastleigh*. **Spencer Conquest**

Key Facts

- Reduced cost alternative to introduce new locos, while still meeting emissions standards.
- Car-bodies of Class 56s, gutted, Modified and rebuilt using a reconditioned engine and new/overhauled equipment.
- Most locos carry a different livery, branded to match.
- Present order for 16, could be extended but considered unlikely.

2023 Fleet

69001	GBR
69002	GBR
69003	GBR
69004	GBR
69005	GBR
69006	GBR
69007	GBR
69008	GBR
69009	GBR
69010	GBR
69011	GBR
69012	GBR
69013	GBR
69014	GBR
69015	GBR
69016	GBR

Rolling Stock Review : 2023-2024

Class 70

For history and pictures of this class, see Modern Locomotives Illustrated Issue 234

Class:	70
Number range:	Freightliner: 70001-70020
	Colas Rail Freight: 70801-70817
Built by:	General Electric, Erie, Pennsylvania§
GE model:	JPH37ACmi
Years introduced:	2009-2017
Wheel arrangement:	Co-Co
Maximum speed:	75mph (121km/h)
Length:	71ft 2½in (21.71m)
Height:	12ft 10in (3.91m)
Width:	8ft 8in (2.64m)
Weight:	129 tonnes
Wheel diameter (new):	42in (1,066mm)
Min curve negotiable:	4 chains (80.46m)
Engine type:	General Electric P616LDA1 'Powerhaul'
Engine output:	3,820hp (2,848kW)
Maximum tractive effort:	122,100lb (544kN)
Continuous tractive effort:	96,000lb (427kN)
Cylinder bore:	7.48in (190mm)
Cylinder stroke:	8.66in (220mm)
Transmission:	Electric
Traction alternator:	GE GTA
Companion alternator:	GE
Traction motor type:	GE AC 5GEB30
No. of traction motors:	6
Brake type:	Air, dynamic
Brake force:	96.7 tonnes
Bogie type:	GE Fabricated
Route availability:	7
Heating type:	Not fitted
Multiple coupling type:	AAR
Fuel tank capacity:	1,320gal (6,000lit)
Lub oil capacity:	212gal (802lit)
Sanding equipment:	Pneumatic

§ 70801 Built by Tulomsas, Eskisehir, Turkey as a demonstrator. 70001-70020 owned by Akiem Rail, 70801-810 are owned by Lombard Leasing and 70811-817 are owned by Beacon Rail.

Above & Below: *Freightliner ordered and took delivery of the first Class 70 high-output diesel-electric from General Electric of the USA in 2009. Classified by GE as JPH37ACmi and in the UK as Class 70. The twin-cab, narrow body design, were the first General Electric locos to be imported to the UK. A fleet of 20 were planned, however, one was lost on delivery, so the fleet stands at 19. Above, No. 70011 is shown from its electrical end with a container train at Leamington Spa. All Freightliner '70s' are painted in the 'Powerhaul' scheme. Below, No. 70017 is seen from the opposite side, with the cooler group at the far end. Both:* **CJM**

Bottom: *The fleet of 19 are based at Freightliner Leeds Midland Road depot, but receive maintenance at Crewe and other Freightliner depots, such as Southampton. The class can be seen throughout the Freightliner system, powering both container and Heavy Haul block consists. No. 70017 is seen at Eastleigh with a container wagon from repair at the nearby works.* **Nathan Williamson**

Key Facts

- One of the Freightliner order (70012) lost on delivery, after falling from a crane into the ship.

- Two follow-on orders placed by Colas Rail Freight, for 17 Class 70/8s.

- With the exception of No. 70801, all locos built by GE in Erie, Pennsylvania, USA, No. 70801 was built by Tulomsas, Eskisehir in Turkey.

Class 70

Right: Between 2013-2017 Colas Rail Freight took delivery of 17 Class 70s. One, No. 70801, had been previously built as a demonstrator of the US design in Turkey, after limited use there, it was then shipped to the UK and upgraded. The others came direct from the US. The fleet operate on general Colas Rail Freight flows, including Network Rail infrastructure trains and contract freight flows, they can usually be found in the Westbury, Bescot, Carlisle and Eastleigh areas. Pennsylvania-built No. 70811, with its radiator and cooler group leading passes Dawlish Warren with a westbound engineers train. **CJM**

Right: The Colas Class 70s are on paper based at Cardiff Canton, however, most maintenance is performed 'in the field' with mobile maintenance crews visiting the locos during lay-over periods. Displaying the more recent diamond cabside badge, No. 70803 is seen from its electrical end passing Worting Junction south of Basingstoke in April 2022. **CJM**

Left: Class 70 front end equipment, applicable to both Freightliner and Colas Rail Freight locos. 1: High level height marker light, 2: Warning horns behind grille, 3: Joint white marker and red tail light, 4: Headlight, (marker, tail and headlights are set in an anti-climber block, designed to stop locos climbing into each other in a heavy front end impact), 5: Windscreen washer filler (behind door), 6: Association of American Railroad (AAR) multiple control jumper socket, 7: Lamp bracket, 8: Air brake pipe (red), 9: Coupling hook and shackle, 10: Main reservoir pipe (red). **CJM**

Below: With the Class 70 came a totally new cab design, as the first operator, Freightliner made a considerable input to the style and equipment positions, involving a number of driver's and the driver's trade union ASLE&F. Looking very different from anything previously seen, the style has the power controller to the right of centre, with the loco and train brake controllers to the left. Loco management computer screens are positioned on either side. **CJM**

2023 Fleet

70/0

70001	FLR
70002	FLR
70003	FLR
70004	FLR
70005	FLR
70006	FLR
70007	FLR
70008	FLR
70009	FLR
70010	FLR
70011	FLR
70013	FLR
70014	FLR
70015	FLR
70016	FLR
70017	FLR
70018	FLR
70019	FLR
70020	FLR

70/8

70801	COL
70802	COL
70803	COL
70804	COL
70805	COL
70806	COL
70807	COL
70808	COL
70809	COL
70810	COL
70811	COL
70812	COL
70813	COL
70814	COL
70815	COL
70816	COL
70817	COL

Rolling Stock Review : 2023-2024

Class 73

For history and pictures of this class, see *Modern Locomotives Illustrated* Issues 198 & 241

Sub-class:	73/1	73/2	73/95
Number range:	73101-73141	73201-73235	73951-73952
Former number range:	E6007-E6049	From Class 73/1 fleet	From Class 73/1 and 73/2 fleet
Southern Region class code:	JB	JB	-
Built by:	English Electric, Vulcan Foundry	Modified by BR Stewarts Lane depot	Modified by RVEL Derby
Years introduced:	1965-1967	1984-1990	2014-2015
Wheel arrangement:	Bo-Bo	Bo-Bo	Bo-Bo
Maximum speed:	90mph (145km/h)	90mph (145km/h)	90mph (145km/h)
Length - buffers retracted:	52ft 6in (16.00m)	52ft 6in (16.00m)	52ft 6in (16.00m)
Length - buffers extended:	53ft 8in (16.35m)	53ft 8in (16.35m)	53ft 8in (16.35m)
Height:	12ft 5⅝in (3.79m)	12ft 5⅝in (3.79m)	12ft 5⅝in (3.79m)
Width:	8ft 8in (2.64m)	8ft 8in (2.64m)	8ft 8in (2.64m)
Weight:	77 tonnes	77 tonnes	76.8 tonnes
Wheelbase:	40ft 9in (12.42m)	40ft 9in (12.42m)	40ft 9in (12.42m)
Bogie wheelbase:	8ft 9in (2.66m)	8ft 9in (2.66m)	8ft 9in (2.66m)
Bogie pivot centres:	32ft 0in (9.75m)	32ft 0in (9.75m)	32ft 0in (9.75m)
Wheel diameter (new):	3ft 4in (1.01m)	3ft 4in (1.01m)	3ft 4in (1.01m)
Min curve negotiable:	4 chains (80.46m)	4 chains (80.46m)	4 chains (80.46m)
Power supply:	660-850V dc third rail	660-850V dc third rail	660-850V dc third rail
Electric output (Nom):	1,600hp (1,193kW)	1,600hp (1,193kW)	1,328hp (990kW)
Electric power at rail (Cont):	1,200hp (895kW)	1,200hp (895kW)	1,100hp (820kW)
Electric power at rail (Max):	3,150hp (2,349kW)	3,150hp (2,349kW)	3,000hp (2,237kW)
Engine type:	English Electric 4SRKT Mk2	English Electric 4SRKT Mk2	2 x Cummins QSK19 of 750hp (560kW)
Engine output:	600hp (447kW)	600hp (447kW)	1,500hp (1,120kW)
Diesel power at rail:	402hp (300kW)	402hp (300kW)	1,200hp (895kW)
Electric tractive effort:	40,000lb (178kN)	40,000lb (178kN)	42,000lb (186kN)
Diesel tractive effort:	36,000lb (160kN)	36,000lb (160kN)	38,000lb (169kN)
Cylinder bore:	10in (250mm)	10in (250mm)	6¼in (159mm)
Cylinder stroke:	12in (300mm)	12in (300mm)	6¼in (159mm)
Main generator type:	EE824-5D	EE824-5D	-
Main alternator type:	-	-	2 x Marathon Magnaplus
Aux generator type:	EE908-5C	EE908-5C	Included in above
Traction motor type:	EE546-1B	EE546-1B	EE546-1B
No. of traction motors:	4	4	4
Gear ratio:	61:19	61:19	61:19
Brake type:	Dual, with high-level air pipes, some modified to air brake only	Dual, with high-level air pipes, all modified to air brake only	Air
Brake force:	31 tonnes	31 tonnes	31 tonnes
Route availability:	6	6	6
Heating type:	Electric - index 66 (electric only)	Electric - index 66 (electric only)	Not fitted
Multiple coupling type:	Electric, 1951-1966 EMU jumpers Diesel - Blue star	Electric, 1951-1966 EMU jumpers Diesel - Blue star	- -
Fuel tank capacity:	310gal (1,409lit)	310gal (1,409lit)	497gal (2,260lit)
Sanding equipment:	Pneumatic	Pneumatic	Pneumatic
Fittings:	Drop-head buck-eye	Drop-head buck-eye	Drop-head buck-eye
Sub-class variations:	Production fleet	Originally dedicated to Gatwick Express operation	RVEL Network Rail rebuild

Below: In 1965-1966 English Electric, Vulcan Foundry, built a production run of 43 electro-diesel locos classified 'JB', 73/1 under TOPS for BR Southern Region, as a follow-on order to six BR built prototypes in 1962. They became the workhorse of the Southern Region for many years. They remain, in much depleted numbers today, mainly working for GB Railfreight. Many locos have been overhauled, refurbished and operate as Class 73/9s. In 2023, 12 Class 73/1 and 73/2s are in service with GB Railfreight. These two sub-classes retain their original English Electric 4SRKT, 600hp (447kW) engine, powering English Electric traction and control equipment. The locos have air, vacuum and EP braking (vacuum removed on some examples). They have duplicate high-level air and control pipes and were able to multiple with 1951-1966 EMU stock (now all withdrawn). Most retain drop-head buck-eye couplings, Pullman rubbing plates and retractable buffers. No. 73119 restored to BR CCE 'Dutch' livery is seen working in multiple with No. 73202, which for many years (until 2023) operated with GTR Southern and is painted in GB Railfreight branded GTR Southern white and green livery. The pair are seen at Wandsworth Road on 15 June 2023 with an engineers train bound for Hoo Junction. **Jamie Squibbs**

44

Rolling Stock Review : 2023-2024

Class 73

73/96
73961-73971
From Class 73/0, 73/1 and 73/2 fleet
-
Modified by Brush-Wabtec
2014-2016
Bo-Bo
90mph (145km/h)
52ft 6in (16.00m)
53ft 8in (16.35m)
12ft 5 5/16 in (3.79m)
8ft 8in (2.64m)
77.5 tonnes
40ft 9in (12.42m)
8ft 9in (2.66m)
32ft 0in (9.75m)
3ft 4in (1.01m)
4 chains (80.46m)
660-850V dc third rail (73961-73965 only)
1,600hp (1,193kW)
1,080hp (805kW)
3,000hp (2,237kW)
MTU V8 R43L 4000
1,600hp (1,193kW)
1,250hp (932kW)
42,000lb (186kN)
34,000lb (151kN)
6.7in (170mm)
8.3in (210mm)
-
Lechmotoren SDV87.53-12
-
EE546-1B
4
61:19
Air
-
31 tonnes
6
Electric - 73961-73965 - index 30
Electric - 73966-73971 - index 70
AAR and Blue star
-
73961-73965 - 310gal (1,409lit)
73966-73971 - 460gal (2,091lit)
Pneumatic
73961-965 - Buck-eye, 73966-971 - Dellner
Brush rebuilds for GBRf

Right Bottom: *Between 2014-2016, GB Railfreight sponsored the rebuild of 11 Class 73s, with new electrical equipment, and a MTU diesel engine of 1,600hp (1,193kW), plus a body upgrade. The rebuilt locos, classified 73/96, consists of five (Nos. 73961-73965) for use on Network Rail services and six on the Caledonian Sleeper services within Scotland. The NR locos are used on engineering and test trains, they are painted in GBRf blue and orange livery and based at Tonbridge. They retain the ability to operate from the third rail supply as well as the MTU engine. They have high level air pipes and multiple control jumper/ receptacle, a high level marker light and Pullman rubbing plate. They have kept drop-head buck-eye couplings and retractable buffers. They are also fitted with an AAR jumper socket and sport standard light clusters.*
Spencer Conquest

Key Facts
- Originally two sub-classes existed, 73/0 - six original locos built at Eastleigh, and 73/1 - 43 modified locos built by English Electric.
- Brush rebuilt 11 as Class 73/9s for GBRf, fitting MTU diesel engines.
- RVEL rebuilt two locos (73951 and 73952) for Network Rail with a major body re-design and total equipment replacement.

Above: *No less than seven GB Railfreight Class 73s are in this view at Tonbridge, where a footbridge gives perfect views of the yard, No. 73107, nearest the camera, shows the radiator end. As can be seen a lot of equipment is carried on the buffer beam of this design, including engine control air pipes, brake pipes, main reservoir pipes, vacuum pipe, as well as low-level blue-star multiple control jumper sockets. On the cab end two sets of dual air brake and main reservoir pipes and multiple jumper cables are carried. A central headlight is fitted above the Pullman rubbing plate.* **CJM**

Below: *Nos. 73212 and 73213 received a number of modifications when operating with Network Rail some years ago, this included the removal of the Pullman rubbing plate and buck-eye couplings, leaving just a hook, main reservoir and brake pipe on the buffer beam. A three section snowplough set was also fitted. The body mounted connections were retained. No. 73213 is seen from the radiator end at Tonbridge. In past times the Class 73/2s were used on Gatwick Express services.* **CJM**

2023 Fleet

73/1			
73101	GBR	73213	GBR
73107	GBR	73235	SWR
73109	GBR		
73110	GBR	**73/9**	
73109	GBR	73951	(S)
73119	GBR	73952	(S)
73128	GBR	73961	GBR
73133	PRI	73962	GBR
73136	GBR	73963	GBR
73138	NRL	73964	GBR
73139	GBR	73965	GBR
73141	GBR	73966	CAL
		73967	CAL
73/2		73968	CAL
73201	GBR	73969	CAL
73202	GBR	73970	CAL
73212	GBR	73971	CAL

Rolling Stock Review : 2023-2024

Class 73

Left: As part of the GB Railfreight/Brush rebuild project for 11 Class 73s, six locos, Nos. 73966-73971 were modified to operate with Mk5 Caledonian sleeper stock. Powering trains within Scotland between Edinburgh Waverley and Aberdeen, Inverness and Fort William. The locos sport a slightly revised body design with a different front end layout. They do not retain active third rail power collection equipment and are thus classified as a diesel loco, the equipment was originally retained, but removed in 2019-2020. The batch are painted in Caledonian Sleeper teal green livery and are maintained at Edinburgh Craigentinny. No. 73969 is seen at Crewe. On these locos only one multiple control jumper/socket is retained on the front end. **Spencer Conquest**

Below Left: Caledonian Sleeper modified Class 73/9 front end layout and equipment. 1: Warning horns, 2: High level marker light, 3: Radio aerial, 4: Multiple control jumper cable, 5: Multiple control jumper socket, 6: AAR Jumper socket, 7: Headlight, 8: Combined white marker light, red tail light, 9: Lamp bracket, 10: Main reservoir pipe (yellow), 11: Air brake pipe (red), 12: ETH jumper socket, 13: ETH jumper cable, 14: Train control jumper socket for Mk5 stock, 15: Train control jumper cable for Mk5 stock, 16: Drop-head Dellner coupling. **CJM**

Right: Class 73 GB Railfreight rebuild program saw the installation of AAR jumper sockets for multiple operation, the jumper is shown installed. This illustration was taken before the retro fitting of Dellner couplings. The two lights on the outer edge of the light cluster box, indicates to ground staff if the train supply is operational (red light) or off (green light), indicating if it is safe to remove the power jumpers. **Antony Christie**

Below: Part of the major Caledonian sleeper Class 73 overhaul, was the total rebuild of the dual position driving cab, maintaining the same basic style, but installing new equipment to operate with the Mk5 stock, updating switches and displays and changing the brake controllers to the MZT Hepos type. The dual position dual handle controllers were retained. **CJM**

Left: In 2014-2015 an in depth rebuild operation was undertaken by Rail Vehicle Engineering Ltd (RVEL) at the RTC Derby, to upgrade two Class 73/1s for Network Rail. This saw the installation of two Cummins QSK19 engines and a new electronic control system. The combined diesel engine output was now 1,500hp (1,120kW). Externally, the locos were heavily rebuilt, with new sides and a modified front end, devoid of all air and jumper connections. New light clusters were installed, with the drop-head buck-eye coupling and Pullman rubbing plate retained. The pair are painted in Network Rail yellow and based at the RTC Derby for the powering of test trains. The structural changes from the original Class 73/1 to the modified examples can be seen if this image is compared to that at the start of the Class 73 section. No. 73951 is seen at Bristol Temple Meads. Both locos were taken out of service in May 2023 and stored at the Severn Valley Railway. **Spencer Conquest**

Rolling Stock Review : 2023-2024

For history and pictures of this class, see *Modern Locomotives Illustrated* Issue 172

Class 86

Sub-class:	86/1, 86/2, 86/4
Number range:	86101, 86259, 86401
Previous number range:	860xx
Former class code:	AL6
Built by:	BR Doncaster
Rebuilt by:	BREL Crewe
Years introduced:	1966
Years modified:	1975
Wheel arrangement:	Bo-Bo
Maximum speed:	100mph (161km/h)
Length:	58ft 6in (17.83m)
Height (pan down):	13ft 0⁹⁄₁₆ in (3.97m)
Width:	8ft 8¼in (2.65m)
Weight:	85 tonnes
Wheelbase:	43ft 6in (13.25m)
Bogie wheelbase:	10ft 9in (3.28m)
Bogie pivot centres:	32ft 9in (9.98m)
Wheel diameter (new):	3ft 9¼in (1.15m)
Min curve negotiable:	4 chains (80.46m)
Power supply:	25kV ac overhead
Traction output (max):	6,100hp (4,549kW)
Traction output (cont):	4,040hp (3,017kW)
Tractive effort:	46,500lb (207kN)
Control system:	HT tap changing
No. of traction motors:	4
Traction motor type:	AEI G282BZ
Gear ratio:	22:65
Brake type:	Dual
Brake force:	40 tonnes
Route availability:	6
Multiple coupling type:	TDM
Heating type:	Electric - index 66
Sanding equipment:	Pneumatic
Sub-class variations:	Fitted with flexicoil suspension, SAB wheels

15 Class 86/6s, owned by Freightliner and stored at Crewe Basford Hall, were exported to Bulgaria in summer 2023, departing the UK through Immingham.

2023 Fleet

86/1
86101 LSL

86/2
86259 PRI

86/4
86401 WCR

Right: *Loco Services Ltd of Crewe own Class 86/1 No. 86101 Sir William A Stanier FRS. It is used with the operators InterCity charter rake and retains TDM push-pull operation. It carries full InterCity Swallow livery. When not in use, it can usually be found stabled at Crewe, where this image was recorded.* **Cliff Beeton**

Below: *Class 86/2 No. 86259 is owned by Les Ross, and powers the electric legs of charter trains, over the West Coast Main Line. The loco sports a version of early BR electric blue with white cab window surrounds and roof, together with small yellow warning ends. No. 86259 is usually stabled and maintained at Rugby. It is dual braked and retains RCH front end jumpers. It is seen at Crewe.* **Spencer Conquest**

Key Facts

- In 2021 Freightliner Class 86s were stored. In 2023 these have found a new home working in Bulgaria.
- No. 86259 is privately owned by Les Ross and authorised for main line operation.
- A number of former Class 86s operate overseas in Hungary and Bulgaria.

For history and pictures of this class, see *Modern Locomotives Illustrated* Issue 199

Class 87

Class:	87
Number range:	87002
Built by:	BREL Crewe
Years introduced:	1973
Wheel arrangement:	Bo-Bo
Maximum speed:	110mph (177km/h)
Length:	58ft 6in (17.83m)
Height (pan down):	13ft 1¼in (3.99m)
Width:	8ft 8¼in (2.65m)
Weight:	83.3 tonnes
Wheelbase:	43ft 6⅛in (13.26m)
Bogie wheelbase:	10ft 9⅛in (3.28m)
Bogie pivot centres:	32ft 9in (9.98m)
Wheel diameter (new):	3ft 9½in (1.16m)
Min curve negotiable:	4 chains (80.47m)
Power supply:	25kV ac overhead
Traction output (max):	7,860hp (5,861kW)
Traction output (con):	5,000hp (3,728kW)
Tractive effort:	58,000lb (258kN)
Control system:	HT tap changing
No of traction motors:	4
Traction motor type:	GEC G412AZ
Gear ratio:	32:73
Brake type:	Air
Brake force:	40 tonnes
Route availability:	6
Multiple coupling type:	TDM
Heating type:	Electric - index 95
Sanding equipment:	Pneumatic

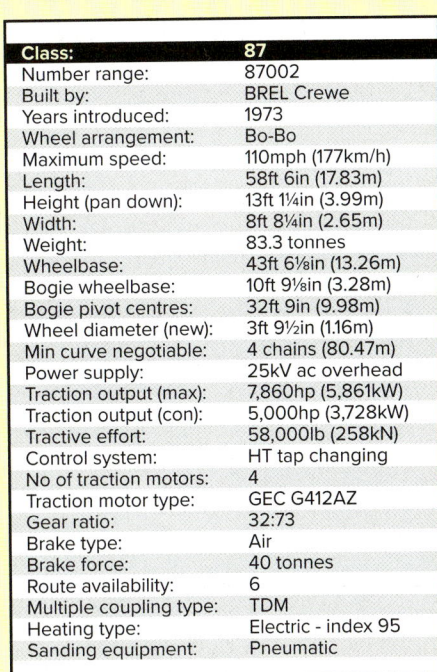

Above: *Class 87s operated over the West Coast route between 1973 and replacement by Pendolino stock, No. 87002 was first preserved by the AC Locomotive Group, restored to main line operation and hired to GB Railfreight to power Caledonian sleeper services. After this work ended in 2019, No. 87002 was sold to Crewe-based Loco Services Ltd and in 2020 was overhauled and restored to InterCity 'Swallow' livery by Arlington Fleet Services at Eastleigh. It is now used for main line charter work on the West Coast Main Line, powering the LSL rake of restored Mk3 stock with a DVT at the remote end. No. 87002 Royal Sovereign is illustrated passing Rugby on 26 August 2022 powering a Euston to Manchester service.* **Antony Christie**

2023 Fleet

87002 LSL

A fleet of 21 locos have been exported to Bulgaria and operate with BZK or Bulmarket.

Key Facts

- Once a fleet of 36 locos, the backbone of West Coast electrified services between 1973 and the early 2000s.
- One loco, No. 87002 *Royal Sovereign* remains operational in the UK, owned by Loco Services Ltd at Crewe and used for charter train duties.
- Long broken up, the final member of the build No. 87101, was a traction development loco in the use of thyristors in traction systems.
- A fleet of 21 locos were exported to Bulgaria for use on freight services.

Rolling Stock Review : 2023-2024   47

Class 88

For history and pictures of this class, see *Modern Locomotives Illustrated* Issue 236

2023 Fleet

88001	DRS
88002	DRS
88003	DRS
88004	DRS
88005	DRS
88006	DRS
88007	DRS
88008	DRS
88009	DRS
88010	DRS

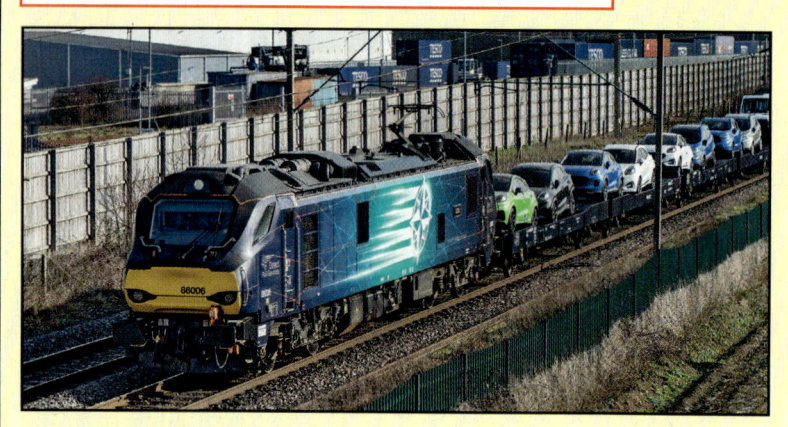

Class:	88
Number range:	88001-88010
Built by:	Stadler, Valencia, Spain
Classification:	UK-Dual-Mode
Year introduced:	2016-2017
Wheel arrangement:	Bo-Bo
Maximum speed:	100mph (161km/h)
Length:	67ft 3in (20.50m)
Height:	12ft 6½in (3.82m)
Width:	8ft 10in (2.69m)
Weight:	86 tonnes
Wheelbase:	48ft 10in (14.90m)
Bogie wheelbase:	9ft 3in (2.80m)
Bogie pivot centres:	37ft 7in (12.10m)
Wheel diameter (new):	43½in (1,100mm)
Min curve negotiable:	4 chains (80.46m)
Power supply:	25kV ac overhead
Traction package:	ABB
Electric output:	5,360hp (4,000kW)
Diesel engine type:	Caterpillar C27 12-cylinder
Engine output:	950hp (708kW)
Starting tractive effort:	71,264lb (317kN)
Continuous tractive effort:	56,200lb (250kN)
Cylinder bore:	5.42in (137.7mm)
Cylinder stroke:	6in (152.4mm)
Main/traction alternator:	ABB AMXL400
Traction motor type:	ABB 4FRA6063 of 600kW at 4,400rpm
No. of traction motors:	4
Brake type:	Air Disc, EP, dynamic regen
Brake force:	73 tonnes
Bogie type:	Stadler Fabricated
Route availability:	7
Axle load:	21.5 tonnes per axle
Heating type:	Electric - 500kW, index 96, Diesel - No
Multiple coupling:	Within class and Class 68 (max 2)
Fuel tank capacity:	395gal (1,800lit)
Sanding equipment:	Pneumatic

Key Facts

- Built by Stadler, these 10 Bi-mode (dual-power) electro-diesel locos are owned by Beacon Rail and operated by DRS. They are based on the body design of the Class 68.

- The diesel can be used as a 'last mile' power source for yard working, but is also strong enough to allow diesel-operation on the main line.

- Class 88s can operate in multiple with Class 68s, under both diesel and electric conditions.

Above & Below: *The UKs most advanced bi-mode locos are the Stadler-built Class 88s, owned by Beacon Rail and leased to Direct Rail Services. The 5,360hp (4,000kW) design, haul heavy freight traffic under the overhead wire, but are also fitted with a 940hp (708kW) Caterpillar C27 12-cylinder diesel engine, used for 'last mile' running, shunting or to power main line light weight duties. Above, sporting DRS livery, No. 88006 Juno is seen from the non-pantograph end. On the left, No. 88010 displays the 'Keeping it cool with less CO2' promotional colours.* **CJM / Antony Christie**

Below Left: *Class 88 front end equipment positions. 1: Cant rail height marker light, 2: Warning horns and air vents, 3: Multiple control jumper sockets, (cable is stowed in the engine compartment, these jumpers are identical to those on the Class 68s and the locos can operate together), 4: Combined head, white marker and red tail light cluster, using LEDs, 5: Emergency lamp bracket, 6: Electric train heat (supply) jumper cable, 7: Electric train heat (supply) jumper socket, 8: Main reservoir pipe, 9: Air brake pipe, 10: Coupling hook and shackle, 11: Adjustable height obstacle deflector plate.* **CJM**

Below Right: *Class 88 driving cab. This is based on the Stadler 'Eurolight' style and is almost as that installed on the Class 68. Extra controls are provided for the operation of the electric side of the loco. The power controller is on the right side and the brake controllers on the left. The cab illustrated is from No. 88003.* **CJM**

Class 90

2023 Fleet

90001	LSL
90002	LSL
90003	FLR
90004	FLR
90005	FLR
90006	FLR
90007	FLR
90008	FLR
90009	FLR
90010	FLR
90011	FLR
90012	FLR
90013	FLR
90014	FLR
90015	FLR
90016	FLR
90017	(S)
90018	(S)
90019	(S)
90020	(S)
90021	(S)
90022	(S)
90023	(S)
90024	(S)
90025	(S)
90026	(S)
90027	(S)
90028	(S)
90029	(S)
90030	(S)
90031	(S)
90032	(S)
90033	(S)
90034	(S)
90035	(S)
90036	(S)
90037	(S)
90038	(S)
90039	(S)
90040	(S)
90041	FLR
90042	FLR
90043	FLR
90044	FLR
90045	FLR
90046	FLR
90047	FLR
90048	FLR
90049	FLR
90050	OLS

For history and pictures of this class, see *Modern Locomotives Illustrated* Issue 210

Right: *Class 90 front end equipment. Applicable to all sub-fleet members, a handful of locos, including Nos. 90001/002 retain buck-eye couplings and a Pullman rubbing plate. 1: Headlight, 2: Combined marker light (white) and tail light (red), 3: Horns behind grille, 4: RCH jumper cable, 5: Air brake pipe (red), 6: Main reservoir pipe (yellow), 7: Electric train supply jumper socket, 8: Electric train supply jumper cable. 9: Coupling hook and shackle.* **CJM**

Class:	90
Number range:	90001-90049 (90050)
Built by:	BREL Crewe
Years introduced:	1987-1990
Wheel arrangement:	Bo-Bo
Maximum speed:	90001-90040 - 110mph (177km/h)
	90041-90050 - 75mph (121km/h)
Length:	61ft 6in (18.74m)
Height (pan down):	13ft 0¼ in (3.96m)
Width:	9ft 0in (2.74m)
Weight:	84.5 tonnes
Wheelbase:	43ft 6in (13.25m)
Bogie wheelbase:	10ft 9in (3.27m)
Bogie pivot centres:	32ft 9in (9.98m)
Wheel diameter (new):	3ft 9½in (1.16m)
Min curve negotiable:	4 chains (80.43m)
Power supply:	25kV ac overhead
Traction output (max):	7,860hp (5,861kW)
Traction output (con):	5,000hp (3,728kW)
Tractive effort:	58,000lb (258kN)
Control system:	Thyristor
No. of traction motors:	4
Traction motor type:	GEC G412CY
Gear ratio:	32:73
Brake type:	Air
Brake force:	40 tonnes
Route availability:	7
Multiple coupling type:	TDM
Heating type (if fitted):	Electric - index 95
	Some Freightliner locos isolated
Sanding equipment:	Pneumatic

Left: *Loco Services Ltd of Crewe, purchased the first two Class 90s, Nos. 90001 and 90002 from Porterbrook after they were displaced from Anglia InterCity operations. The pair now carry near original style InterCity livery and operate LSL charter services, mainly over the West Coast Main Line, formed of the operators Mk3 passenger formation and a remote Mk3 Driving Trailer. No. 90001 Royal Scot is seen at Crewe.* **Antony Christie**

Right Middle: *In summer 2023, 24 Class 90s were on the books of DB-Cargo and were all stored at Crewe due to the high cost of using electric locos. A fleet of 23 are operated by Freightliner. Based at Crewe, powering long distance container traffic over the West Coast route and via North London to Ipswich. They currently display either Freightliner green, 'Powerhaul' or the latest Genesee & Wyoming orange and black colours. Originally the locos were fitted with buck-eye couplings and rubbing plates, but these have now been removed. Viewed from their four grille side, immaculate Nos. 90014 and 90006 head north near Kenton on 14 July 2022.* **CJM**

Key Facts

- Once the backbone of West Coast passenger services and then the Great Anglia route, most are now operated by freight companies.
- DB and Freightliner operate 47 locos on long distance services (all DB locos into store summer 2023).
- Two locos, Nos. 90001/90002 are owned by Locomotive Services Ltd at Crewe for charter train use.

Right: *Several Freightliner Class 90s sport the original green livery, either in the standard style or in the 'Powerhaul' design. 'Powerhaul'-liveried No. 90042 pilots a Genesee & Wyoming-liveried No. 90048 at Carlisle. To improve coupling and reduce buffer locking fixed oval buffers are now installed.* **Antony Christie**

Rolling Stock Review : 2023-2024

Class 91

For history and pictures of this class, see *Modern Locomotives Illustrated* Issue 229

Class:	91
Number range:	91101-91130
Built by:	BREL Crewe
Rebuilt:	Adtranz/Bombardier Doncaster
Years introduced:	1988-1991
Years refurbished to 91/1:	2000-2002
Wheel arrangement:	Bo-Bo
Maximum operating speed:	125mph (201km/h)
	110mph (177km/h) slab end leading
Maximum design speed:	140mph (225km/h)
Length:	63ft 8in (19.40m)
Height (pan down):	12ft 4in (3.76m)
Width:	9ft 0in (2.74m)
Weight:	84 tonnes
Wheelbase:	45ft 4½in (13.83m)
Bogie wheelbase:	10ft 1⅞in (3.09m)
Bogie pivot centres:	34ft 5½in (10.50m)
Wheel diameter (new):	3ft 3½in (1.00m)
Min curve negotiable:	4 chains (80.49m)
Power supply:	25kV ac overhead
Traction output (max):	6,300hp (4,700kW)
Traction output (con):	6,090hp (4,531kW)
Tractive effort:	43,000lb (190kN)
Control system:	Thyristor
No. of traction motors:	4
Traction motor type:	GEC G426AZ
Gear ratio:	1.74:1
Brake type:	Air
Brake force:	45 tonnes
Route availability:	7
Heating type:	Electric - index 95
Multiple coupling type:	TDM
Sanding equipment:	Pneumatic
Special fittings:	Drop-head buck-eye couplings

Key Facts

- Operated as key motive power on East Coast between London Kings' Cross and Leeds, Newcastle, Edinburgh and Glasgow with Mk4 stock, until replaced by 'Azuma' fleet in 2019-2020.
- 12 locos remain in service with LNER for the foreseeable future.
- Several locos carry impressive promotional 'pictogram' graphics.
- Locos have a driving cab at the slab end, but are restricted to 110mph (177km/h) if the slab end is leading.
- Two '91s' (91117/120) are on the books of Europhoenix, but stored.

2023 Fleet

91101	LNE
91105	LNE
91106	LNE
91107	LNE
91109	LNE
91110	LNE
91111	LNE
91114	LNE
91117	EPX
91119	LNE
91120	EPX
91124	LNE
91127	LNE
91130	LNE

Above & Below: *Introduced between 1988-1991 for the East Coast electrification project, the Class 91s were the backbone of East Coast services for over 25 years, replaced by Class 800/801 'Azuma' stock from 2019. Twelve of the class and a number of Mk4 sets remain with LNER for the foreseeable future. The locos have one raked and one slab cab end, forming part of semi-fixed Mk4 formation, the loco is usually coupled at the north end, with a DVT at the south. Above, No. 91111 in For the Fallen 'Poppy' livery is seen at King's Cross. From summer 2022, an overhaul program commenced at Wabtec Doncaster, seeing all non-pictogram locos emerge in a Oxblood and white livery, based on the original InterCity style. No. 91106 is seen at Doncaster.* **Antony Christie / CJM**

Left: *A full driving cab is provided at both ends of the Class 91, when driving from the main 'raked' end the top speed is 125mph (201km/h), however, if for any reason the 'slab' end is leading, the maximum authorised speed is 110mph (177km/h). The cab is styled on the Class 90, which was under production at the same time. The power controller is on the right side and the brake controller on the left.* **CJM**

For history and pictures of this class, see Modern Locomotives Illustrated Issue 213

Class 92

Key Facts

- Introduced by UK Railfreight, SNCF (French Railways) and Eurostar for Channel Tunnel duties.
- Only main line electric locos authorised to operate through the Channel Tunnel.
- Due to a downturn in freight and the abandonment of overnight passenger services through the Channel Tunnel, many locos became spare.
- Thirteen of the DB locos now operate in Eastern Europe.
- A batch of GBRf locos are modified with drop-head Dellner couplings to operate Mk5 stock on the Caledonian Sleeper services between London and Glasgow/Edinburgh.

Class:	92
Number range:	92001-92046
Built by:	Brush Traction
Years introduced:	1993-1996
Wheel arrangement:	Co-Co
Maximum speed:	87mph (140km/h)
Length:	70ft 1in (21.34m)
Height (pan down):	13ft 0in (3.96m)
Width:	8ft 8in (2.67m)
Weight:	126 tonnes
Wheelbase:	56ft 6in (17.22m)
Bogie wheelbase:	14ft 1in (4.29m)
Bogie pivot centres:	41ft 10½in (12.75m)
Wheel diameter (new):	3ft 7in (1.07m)
Min curve negotiable:	6 chains (120.7m)
Power supply:	Overhead at 25kV ac or Third rail at 750V dc
Traction output (max):	6,760hp (5,040kW) - overhead ac power supply 5,360hp (4,000kW) - third rail dc power supply
Tractive effort:	Normal - 81,000lb (360kN) Boost - 90,000lb (400kN)
Control system:	Asynchronous 3-phase
No. of traction motors:	6
Traction motor type:	ABB 6FRA 7059B
Brake type:	Air, rheostatic and regenerative
Brake force:	63 tonnes
Coupling:	Screw Caledonian Sleeper locos - Dellner T12
Route availability:	7
Multiple coupling type:	TDM
Heating type:	Electric – ac supply - index 180 § dc supply - index 70

§ Caledonian Sleeper locos have higher HEP output for hotel power.
92009-011/015/016/018/019/023/031/032/036/038/041-044 fitted with TVM430 for operation over HS1.
92006/010/014/018/020/023/028/032/033/038/043 fitted with drop-head Dellner couplings.

Above, Below & Inset: *The Brush-built Class 92s, were introduced between 1993-1995 for use through the Channel Tunnel, as well as on UK domestic duties. Designed for dual AC/DC operation, using the overhead (AC) system or third rail (DC) system. The fleet was originally operated by Railfreight, Eurostar UK and French Railways (SNCF). The Railfreight locos now operated by DB-Cargo, sees just a few remaining operational in the UK, a batch of 13 have been exported to Eastern Europe, with some sold to foreign administrations. A handful are modified to operate over 'HS1' between the Channel Tunnel and London. In 2023, a small number retain the ability to operate from the third rail. Above, displaying EWS-branded Railfreight grey, No. 92041 is shown operating over HS1. The two bodysides of the '92' are near identical, with two grille panels behind the cab door on the non-driving side. In 2023, 16 members of the fleet were on the books of GB Railfreight, not all are operational. The GBRf '92s' power either domestic freight, international freight or the London to Edinburgh/Glasgow sections of the Caledonian Sleeper services. Liveries of the GBRf fleet are varied, some carry full GBRf colours, others carry the 'rounded' livery style by the cab doors, while others carry Caledonian Sleeper teal green, as displayed on No. 92038 below at London Euston. To operate with the CAF-built Mk5 sleeper stock, drop-head Dellner couplings and a revised jumper connection have been fitted to 11 locos. The inset image shows the drop-head Dellner in its raised (coupling) position.* **Jamie Squibbs / Spencer Conquest / Robin Ralston**

Rolling Stock Review : 2023-2024

Class 92

Above: *No. 92032 a GBRf loco which still retains conventional draw-gear, is seen at Edinburgh Waverley. The loco displays full GB Railfreight blue and orange livery.*
Adrian Paul

Right: *Class 92 driving cab, this is based on the general style of the Class 60, with many additional fittings. The '92' cab is one of the most complex on any loco, able to operate using conventional, or Channel Tunnel signalling and work from both 25kV AC overhead or 750V DC third rail supplies. The power controller is to the right of the driver and the brake controllers to the left.*
CJM

2023 Fleet

Number	Operator
92001	EXP
92002	EXP
92003	EXP
92004	DBC
92005	EXP
92006	GBR
92007	DBC
92008	DBC
92009	DBC
92010	GBR
92011	DBC
92012	EXP
92013	DBC
92014	GBR
92015	DBC
92016	DBC
92017	DBC
92018	GBR
92019	DBC
92020	GBR
92021	GBR
92022	EXP
92023	GBR
92024	EXP
92025	EXP
92026	EXP
92027	EXP
92028	GBR
92029	DBC
92030	EXP
92031	DBC
92032	GBR
92033	GBR
92034	EXP
92035	DBC
92036	DBC
92037	DBC
92038	GBR
92039	EXP
92040	GBR
92041	DBC
92042	DBC
92043	GBR
92044	GBR
92045	GBR
92046	GBR

Class 99

After the Stadler Class 93s were ordered by Rail Operations Group/Star Capita in December 2018. GB Railfreight announced a new loco build in March 2022, funded by Beacon Rail, when 30 hi-tech bi-mode Co-Co locos for delivery from 2025 were announced, to be built by Stadler at their Spanish facility. The GBRf fleet will be classified 99 and used to reduce the level of diesel (Class 66) operation of trains working under the wires. The product is based on the Stadler Eurolight platform. Technical details might change as production progresses.

Artist's impression of how the Class 99 might look.
Stadler

Class:	99 Stadler Euro Dual
Number range:	99001-99050
Built by:	Stadler, Valencia, Spain
Years introduced:	2024-2026
Wheel arrangement:	Co-Co
Maximum speed:	75mph (121km/h)
Length:	70ft 1in (21.34m)
Height (pan down):	12ft 2in (3.86m)
Width:	8ft 7in (2.65m)
Weight:	113 tonnes
Wheelbase:	tba
Bogie wheelbase:	tba
Bogie pivot centres:	tba
Wheel diameter (new):	43.3in (1.10m)
Min curve negotiable:	tba
Power supply:	Overhead at 25kV ac
Diesel engine:	Cummins QSK50 V12
Traction output: supply	8,200hp (6.120kW) - ac power 2,415hp (1,800kW) - diesel
Tractive effort:	Max - 112,404lb (500kN) Cont - 96,668lb (430kN)
No. of traction motors:	6
Traction motor type:	ABB AMXL 400 Asynchronous
Brake type:	Air, rheostatic and regenerative
Brake force:	tba tonnes
Coupling:	Screw
Route availability:	7
Multiple coupling type:	Stadler
Heating type:	Not fitted
Owner:	Beacon Rail

Class 93

93001	ROG
93002	ROG
93003	ROG
93004	ROG
93005	ROG
93006	ROG
93007	ROG
93008	ROG
93009	ROG
93010	ROG
93011	ROG
93012	ROG
93013	ROG
93014	ROG
93015	ROG
93016	ROG
93017	ROG
93018	ROG
93019	ROG
93020	ROG
93021	ROG
93022	ROG
93023	ROG
93024	ROG
93025	ROG
93026	ROG
93027	ROG
93028	ROG
93029	ROG
93030	ROG

Class:	93
Number range:	93001-93030
Built by:	Stadler, Spain
Years introduced:	2023-2025
Wheel arrangement:	Bo-Bo
Maximum speed:	110mph (177km/h)
Length:	20.50m
Height (pan down):	12ft 6½in (3.82m)
Width:	2.69m
Weight:	86 tonnes
Wheel diameter (new):	42½in (1.08m)
Power supply:	Overhead at 25kV ac, diesel or battery
Engine:	Caterpillar C32 of 1,200hp (900kW) V12
Battery:	Total: 540hp (400kW) 2 x 200kW using lithium-titanium oxide chemistry
Traction output (max):	Electric 5,364hp (4,000kW) Boost: 6,400hp (4.8kW) [max 30min] Diesel: 1,700hp (1,300Kw) with battery boost
Tractive effort:	65,195lb (290kN)
No. of traction motors:	4
Traction motor type:	ABB AMXL 400
Brake type:	Air and regenerative able to recharge battery
Coupling:	Schackle, Voith 132 (drop-head)
Fuel capacity:	tbc
Route availability:	7
Multiple coupling type:	Stadler
Heating type:	Electric - index 96
Owner/Operator:	Rail Operations Group

Above: Class 93 front end detail, showing the factory fitted drop head Dellner coupling, compatible with most modern multiple unit trains, in the raised position a standard shackle coupling is provided, dual air and main reservoir pipes are fitted, with standard electric train supply connections with a heating index of 96. Jumper and multiple control connections are mounted below the cab window.

Right: Class 93 driving cab, based on the standard Stadler 'Eurolight' layout and similar to the Class 68 and 88.

Below: The Stadler-built Class 93 is one of the most complex dual power (electro-diesel) plus battery powered locos ever built, able to operate from 25kV ac overhead, its on board diesel-electric engine or from battery power. No. 93001, in Rail Operations Group green livery is shown from its No. 1 (pantograph) end, during commissioning trials at Worksop in July 2023. All: **Richard Tuplin**

Rolling Stock Review : 2023-2024

SUBSCRIBE

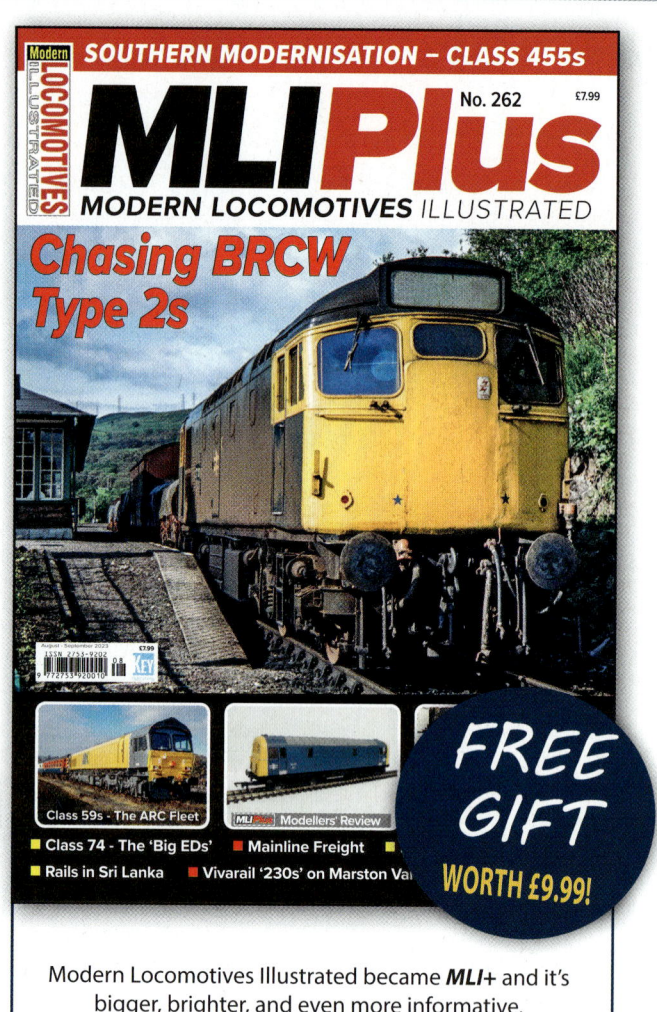

Modern Railways has earned its reputation in the industry as an established and highly respected railway journal.

shop.keypublishing.com/mrsubs

Modern Locomotives Illustrated became **MLI+** and it's bigger, brighter, and even more informative.

shop.keypublishing.com/bawsubs

ORDER DIRECT FROM OUR SHOP...
shop.keypublish

OR CALL +44 (0)1780 480404

(Lines open 9.00-5.30, Monday-Friday GMT)

KEY Publishing

TODAY

SAVE UP TO £41 WHEN YOU SUBSCRIBE!

Hornby Magazine

FREE GIFT WORTH £29.00!

Hornby Magazine takes a unique approach to model railways with both the relatively inexperienced and the seasoned modeller in mind.

shop.keypublishing.com/mlisubs

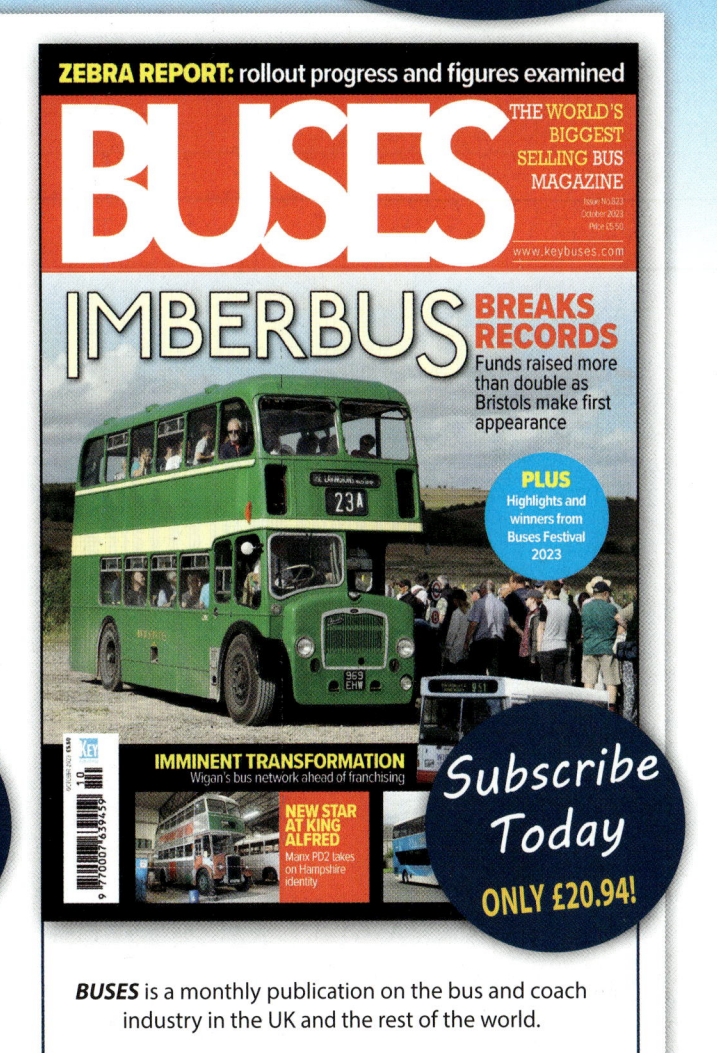

BUSES

Subscribe Today ONLY £20.94!

BUSES is a monthly publication on the bus and coach industry in the UK and the rest of the world.

shop.keypublishing.com/bmsubs

ing.com

Class 139

Class:	139
Vehicle type:	DMS (PPM-60)
Vehicle number range:	39001-39002
Set number range:	139001-139002
TOPS classification:	139
Introduced:	2007-2008
Max speed:	20mph (32km/h)
Built by:	Main Road Sheet Metal, Leyland, for Parry People Movers
Vehicle length (over body):	28ft 6in (8.7m)
Height:	12ft 3in (3.77m)
Width:	7ft 10in (2.81m)
Seating:	20S
Internal layout:	1+1, longitudinal
Gangway:	Not fitted
Toilets:	Not fitted
Weight:	12.5 tonnes
Brake type:	Regen using flywheel & air
Primary power:	Ford MVH-420 engine
Horsepower (total):	86hp (64kW)
Flywheel (stored power):	500kg/1m dia 1,000-1,500rpm wheel
Transmission:	Tandler bevel box with Linde hydrostatic transmission and spiral bevel gearbox
Coupling type:	Emergency only
Multiple restriction:	Not equipped
Door type:	Double-leaf folding
Body structure:	Steel
Owner/Operator:	West Midlands Railway / Pre-Metro Operations
Note:	Only certified to operate Stourbridge Junction to Stourbridge Town

Top: Two Parry People Mover (PPM) ultra light weight single-vehicle twin-cabbed vehicles entered service on the Stourbridge Junction to Stourbridge Town 'shuttle' in 2007-2008. They are based at a depot adjacent to Stourbridge Junction station. One vehicle operates at a time. The pair are classified '139' under TOPS and in 2023 carry WMR mauve/ gold livery. No. 139002 departs from Stourbridge Junction on its three minute journey. **Antony Christie**

Above: Seating for 20 standard class passengers, in a mix of single and longitudinal style are provided, plus a large standing area. The cab is a part of the saloon, with in recent years a plastic hinged door seperating the area. **Adrian Paul**

Above Left: Class 139 front end equipment. 1: Destination indicator, 2: Headlight, 3: Marker or running light, 4: Red tail light, 5: Emergency hauling 'lug'. These vehicles do not conform to main line running standards and must not operate off the Stourbridge Junction to Stourbridge Town line. **CJM**

Below: A driving position is located on the unconventional right side of the vehicle at each end. A combined hydrostatic power/brake controller is located on the right, with a pneumatic brake valve on the left. Vehicles are not fitted with AWS or TPWS, as the top speed is only 20mph (32km/h), lower than the threshold figure. **CJM**

2023 Fleet
139001 WMR
139002 WMR

Key Facts

- Two double-ended, ultra light-weight vehicles, currently work for West Midlands Railway.
- Certified to operate between Stourbridge Junction and Stourbridge Town only.
- '139s' do not confirm to heavy rail crash worthiness.
- Based at special shed building adjacent to Stourbridge Junction station.
- Only one vehicle operates at a time.
- Not fitted with conventional draw-gear, AWS or TPWS.

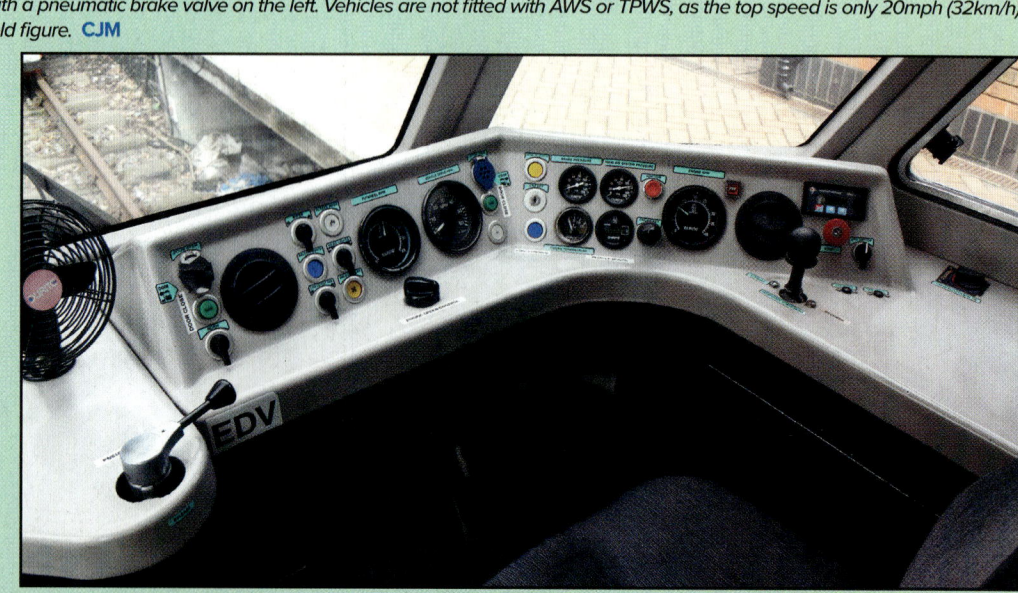

'Sprinter' — Class 150

2023 Fleet

150/0
150001	NOR		
150002	NOR		
150003	NOR		
150004	NOR		
150005	NOR		
150006	NOR		

150/1
150101	NOR
150102	NOR
150103	NOR
150104	NOR
150105	NOR
150106	NOR
150107	NOR
150108	NOR
150109	NOR
150110	NOR
150111	NOR
150113	NOR
150114	NOR
150115	NOR
150118	NOR
150119	NOR
150120	NOR
150121	NOR
150122	NOR
150123	NOR
150124	NOR
150125	NOR
150126	NOR
150127	NOR
150128	NOR
150129	NOR
150130	NOR
150131	NOR
150132	NOR
150133	NOR
150134	NOR
150135	NOR
150136	NOR
150137	WMR
150138	NOR
150139	NOR
150140	NOR
150141	WMR
150142	NOR
150143	NOR
150144	NOR
150145	NOR
150146	NOR
150148	NOR
150149	NOR
150150	NOR

150/2
150201	NOR	150243	GWR
150202	OLS	150244	GWR
150203	NOR	150245	GWR
150204	NOR	150246	GWR
150205	NOR	150247	GWR
150206	NOR	150248	GWR
150207	GWR	150249	GWR
150208	TFW	150250	TFW
150210	NOR	150251	TFW
150211	NOR	150252	TFW
150213	TFW	150253	TFW
150214	NOR	150254	TFW
150215	NOR	150255	GWR
150216	GWR	150256	TFW
150217	TFW	150257	GWR
150218	NOR	150258	TFW
150219	GWR	150259	TFW
150220	NOR	150260	TFW
150221	GWR	150261	GWR
150222	NOR	150262	TFW
150223	NOR	150263	GWR
150225	NOR	150264	GWR
150226	NOR	150265	GWR
150227	TFW	150266	GWR
150228	NOR	150267	TFW
150229	TFW	150268	NOR
150230	TFW	150269	NOR
150231	TFW	150270	NOR
150232	GWR	150271	NOR
150233	GWR	150272	NOR
150234	GWR	150273	NOR
150235	TFW	150274	NOR
150236	TFW	150275	NOR
150237	TFW	150276	NOR
150238	GWR	150277	NOR
150239	GWR	150278	TFW
150240	TFW	150279	TFW
150241	TFW	150280	TFW
150242	TFW	150281	TFW
		150282	TFW
		150283	TFW
		150284	TFW
		150285	TFW

For history and pictures of this class, see Modern Locomotives Illustrated Issue 211

Class:	150/0	150/1	150/2
Number range:	150001-150006§	150101-150150	150201-150285
Built by:	BREL York	BREL York	BREL York
Introduced:	1984	1985-1986	1986-1987
Max speed:	75mph (121km/h)	75mph (121km/h)	75mph (121km/h)
Formation:	DMSL+MS+DMS	DMSL+DMS	DMSL+DMS
Vehicle numbers:	DMSL - 55200-55201 MS - 55400-55401 DMS - 55300-55301	DMSL - 52101-52150 DMS - 57101-57150	DMSL - 52201-52285 DMS - 57201-57285
Vehicle length:	DMSL - 65ft 9¾in (20.06m) MS - 66ft 2½in (20.18m) DMS - 65ft 9¾in (20.06m)	DMSL - 65ft 9¾in (20.06m) DMS - 65ft 9¾in (20.06m)	DMSL - 65ft 9¾in (20.06m) DMS - 65ft 9¾in (20.06m)
Height:	12ft 4½in (3.77m)	12ft 4½in (3.77m)	12ft 4½in (3.77m)
Width:	9ft 3⅛in (2.82m)	9ft 3⅛in (2.82m)	9ft 3⅛in (2.82m)
Weight:	Total - 99.3 tonnes DMSL - 35.4 tonnes MS - 34.1 tonnes DMS - 29.8 tonnes	Total - 76.4 tonnes DMSL - 38.3 tonnes DMS - 38.1 tonnes	Total - 74 tonnes DMSL - 37.5 tonnes DMS - 36.5 tonnes
Internal layout:	2+3	2+3	2+3 or 2+2
Seating:	Total - 240S DMSL - 72S MS - 92S DMS - 76S	Total - 124-144S DMSL - 59-71S DMS - 65-73S	Total - 116-149S DMSL - 60-73S DMS - 56-76S
Gangway:	Within set	Within set	Throughout
Toilets:	DMSL - 1	DMSL - 1	DMSL - 1
Bogie type:	Powered - BX8P Trailer - BX8T	Powered - BP38 Trailer - BT38	Powered - BP38 Trailer - BT38
Power unit:	1 x NT855R4 of 285hp (213kW) per vehicle	1 x NT855R5 of 285hp (213kW) per vehicle	1 x NT855R5 of 285hp (213kW) per vehicle
Transmission:	Hydraulic	Hydraulic	Hydraulic
Transmission type:	Voith T211r	Voith T211r	Voith T211r
Horsepower:	855hp (639kW)	570hp (426kW)	570hp (426kW)
Brake type:	Air EP	Air EP	Air EP
Coupling type:	Outer - BSI Between cars - Bar	Outer - BSI Between cars - Bar	Outer - BSI Between cars - Bar
Multiple restriction:	Class 15x, 170, 172	Class 15x, 170, 172	Class 15x, 170, 172
Door type:	Bi-parting sliding (cab door - slide)	Bi-parting sliding (cab door - slam)	Bi-parting sliding (cab door - slide)
Body structure:	Steel	Steel	Steel
Operator:	Northern Rail	Northern Rail, WMR	GWR, TfW, Northern Rail
Owner:	Angel Trains	Angel Trains	Angel Trains, Porterbrook
Notes:	Original prototype sets § Sets 150003-150006, formed of a Class 150/1 plus one 150/2 vehicle.	Non-corridor sets	Corridor sets

Right: *The first of the BR 'Sprinter' family emerged from BREL York in 1984, with two prototype three-car sets, built as development trains to compare with the Metro-Cammell Class 151s. BREL won the production order, with a fleet of 50 two-car sets. The two prototype units were the only three-car formed Class 150s, they are currently working with Northern. They differ from production Class 150/1s by having power operated crew doors. No. 150002, which for a period operated as a Class 154 during development work for Class 158s, has a slightly different roof profile. The set is seen at Sheffield.* **Antony Christie**

Key Facts

- Largest fleet of second generation 'Sprinter' DMU stock.
- Three sub-classes - 150/0 three-car prototype sets, Class 150/1 - two-car non-corridor sets, Class 150/2 - two-car corridor sets.
- A large number of different interior designs can be found.
- Class 150/1s fitted with hinged crew access doors.
- Four 'extra' 150/0 sets operate in Manchester-Leeds area, formed of a Class 150/1 with a Class 150/2 vehicle coupled between, operating alongside Class 150/0 sets.

Right: *Between 1985-1986, BREL York built a fleet of 50 production two-car 'Sprinter' sets, these were non-gangwayed, but similar to the prototype Class 150/0s. Classified as 150/1, the sets incorporated a hinged crew door. Today, the sub-class are operated by Northern and carry Northern white and mauve livery. Set No. 150142 is seen at Southport from its DMSL end. Front end equipment consists of a BSI coupling, incorporating air and electrical connections. Two standard light clusters, housing a marker, tail and head light are mounted on each side. A destination indicator is located in the upper section of the middle front window.* **CJM**

Rolling Stock Review : 2023-2024

Class 150   'Sprinter'

Above: When introduced, the Class 150/1s were fitted with 2+3 seating, for medium-distance regional services. Today, after more than 30 years in service and several overhauls and refurbishments, revised interiors with a mix of 2+2 and 2+3 seating is found. A universal access toilet, as well as wheelchair space is also provided. The interior of a Class 150/1 DMS is shown. Class 150/1s have straight glass dividers at door positions and most have loop top seat handles. **CJM**

Above: A fleet of 85 Class 150/2 sets, built with end corridors, emerged from BREL York in 1986-87. Originally, seating was in the 2+3 layout, with one toilet, at the inner end of the DMSL (52xxx) vehicle. Angled pillars are incorporated by door pockets, with triangular glazed screens, an above seat light luggage rack are provided. Today, sets are refurbished, with the fleets main operator Northern retaining the 2+3 layout, but with revised seats, decor, lighting and a universal toilets. Set No. 150273 is seen at Mexborough. **Antony Christie**

Left & Below: Class 150/2 front end detail. 1: Destination indicator, 2: Front gangway door, 3: Lamp bracket, 4: Headlight, 5: Combined white marker and red tail light, 6: BSI coupling, 7: Electrical connection box with roller cover, 8: Emergency jumper connection socket, 9: Obstacle deflector plate, 10: Air warning horns. 11: Forward facing camera, illustrated below. Both: **CJM**

Due to the withdrawal of the Class 230 Vivarail sets from the West Midlands Railway Marston Vale Line at the end of 2022, other rolling stock was sought in early 2023. With restricted length platforms the choice was limited to Class 150s and three former Northern sets Nos. 150137, 150139 and 150141 have been placed on long-term loan from autumn 2023. Before entry into service, sets will be internally upgraded and given a revised WMR Marston Vale white and green livery, as illustrated in the graphic. **WMR**

Above & Inset: During 2023, Transport for Wales operated 36 Class 150/2 sets, plus two on sub-lease from Northern, allocated to Cardiff Canton depot working Valley Line and longer distance services. The fleet now carries the latest TfW white and red livery and are in their declining years in traffic, soon to be replaced by Class 197 and 398 'Tram-Train' sets. All TfW '150s' have 2+2 low-density seating, using individual seating with hinged armrests, finished in a grey fleck moquette, with a red grab handle at the top of each seat. The sets a universal access toilet. Set No. 150283 is seen at Radyr. The inset image shows the 2+2 interior layout, with a mix of group and airline style seating. Charging plugs and USB sockets are provided. Both: **CJM**

'Sprinter' — Class 150

Above, Below, Right & Insets: Exeter depot is the base for a fleet of 20 Class 150/2 sets. These operate local services on the Exeter to Barnstaple, Okehampton, Exmouth and Paignton routes, as well as throughout the GWR western area. All but two sets carry the GWR branded green livery, the exceptions are Nos. 150219 and 150238 which still carry First Great Western blue and pink. Interiors have all been upgraded and now incorporate a universal access toilet. Set No. 150233 is seen in the main image passing Dawlish Warren, led by its 52xxx vehicle. These sets have a medium life expectancy with GWR (until at least summer 2027). The inset left view shows the Robust Train Positioning System (RTPS) equipment fitted to the front of set 150263 for trials. The inset right shows the driving side body of the DMSL, with a blanked-out window at the inner end, adjacent to the universal toilet compartment. Below left, on of the GW blue-liveried pair No. 150238 is seen from its DMSL coach. The image below right shows the interior with 2+2 Chapman seating. Door screens are of the angled style. Ceiling mounted passenger information screens are provided. All: **CJM**

Class 950 No. 950001

During the mid-1980s, the Director of Civil Engineering, based at the Railway Technical Centre (RTC), Derby sought to obtain a new two-car test train for track inspection and assessment work.

Funding was released for the building of two extra, heavily modified, body shells as part of the Class 150/1 build at BREL York. These body shells were then fitted out at the RTC Derby.

Emerging in 1987, the train was originally described as a Class 180, but this was never officially carried. It later became Class 950, No. 950001. It is formed of vehicles Nos. 999600 and 999601 and today carries branded Network Rail track inspection yellow livery.

It is based at the RTC Derby and operates throughout the railway system on England, Wales and Scotland on a programmed basis.

Over the years, the front ends have been modified many times to carry different forms of lighting and recording equipment. Different cabling is also found on the cab ends. Currently, the train is operated by Loram.

Above & Left: Led by vehicle No. 999600, set No. 950001, which only displays its set number on this end, is seen calling at Bangor on 28 March 2023. On the outside, one side of each vehicle has normal windows, the other is panelled by equipment positioning. On the exterior, Network Rail decals with pictogram 'Track Recording Unit' branding is carried. The cabs are virtually identical to the production Class 150/1 sets. The power controller is on the right, with brake controller to the left. On the right, is a Radio Electronic Token Block transmitter/receiver. Both: **CJM**

Class 153 'Super Sprinter'

For history and pictures of this class, see *Modern Locomotives Illustrated* Issue 211

Class:	153
Number range:	153301-153385
Original build:	Leyland Bus, Workington (as Class 155)
Rebuilt (single car) by:	Hunslet-Barclay, Kilmarnock
Introduced:	As 153 - 1991-1992
	Originally as Class 155 2-car sets - 1987-1988
Max speed:	75mph (121km/h)
Formation:	DMSL
Vehicle numbers:	52301-52335, 57351-57387
Vehicle length:	76ft 5 in (23.29m)
Height:	12ft 3⅜in (3.75m)
Width:	8ft 10in (2.69m)
Weight:	41.2 tonnes
Internal layout:	2+2
Seating:	66 or 72S depending on layout
Gangway:	Throughout
Toilets:	1 (isolated on some sets)
Bogie type:	Powered - BREL P3-10
	Trailer - BREL BT38
Power unit:	1 x Cummins NT855R5
Transmission:	Hydraulic
Transmission type:	Voith T211r
Horsepower (total):	285hp (213kW)
Brake type:	Air EP
Coupling type:	BSI
Multiple restriction:	Class 15x, 170, 172 series
Door type:	Single-leaf sliding plug
Body structure:	Aluminium alloy on steel frame
Operator:	Transport for Wales and Scotrail
Owner:	Angel Trains, Porterbrook Leasing, TfW
	Many vehicles off-lease and stored

Above: The Class 153s sport two different front end styles, following conversion from Class 155 two-car sets. The original end is shown left, with the 'new' (or small cab end) right. Equipment positions are. 1: Destination indicator, 2: Gangway door, 3: Lamp bracket, 4: Headlight, 5: Combined white marker and red tail light, 6: BSI coupling, 7: Electric connection box, 8: Warning horns, 9: Emergency jumper socket. **CJM / Antony Christie**

Below & Lower: A fleet of 70 single car Class 153s were introduced in 1991-1992 by splitting the original fleet of two-car Class 155s and building a new (small) cab in the previous vestibule area. The conversion allowed Regional Railways to have a fleet for use on rural branch lines and secondary routes. Today, the fleet is in rapid decline, with in 2023 only Transport for Wales and ScotRail still operating sets. To allow TfW use until new trains are delivered, some '153s' have been upgraded to meet regulations for passenger with reduced mobility, with an universal toilet installed and extra space for a wheelchair with an adjacent call point. Unmodified sets have been renumbered in the 1539xx series. Seating is in the 2+2 style. Modified sets have had their toilet compartment moved to the opposite end of the vehicle to allow access. In the two views below, both modified and unmodified vehicles are shown, the upper illustration shows modified No. 153325, showing plated windows at both ends adjacent to the original an present toilet positions. The image below of No. 153913, an unmodified sets shows window plating at one end. The vehicles carry standard TfW grey, black and red livery. Both: **CJM**

Above: Transport for Wales Class 153 interior, showing the 2+2 layout using Chapham seats. Seats are a mix of group and airline, with airline seats facing the middle of the coach. On Class 1539xx sets, the toilet compartment should be locked out of use. **CJM**

Key Facts

- Built as two-car Class 155s for BR Provincial sector.
- Rebuilt as single cars by Hunslet Barclay for use on rural lines.
- Some vehicles modified to meet Passengers with Restricted Mobility restrictions, with a universal toilet compartment at the opposite end of the coach to where it was originally fitted.
- The two driving cab ends are of different design, original has low-level light clusters, 'new' has mid- height light clusters. Different cab layouts.

Left: The 'new' Class 153 small driving cab, built into the former vestibule position and providing very cramped accommodation for the driver. The brake controller is to the left and the power controller to the right. **CJM**

2023 Fleet

153303	TFW
153305	ASR
153906	TFW
153307	TFW*
153909	TFW
153910	TFW
153311	NRL
153312	TFW
153913	TFW
153914	TFW
153318	TFW
153320	TFW
153921	TFW
153922	TFW
153323	TFW
153325	TFW
153926	TFW
153327	TFW
153328	TFW*
153329	TFW
153331	TFW*
153333	TFW
153935	TFW
153352	TFW*
153353	TFW
153361	TFW
153362	TFW
153367	TFW
153968	TFW
153369	TFW
153370	ASR
153371	ASR
153972	TFW
153373	ASR
153376	NRL
153377	ASR
153380	ASR
153982	TFW
153385	NRL

* For TfW 'Active Travel'

Rolling Stock Review : 2023-2024

'Super Sprinter' — Class 153

Above & Left: In 2019-2021 ScotRail and Brodie Engineering of Kilmarnock transformed five Class 153s Nos. 153305/370/373/377/380, with a major engineering upgrade to form 'Active Travel' vehicles, including provision for cycles and large sporting equipment. Introduced in late 2021, the coaches operate on the West Highland Line from Glasgow to Oban. The former Great Western sets had a full internal rebuild, including installation of a free wi-fi system, at-seat power sockets and a universal retention toilet. In addition, to providing custom-designed cycle and luggage racks, 20 'high class' seats in the 2+2 layout, set out in groups around tables are located at one end. Above, No. 153373 is seen at Glasgow Queen Street with its luggage/bike space at the near end. On the left, the 2+2 group seating is seen, which includes route maps set into the table tops, carpets on the floor and above seat light luggage racks. Both: **Adrian Paul**

Left, Left Below & Below: In 2021-2023, Network Rail obtained several Class 153s, Nos. 153311, 153376, 153379, 153384 and 153385 for conversion as Video Inspection Units (VIUs). The vehicles are fitted with video surveillance equipment on the ends and underslung, with the results recorded in equipment mounted in the former passenger area. The VIUs can operate just with a driver and no technical staff on board or can be attached to a service train, as long as it is fitted with BSI couplings. On the left, No. 153385 is seen from its small cab end, fitted with forward facing recording equipment. This vehicle retains its former East Midlands Trains livery now with black Network Rail branding. Below left is VIU2, No. 153376, captured south of York on 30 May 2023, which in 2022 was upgraded and repainted into a new Inspection Train sky blue livery at the RTC Derby. The vehicle carries a number of forward facing cameras and can if needed by fitted with, downward facing cameras, as shown in the illustration of the No. 1 front end below right.
Spencer Conquest / Phil Pecious / CJM

Rolling Stock Review : 2023-2024

VISIT OUR ONLINE SHOP
TO VIEW OUR FULL RANGE OF RAIL BOOKS

Key Shop

shop.keypublishing.com/books

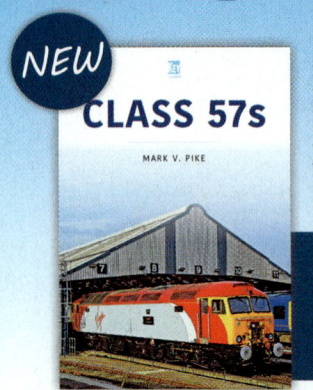

NEW

CLASS 57s
MARK V. PIKE

Known irreverently as 'bodysnatchers', the Class 57s have filled a gap in available traction for the last two decades and the fleet is intact today, continuing to see use with various TOCs and future-proofed by overhaul. All 33 locos are depicted in this volume, with over 200 images in number order from 57001–57012, 57301–57316 and 57601–57605, showing a variety of duties with which they have been entrusted in the last 20 years.

ONLY £16.99
SUBSCRIBERS don't forget to use your £2 OFF DISCOUNT CODE!

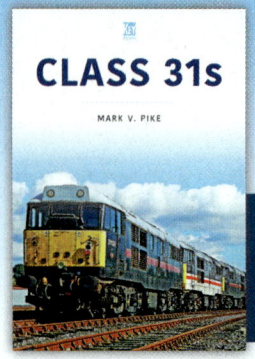

CLASS 31s
MARK V. PIKE

Perhaps overshadowed by more 'exotic' locos, the Class 31s nonetheless played a significant part in British Railways' Modernisation Plan.

This volume includes more than 200 images showing these workhorses in action, mostly in the south of England.

ONLY £16.99
SUBSCRIBERS don't forget to use your £2 OFF DISCOUNT CODE!

RAIL FREIGHT
LONDON AND THE SOUTH EAST
PAUL SHANNON

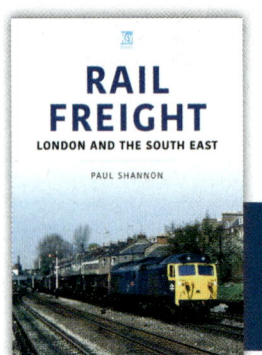

Illustrated with more than 160 photographs, many of which are previously unpublished, this volume looks at the changing face of rail freight in London and the South East of England. It details the transformation in traction, rolling stock and railway infrastructure over four decades.

ONLY £16.99
SUBSCRIBERS don't forget to use your £2 OFF DISCOUNT CODE!

ELECTROSTAR
CAPITAL COMMUTER
IAN BUCK

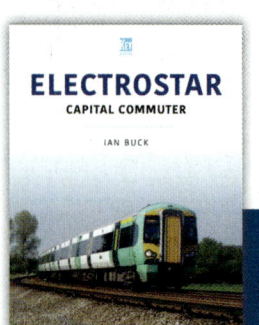

With nearly 200 previously unpublished images, this book gives an overview of the routes the Electrostar has worked, or is still working, as well as the different companies that these unsung heroes of the everyday railway have served.

ONLY £15.99
SUBSCRIBERS don't forget to use your £2 OFF DISCOUNT CODE!

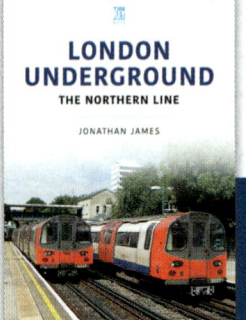

LONDON UNDERGROUND
THE NORTHERN LINE
JONATHAN JAMES

This book charts the progression of the Northern line, which dates back more than 125 years, and illustrates how three railway companies evolved and then merged to become one vast operation integrated into London's transport system.

ONLY £17.99
SUBSCRIBERS don't forget to use your £2 OFF DISCOUNT CODE!

HERITAGE DMUs
THE FINAL YEARS
IAN BUCK

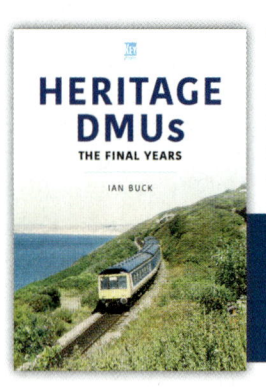

This book provides an overview of the final operations that the heritage DMUs were involved in. These unsung heroes of the 1955 modernisation plan certainly gave their worth and, thankfully, many can still be seen today on the railway preservation scene.

ONLY £16.99
SUBSCRIBERS don't forget to use your £2 OFF DISCOUNT CODE!

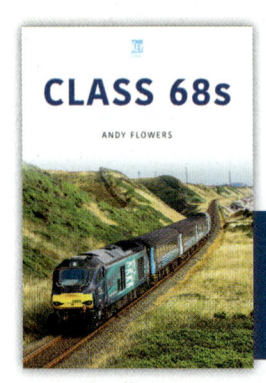

CLASS 68s
ANDY FLOWERS

A detailed technical specification is included in this book, together with a review of DRS's chartered services of recent years, including on railtours. Illustrated with more than 100 photographs, this visual guide presents a comprehensive overview of the role of today's Class 68.

ONLY £16.99
SUBSCRIBERS don't forget to use your £2 OFF DISCOUNT CODE!

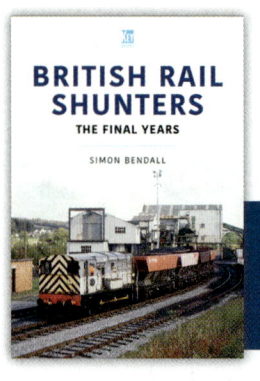

BRITISH RAIL SHUNTERS
THE FINAL YEARS
SIMON BENDALL

This book delves into a period of substantial reorganisation with a pictorial look at the shunter fleet in its final years of BR ownership between 1990 and 1996, tracing the changes in operators, the emergence of new colour schemes and alterations in deployment.

ONLY £16.99
SUBSCRIBERS don't forget to use your £2 OFF DISCOUNT CODE!

FREE P&P* when you order online at…
shop.keypublishing.com/books
Call +44 (0)1780 480404 (Monday to Friday 9am-5.30pm GMT)

*Free 2nd class P&P on all UK & BFPO orders. Overseas charges apply.

'Super Sprinter' — Class 155

Class:	155
Number range:	155341-155347
Built by:	Leyland Bus, Workington
Introduced:	1988
Max speed:	75mph (121km/h)
Formation:	DMSL+DMS
Vehicle numbers:	DMSL - 52341-52347
	DMS - 57341-57347
Vehicle length:	76ft 5in (23.29m)
Height:	12ft 3⅜in (3.75m)
Width:	8ft 10in (2.69m)
Weight:	Total - 77.8 tonnes
	DMSL - 39.2 tonnes
	DMS - 38.6 tonnes
Internal layout:	2+2
Seating:	Total - 140S
	DMSL - 64S, DMS - 76S
Gangway:	Throughout
Toilets:	DMSL - 1
Bogie type:	Powered - BREL P3-10
	Trailer - BREL BT38
Power unit:	1 x Cummins NT855R5 of 285hp (213 kW) per vehicle
Transmission:	Hydraulic
Transmission type:	Voith T211r
Horsepower (total):	570hp (426kW)
Brake type:	Air EP
Coupling type:	BSI
Multiple restriction:	Class 15x, 170, 172 series
Door type:	Single-leaf sliding plug
Body structure:	Aluminium alloy on steel frame
Operator:	Northern
Owner:	Porterbrook Leasing

Above: Seven, two-car Class 155s operate for Northern, based at Leeds Neville Hill depot. They work on West Yorkshire services, to locations such as York, Scarborough and Hull. These standard class only sets have 2+2 seating in a mix of group and airline styles, seating is for 140. A universal toilet is located in the DMSL (52xxx) vehicle, which also houses a wheelchair space and call point. All sets are painted in Northern livery. Set No. 155345 is seen arriving at Hull on the York-Hull corridor. **CJM**

Left: Class 155 front end layout. 1: Route/destination indicator, 2: Gangway door, 3: Lamp bracket, 4: Forward facing camera, 5: Headlight, 6: Combined marker light (white and tail light (red), 7: Air horns, 8: BSI coupling, 9: Electrical connection box with rotary cover, 10: Obstacle deflector plate, 11: Emergency jumper socket. **Adrian Paul**

Right Lower: When introduced in 1988, these seven units were funded by West Yorkshire Passenger Transport Executive. The internally fitted makers plate still reflects this original ownership. **Adrian Paul**

Below: The Class 155 interior is based on the 2+2 layout using Chapman high back seats. Seats are arranged in a mix of airline and group layout, Above seat light luggage rakes are installed and as these sets are not air conditioned, opening hopper windows are provided in every window. Passenger access to the seating area is by way of a single leaf sliding plug door at vehicle ends, feeding a transverse walkway from which a door leads into the passenger saloon. **Adrian Paul**

Key Facts

- Seven, two-car Class 155s operate with Northern on West Yorkshire services. When introduced a fleet of 42 two-car sets were built, 35 were later modified as Class 153 'Bubble' cars.
- Sets were known as 'Super Sprinter' units when introduced.
- Sets now sport the latest Northern white livery have Chapman seting and sport revised front lamp clusters.

2023 Fleet

155341	NOR
155342	NOR
155343	NOR
155344	NOR
155345	NOR
155346	NOR
155347	NOR

Rolling Stock Review : 2023-2024

Class 156   'Super Sprinter'

For history and pictures of this class, see Modern Locomotives Illustrated issue 211

Class:	156
Number range:	156401-156514, 156902-156922
Built by:	Metro-Cammell, Birmingham
Introduced:	1987-1989
Max speed:	75mph (121km/h)
Formation:	DMSL+DMS
Vehicle numbers:	DMSL - 52401-52514
	DMS - 57401-57514
Vehicle length:	75ft 6in (23.03m)
Height:	12ft 6in (3.81m)
Width:	8ft 11in (2.73m)
Weight:	Total - 74.7 tonnes
	DMSL - 38.6 tonnes
	DMS - 36.1 tonnes
Internal layout:	2+2
Seating:	Total - 140-152S
	DMSL - 68-74S
	DMS - 72-78S
Gangway:	Throughout
Toilets:	DMSL - 1
Bogie type:	Powered - BREL P3-10
	Trailer - BREL BT38
Power unit:	1 x Cummins NT855R5 of 285hp (213kW) per vehicle
Transmission:	Hydraulic
Transmission type:	Voith T211r
Horsepower (total):	570hp (426kW)
Brake type:	Air EP
Coupling type:	BSI
Multiple restriction:	Class 15x, 170, 172 series
Door type:	Single-leaf sliding
Body structure:	Steel
Operator:	Northern, ScotRail
Owner:	Porterbrook Leasing (38), Angel Trains (75), Brodie Rail (1)

Right: *Class 156 front end equipment positions. 1: Route/destination indicator, 2: Gangway door, 3: Headlight, 4: Combined white marker and red tail light, 5: BSI coupling, 6: Warning horns (behind), 7: Electric connection box with rotary cover, 8: Emergency jumper socket, 9: Obstacle deflector plough plate. Forward facing cameras are progressively being fitted in the non-driving front window.* **Antony Christie**

Left: *Scottish Railways operated a fleet of 43 Class 156s, allocated to Corkerhill depot, Glasgow, In mid-2023. All are painted in the current Scottish Railways blue and white livery. The Scottish fleet is owned by Angel Trains, except set No. 156478, this set was officially written-off by its owners due to flood damage, it was purchased by Brodie Engineering of Kilmarnock, who rebuilt the set and now lease it to Scottish Railways. Set No. 156501 is seen stabled in the bay at Carlisle with a Glassgow via Annan service.* **CJM**

Below: *In mid-2023, Northern operated a fleet of 51 Class 156s, allocated to Heaton (Newcastle) or Newton Heath (Manchester) depots. The fleet work longer-distance services alongside Class 158s, all have been refurbished in recent years to incorporate a universal style toilet in the 52xxx vehicle. Sets carry the standard Northern white and blue livery. A small number of sets are named, these are applied in 'stick-on' plates at cant rail height just inward of the end passenger doors. No. 156469, with the name* The Royal Northumberland Fusiliers (The Fighting Fifth) *is shown departing from Carlisle with a working to Newcastle on 4 June 2022.* **CJM**

Rolling Stock Review : 2023-2024

'Super Sprinter' — Class 156

2023 Fleet

156401	NOR	156458	ASR
156402	NOR	156459	NOR
156403	NOR	156460	NOR
156404	NOR	156461	NOR
156405	NOR	156462	ASR
156406	NOR	156463	NOR
156907	OLS	156464	NOR
156408	NOR	156465	NOR
156909	OLS	156466	NOR
156410	NOR	156467	ASR
156411	NOR	156468	NOR
156412	NOR	156469	NOR
156413	NOR	156470	EMR
156414	NOR	156471	NOR
156415	NOR	156472	NOR
156916	OLS	156473	EMR
156417	NOR	156474	ASR
156918	OLS	156475	NOR
156419	NOR	156476	ASR
156420	NOR	156477	ASR
156421	NOR	156478	ASR
156422	NOR	156479	NOR
156423	NOR	156480	NOR
156424	NOR	156481	NOR
156425	NOR	156482	NOR
156426	NOR	156483	NOR
156427	NOR	156484	NOR
156428	NOR	156485	NOR
156429	NOR	156486	NOR
156430	ASR	156487	NOR
156431	ASR	156488	NOR
156432	ASR	156489	NOR
156433	ASR	156490	NOR
156434	ASR	156491	NOR
156435	ASR	156492	ASR
156436	ASR	156493	ASR
156437	ASR	156494	ASR
156438	NOR	156495	ASR
156439	ASR	156496	NOR
156440	NOR	156497	EMR
156441	NOR	156498	EMR
156442	ASR	156499	ASR
156443	NOR	156500	ASR
156444	NOR	156501	ASR
156445	ASR	156502	ASR
156446	ASR	156503	ASR
156447	NOR	156504	ASR
156448	NOR	156505	ASR
156449	NOR	156506	ASR
156450	ASR	156507	ASR
156451	NOR	156508	ASR
156452	NOR	156509	ASR
156453	ASR	156510	ASR
156454	NOR	156511	ASR
156455	NOR	156512	ASR
156456	ASR	156513	ASR
156457	ASR	156514	ASR

Right: Northern Class 156, interior, showing set No. 156460 with Chapman seating. Seats are in the low-density 2+2 style with a mix of group and airline layouts. hinged arm rests are available, fold down tables on airline style seats. Charging points are available by most seats, above seat luggage racks are provided and a modern automated passenger information display. **CJM**

Right Lower: In 2023, the 18 Class 156s operated by East Midlands Railway on 'Regional' services, were transferred to Northern, many in 2023-2024 still retain EMR livery. This batch sports 2+2 seating in a mix of airline and group. Many sets carry aubergine and white colours. Set No. 156406, is seen from its 52xxx DMSL vehicle. The '2' on the front end indicates the 52xxx coach. **Cliff Beeton**

Left: Although many of the former East Midlands Class 156s re-branded in aubergine and while colours before transfer away to Northern, many interiors still sport the earlier Stagecoach colours, but still look very neat, except for carpets which look very worn. This view shows the interior of set 156405, showing the comfortable Chapman style seats, set out in a mix of group and airline styles. Northern are scheduled to fully refurbish the inherited '156' interiors. **CJM**

Key Facts

- Metro-Cammell built 'Super Sprinter' sets, currently operated by East Midlands, Northern and ScotRail.
- All units modified with universal access toilets and a wheelchair space.
- Many different interior designs and seating can be found.
- Some ScotRail sets fitted with Radio Electronic Token Block (RETB) equipment for single line operation.

Class 158

For history and pictures of this class, see Modern Locomotives Illustrated Issue 190

Class:	158/0 (2-car)	158/0 (3-car)	158/8	158/9
Number range:	158701-158872 (not 157752-759/798)	158752-158759/158798	158880-158890	158901-158910
Built by:	BREL Derby	BREL Derby	BREL Derby	BREL Derby
Rebuilt by:			Wabtec, Doncaster	
Introduced:	1990-1992	1990-1992, as 3-car - 2000	1991, as 158/8 - 2007	1991
Max speed:	90mph (145km/h)	90mph (145km/h)	90mph (145km/h)	90mph (145km/h)
Formation:	DMSL(A) + DMSL(B) or DMSL(A) + DMCL	DMSL(A)+MSL+DMSL(B)	DMCL+DMSL	DMSL+DMS
Vehicle numbers:	DMSL(A) - 57701-57872 (series) DMSL(B) or DMCL - 52701-52872 (series)	DMSL(A) - 52752-52759/52798 MSL - 58702-58716 DMSL(B) - 57752-57759/57798	DMCL - 52737-52814 series DMSL - 57737-57814 series	DMSL - 52901-52910 DMS - 57901-57910
Vehicle length:	76ft 1¾in (23.20m)	76ft 1¾in (23.20m)	76ft 1¾in (23.20m)	76ft 1¾in (23.20m)
Height:	12ft 6in (3.81m)	12ft 6in (3.81m)	12ft 6in (3.81m)	12ft 6in (3.81m)
Width:	9ft 3¼in (2.82m)	9ft 3¼in (2.82m)	9ft 3¼in (2.82m)	9ft 3¼in (2.82m)
Weight:	Total - 77 tonnes DMSL - 38.5 tonnes DMCL - 38.5 tonnes	Total - 115.5 tonnes DMSL - 38.5 tonnes MSL - 38.5 tonnes DMSL - 38.5 tonnes	Total - 77 tonnes DMCL - 38.5 tonnes DMSL - 38.5 tonnes	Total - 77 tonnes DMSL - 38.5 tonnes DMS - 38.5 tonnes
Internal layout:	2+2S, 2+2F	2+2	2+2S, 2+1F	2+2
Seating:	Total - 13-15F/110-127S DMSL - 64-74S DMCL - 13-15F / 46-53S depending on operator	Total - 204S DMSL(A) - 68S MSL - 66S + 3 tip up DMSL(B) - 70S	Total - 13F/114S DMCL - 13F/44S DMSL - 70S	Total - 142S DMSL - 70S DMSL - 72S
Gangway:	Throughout	Throughout	Throughout	Throughout
Toilets:	DMSL, DMCL 1	DMSL, MSL - 1	DMCL, DMSL - 1	DMSL - 1
Bogie type:	Powered - BREL P4 Trailer - BREL T4	Powered - BREL P4 Trailer - BREL T4	Powered - BREL P4 Trailer - BREL T4	Powered - BREL P4 Trailer - BREL T4
Power unit:	158701-814 1 x NT855R of 350hp (260kW) per vehicle 158815-862 1 x Perkins 2006-TWH of 350hp (260kW) per vehicle 158863-872 1 x NT855R of 400hp (300kW) per vehicle	1 x NT855R of 350hp per vehicle	1 x NT855R of 350hp per vehicle	1 x NT855R of 350hp per vehicle
Transmission:	Hydraulic	Hydraulic	Hydraulic	Hydraulic
Transmission type:	Voith T211r	Voith T211r	Voith T211r	Voith T211r
Horsepower (total):	158701-158862 - 700hp (522kW) 158863-158872 - 800hp (597kW)	1050hp (780kW)	700hp (522kW)	700hp (522kW)
Brake type:	Air EP	Air EP	Air EP	Air EP
Coupling type:	BSI	BSI	BSI	BSI
Multiple restriction:	Class 15x, 170, 172	Class 15x, 170, 172	Class 15x, 170, 172	Class 15x, 170, 172
Door type:	Bi-parting swing plug	Bi-parting swing plug	Bi-parting swing plug	Bi-parting swing plug
Body structure:	Aluminium	Aluminium	Aluminium	Aluminium
Operator:	ScotRail, Northern, Transport for Wales, GWR, East Midlands Railways	Northern, GWR	South Western Railway	Northern
Owner:	Porterbrook Leasing / Angel Trains	Porterbrook Leasing	Porterbrook Leasing	Eversholt Leasing

Above & Inset: *The Welsh rail operation, controlled by the Welsh Government and trading as Transport for Wales (TfW), uses the Wales and Borders operating title. The business is undergoing a major transition in terms of rolling stock with a near complete fleet replacement. In 2023, 24 two-car Class 158s are in use, these carry the TfW white livery, with black window surrounds and contrasting red passenger doors. Sets Nos. 158835 and 158829 arrive at Gobowen on 12 September 2022 with a Holyhead to Birmingham International service, with its 52xxx coach leading. The TfW Class 158s are fitted with the European Rail Traffic Management System (ERTMS) for operation over the Cambrian routes. All sets are refurbished and include Grammer seating for 132 standard class passengers with seating finished in a dark grey moquette with white and red fleck, with red fittings, as illustrated left.* **CJM / Antony Christie**

Rolling Stock Review : 2023-2024

Class 158

158/0
158950-158959
BREL Derby

1992 (as 3-car 2008)
90mph (145km/h)
DMSL+DMSL+DMSL

DMSL - 527xx and 577xx series
DMSL - 527xx and 577xx series
DMSL - 527xx and 577xx series
76ft 1¾in (23.20m)
12ft 6in (3.81m)
9ft 3¼in (2.82m)
Total - 115.5 tonnes
DMSL - 38.5 tonnes
DMSL - 38.5 tonnes
DMSL - 38.5 tonnes
2+2
Total - 200S - 202S
DMSL - 66S
DMSL - 68S
DMSL - 66S or 68
Throughout
DMSL - 1
Powered - BREL P4
Trailer - BREL T4
1 x NT855R of 350hp per vehicle

Hydraulic
Voith T211r
1050hp (780kW)

Air EP
BSI
Class 15x, 170, 172
Bi-parting swing plug
Aluminium
GWR

Porterbrook Leasing

Left: *Class 158 front end layout. 1: Destination/route indicator (electronic type), 2: Gangway door, 3: White marker light, 4: Headlight, 5: Red tail light, 6: BSI coupling, 7: Warning horns, recessed by side and rear of coupling, 8: Electrical connector box with roller cover, 9: Emergency jumper socket (behind locked door). Set No. 158855 is from the Northern allocation and is seen at Sheffield. Some revisions in equipment style may be found, including the increased use of revised light clusters with a combined marker/tail light. Some sets have a curved obstacle deflector (plough).* **CJM**

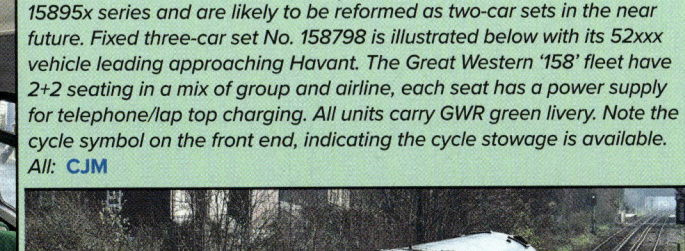

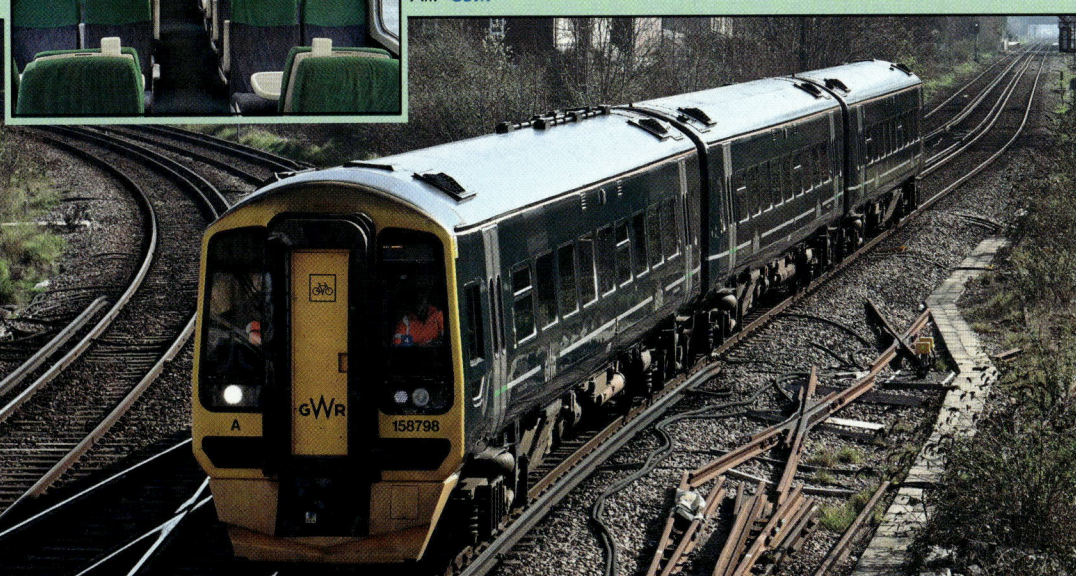

Right, Below & inset: *Great Western Railway, operates a fleet of two- and three-car '158s', based at Bristol St Philips Marsh and Exeter depots. One three-car is formed DMSL+MSL+DMSL and the other three-car sets operate with three driving cars. In the above image, two-car set No. 158767 is seen heading west at Dawlish Warren. The three driving car sets, are numbered in the 15895x series and are likely to be reformed as two-car sets in the near future. Fixed three-car set No. 158798 is illustrated below with its 52xxx vehicle leading approaching Havant. The Great Western '158' fleet have 2+2 seating in a mix of group and airline, each seat has a power supply for telephone/lap top charging. All units carry GWR green livery. Note the cycle symbol on the front end, indicating the cycle stowage is available. All:* **CJM**

2023 Fleet

158/0
158701	ASR	158815	NOR
158702	ASR	158816	NOR
158703	ASR	158817	NOR
158704	ASR	158818	TFW
158705	ASR	158819	TFW
158706	ASR	158820	TFW
158707	ASR	158821	TFW
158708	ASR	158822	TFW
158709	ASR	158823	TFW
158710	ASR	158824	TFW
158711	ASR	158825	TFW
158712	ASR	158826	TFW
158713	ASR	158827	TFW
158714	ASR	158828	TFW
158715	ASR	158829	TFW
158716	ASR	158830	TFW
158717	ASR	158831	TFW
158718	ASR	158832	TFW
158719	ASR	158833	TFW
158720	ASR	158834	TFW
158721	ASR	158835	TFW
158722	ASR	158836	TFW
158723	ASR	158837	TFW
158724	ASR	158838	TFW
158725	ASR	158839	TFW
158726	ASR	158840	TFW
158727	ASR	158841	TFW
158728	ASR	158842	NOR
158729	ASR	158843	NOR
158730	ASR	158844	NOR
158731	ASR	158845	NOR
158732	ASR	158846	EMR
158733	ASR	158847	EMR
158734	ASR	158848	NOR
158735	ASR	158849	NOR
158736	ASR	158850	NOR
158737	ASR	158851	NOR
158738	ASR	158852	EMR
158739	ASR	158853	NOR
158740	ASR	158854	NOR
158741	ASR	158855	NOR
158745	GWR	158856	EMR
158747	GWR	158857	EMR
158748	GWR	158858	EMR
158749	GWR	158859	NOR
158752	NOR	158860	NOR
158753	NOR	158861	EMR
158754	NOR	158862	EMR
158755	NOR	158863	NOR
158756	NOR	158864	EMR
158757	NOR	158865	EMR
158758	NOR	158866	EMR
158759	NOR	158867	NOR
158760	GWR	158868	NOR
158762	GWR	158869	NOR
158765	GWR	158870	NOR
158766	GWR	158871	NOR
158767	GWR	158872	NOR
158768	GWR	**158/8**	
158769	GWR	158880	SWR
158770	EMR	158881	SWR
158771	EMR	158882	SWR
158773	EMR	158883	SWR
158774	EMR	158884	SWR
158777	EMR	158885	SWR
158780	EMR	158886	SWR
158782	NOR	158887	SWR
158783	EMR	158888	SWR
158784	NOR	158889	SWR
158785	EMR	158890	SWR
158786	NOR		
158787	NOR	**158/9**	
158788	NOR	158901	NOR
158789	NOR	158902	NOR
158790	NOR	158903	NOR
158791	NOR	158904	NOR
158792	NOR	158905	NOR
158793	NOR	158906	NOR
158794	NOR	158907	NOR
158795	NOR	158908	NOR
158796	NOR	158909	NOR
158797	NOR	158910	NOR
158798	GWR		
158799	EMR	**158/95**	
158806	EMR	158950	GWR
158810	EMR	158951	GWR
158812	EMR	158958	GWR
158813	EMR	158959	GWR

67

Class 158

Above & Inset: East Midlands Railway operate a fleet of 26 two-car '158s' on longer distance Regional services. They are based at Nottingham Eastcroft, in 2022-2023 being re-branded in the latest EMR Regional aubergine and white colours, but in mid 2023 a number were still carrying the red, orange, white and blue livery. No. 158788 is illustrated passing Dore & Totley with a Liverpool to Norwich service, seen from its 52xxx coach. The inset shows the interior with Grammer seating. Both: **CJM**

Above, Left & Lower Left: In 2023 Northern operates a fleet of 53 Class 158s, running in both two- and three-car formations. The 45 two-car sets are based at Leeds Neville Hill or Heaton, Newcastle. The eight three-car sets (Nos. 158752-158759) are allocated to Leeds Neville Hill. The three-car sets are formed with a Motor Standard Lavatory (MSL) between the two driving cars. In the upper left view, three-car set No. 158757 is seen at Selby on a York to Hull service, while above is MSL No. 58716 from set No. 158752. Below left, two-car set No. 158845 is seen departing from Carlisle bound for Newcastle. All Northern sets carry white and blue livery, all have been refurbished in recent years to meet passengers with restricted mobility standards. A standard Northern interior using 2+2 seating in the group and airline styles is used.
CJM (2) / Antony Christie

Rolling Stock Review : 2023-2024

Class 158

Above: The original Class 158s were introduced from 1990, when the high-specification two and three car sets emerged from BREL Derby for BR Provincial, (later Regional Railways), for use on longer-distance or 'Express' routes. The first batch was allocated to Scotland, where many remain today. Now painted in Scottish Railways colours. Sets Nos. 158708 and 158712 are seen approaching Haymarket. The Scottish sets sport revised two-light head/combined marker/tail lights and are fitted with a curved snowplough/deflector plate. **CJM**

Right: The interior of the Class 158s is set out in the 2+2 style. If first class seating is provided this is in the area directly inward of the driving cab in DMC vehicle. Since introduction, sets have undergone several major refurbishments and the Scottish sets now have Grammer seating, a quality passenger information system, extra luggage stands and a cycle stowage area. Seating is finished in Scottish Railways moquette. **CJM**

Key Facts

- The most successful second generation DMU design.
- Ordered originally by BR Provincial for longer-distance domestic services.
- Currently several different seat styles can be found, all based on the 2+2 low-density layout.
- Some revisions to front ends can be found, including obstacle deflectors, light clusters styles, destination indicators and gangway doors.
- A small number of three-car sets with a MSL were built, some have since been modified as Class 159/1s.
- Four three-car units (formed of three driving cars), operate on Great Western.

Arcola Energy are planning to convert two Scottish Railways Class 158s to operate on hydrogen for use on the Far North routes. Prototype sets should be operational in 2025 and if adopted fleet conversion, entry into passenger service could be in 2026.

Above & Inset: South Western Railway operates a fleet of 11 two-car Class 158s (Nos. 158880-158890), based at Salisbury. These work alongside '159s' on the Waterloo-Salisbury-Exeter route, as well as the Salisbury-Romsey 'shuttle'. Interiors are in the low-density 2+2 style, with a mix of group and airline styles. A small first class area, seating 13, is located behind the cab in the DMCL. Set No. 158888 is shown in the current white and dark blue livery. Inset left, shows the standard class interior. **CJM / Antony Christie**

Rolling Stock Review : 2023-2024 — 69

Class 159

For history and pictures of this class, see *Modern Locomotives Illustrated* Issue 190

Sub-class:	159/0	159/1
Number range:	159001-159022	159101-159108
Former number range:	-	158800-158814 range
Built by:	BREL Derby, fitted out at Rosyth Dockyard	Originally: BREL Derby Rebuilt: Wabtec, Doncaster
Introduced:	1992-1993	As 158 - Originally 1991 As 159 - 2006
Max speed:	90mph (145km/h)	90mph (145km/h)
Formation:	DMCL+MSL+DMSL	DMCL+MSL+DMSL
Vehicle numbers:	DMCL - 52873-52894 MSL - 58718-58739 DMSL - 57873-57894	DMCL - 52800-52811 range MSL - 58701-58717 range DMSL - 57800-57811 range
Vehicle length:	76ft 1¾in (23.20m)	76ft 1¾in (23.20m)
Height:	12ft 6in (3.81m)	12ft 6in (3.81m)
Width:	9ft 3¼in (2.82m)	9ft 3¼in (2.82m)
Weight:	Total - 114.3 tonnes DMCL - 38.5 tonnes MSL - 38 tonnes DMSL - 37.8 tonnes	Total - 114.3 tonnes DMCL - 38.5 tonnes MSL - 38 tonnes DMSL - 37.8 tonnes
Internal layout:	2+1F/2+2S	2+1F/2+2S
Seating:	Total - 23F/170S DMCL - 23F/28S MSL - 70S DMSL - 72S	Total - 24F/170S DMCL - 24F/28S MSL - 70S DMSL - 72S
Gangway:	Throughout	Throughout
Toilets:	DMCL, DMSL, MSL - 1	DMCL, DMSL, MSL - 1
Bogie type:	Powered - BREL P4-4 Trailer - BREL T4-4	Powered - BREL P4-4 Trailer - BREL T4-4
Power unit:	1 x Cummins NTA855R of 400hp (300kW) per vehicle	1 x Cummins NTA855R of 350hp (260kW) per vehicle
Transmission:	Hydraulic	Hydraulic
Transmission type:	Voith T211r	Voith T211r
Horsepower (total):	1,200hp (900kW)	1,050hp (780kW)
Brake type:	Air EP	Air EP
Coupling type:	BSI	BSI
Multiple restriction:	Class 15x, 170, 172	Class 15x, 170, 172
Door type:	Bi-parting swing plug	Bi-parting swing plug
Body structure:	Aluminium	Aluminium
Operator:	South Western Railway	South Western Railway
Owner:	Porterbrook Leasing	Porterbrook Leasing

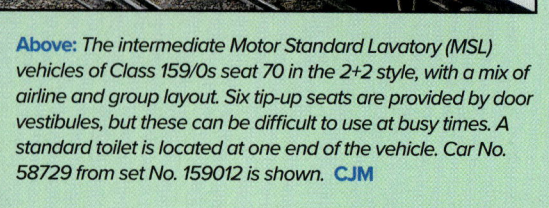

Right: *Class 159 front end equipment, 1: Destination indicator, 2: Gangway door, 3: Headlight, 4: Combined white marker and red tail light, 5: BSI coupling with roller cover electricial connection below, 6: Emergency jumper socket (behind cover), 7: Warning horns.* **CJM**

Left & Bottom: *The 22 Class 159s, a follow-on to the Class 158 build, were ordered by Network SouthEast to replace loco-hauled stock on the Waterloo-Salisbury-Exeter line. Built at BREL Derby, but fitted out at MoD Rosyth, the 159/0s are considered the 'Rolls Royce' of second generation DMUs. Based at Salisbury and operated by South Western Railway, the '159/0' sets are refurbished, seat 23 first and 170 standard class passengers in the 2+1 and 2+2 styles. In the view left, set No. 159016 is seen from its DMS vehicle, sporting SWR light grey and blue colours. The original 22 Class 159/0s were built with the high-output Cummins NTA855R engines set to deliver 400hp (300kW) per vehicle, used due to the undulating nature of the route. In the view left bottom, No. 159021 is seen at Waterloo with its Driving Motor Composite coach, showing the yellow surround to the leading pair of passenger doors and a yellow band at cant rail height, indicating the first class seating area. In usual operation, the first class coach is formed at the London end of formations.* Both: **CJM**

Above: *The intermediate Motor Standard Lavatory (MSL) vehicles of Class 159/0s seat 70 in the 2+2 style, with a mix of airline and group layout. Six tip-up seats are provided by door vestibules, but these can be difficult to use at busy times. A standard toilet is located at one end of the vehicle. Car No. 58729 from set No. 159012 is shown.* **CJM**

Class 159

159/0
159001	SWR
159002	SWR
159003	SWR
159004	SWR
159005	SWR
159006	SWR
159007	SWR
159008	SWR
159009	SWR
159010	SWR
159011	SWR
159012	SWR
159013	SWR
159014	SWR
159015	SWR
159016	SWR
159017	SWR
159018	SWR
159019	SWR
159020	SWR
159021	SWR
159022	SWR

159/1
159101	SWR
159103	SWR
159104	SWR
159105	SWR
159106	SWR
159107	SWR
159108	SWR

Right Top: *In 2006, extra stock was required to work an increased Waterloo-Salisbury-Exeter service, to fulfill this, eight Class 159/1 three-car sets were converted from Class 158s by Wabtec, Doncaster. The rebuild brought the '158s' up to '159' standards. Slight detail difference in seating, resulted in one extra first class seat per train. The 159/1s are fitted with standard 350hp (260kW) engines. Displaying South West 'white' livery from the Stagecoach era, but with SWR branding, a livery carried by all 159/1s, No. 159108 is shown from its DMS end at Worting Junction.* **CJM**

Right Middle: *Class 159/1 No. 159105 is seen from its first class DMC end. Identifiable by a '1' applied to the blue bodywork inwards of the passenger doors.* **CJM**

Right: *Class 159/1 intermediate Motor Standard Lavatory (MSL) in white/blue livery with South Western Railway branding. This MSL is from set No. 159106.* **Antony Christie**

Key Facts

- Upgraded version of Class 158 for the Network SouthEast Waterloo-Salisbury-Exeter route.
- Original 22 sets built classified as 159/0. In 2006, eight Class 158s were converted as 159/1.
- Fitted with modified vestibule door controls to allow the vestibule at the driving end to be a dedicated crew area.
- Class 159/0s fitted with higher output engines of 400hp (300kW).
- In 2023, Class 159/0s carry full SWR livery, 159/1s retain older branded Stagecoach colours.

Below Left & Below Right: *Class 159 interiors. On the left is the first class, based on the 2+1 style using a group layout. On the right is a standard class area, using the 2+2 style. Through the train there is a mix of airline and group, this shows the blue moquette with leather headrests. All sets have ceiling hung Passenger Information System (PIS) displays, providing line of route information or urgent travel messages.* **Both: Antony Christie**

Class 165 'Networker Turbo'

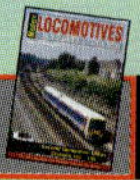

For history and pictures of this class, see Modern Locomotives Illustrated Issue 216

Sub-class:	165/0	165/1
Number range:	165001-165039	165101-165137
Built by:	BREL/ABB York	BREL/ABB York
Introduced:	1990-1992	1992-1993
Max speed:	75mph (121km/h)	75mph (121km/h)
Formation:	165001-165028 - DMSL+DMS 165029-165039 - DMSL+MS+DMS	165101-165117 - DMSL+MS+DMS 165118-165137 - DMSL+DMS
Vehicle numbers:	165001-165028: DMSL - 58801-58822, 58873-58878 DMS - 58834-58855, 58867-58872 165029-165039: DMSL - 58823-58833 MS - 55404-55414 DMS - 58856-58866	165101-165117: DMSL - 58953-58969 MS - 55415-55431 DMS - 58916-58932 165118-165137: DMSL - 58879-58898 DMS - 58933-58952
Vehicle length:	DMSL / DMS - 75ft 2½in (22.92m) MS - 74ft 6½in (22.72m)	DMSL / DMS - 75ft 2½in (22.92m) MS - 74ft 6½in (22.72m)
Height:	12ft 5¼in (3.79m)	12ft 5¼in (3.79m)
Width:	9ft 2½in (2.81m)	9ft 2½in (2.81m)
Weight:	165001-165028 - Total - 79.5 tonnes DMSL - 40.1 tonnes DMS - 39.4 tonnes 165029-165039 - Total - 116.5 tonnes DMSL - 40.1 tonnes MS - 37 tonnes DMS - 39.4 tonnes	165101-165117 - Total - 112 tonnes DMSL - 38 tonnes MS - 37 tonnes DMS - 37 tonnes 165118-165137 - Total - 75 tonnes DMSL - 38 tonnes DMS - 37 tonnes
Internal layout:	2+2/2+3	2+2F/2+3S
Seating:	165001-165028 - Total - 171S + 7 tip up DMSL - 77S + 7 tip up DMS - 94S 165029-165039 - Total - 277S + 7 tip up DMSL - 77S + 7 tip up MS - 106S DMS - 94S	165101-165117 - Total - 257S DMSL - 67S MS - 106S DMS - 84S 165118-165137 - Total - 157S DMSL - 73S DMS - 84S
Gangway:	Within set	Within set
Toilets:	DMSL - 1	DMSL - 1
Bogie type:	Powered - BREL P3-17 Trailer - BREL T3-17	Powered - BREL P3-17 Trailer - BREL T3-17
Power unit:	1 x Perkins 2006TWH of 350hp (260kW) per vehicle	1 x Perkins 2006TWH of 350hp (260kW) per vehicle
Transmission:	Hydraulic	Hydraulic
Transmission type:	Voith T211r	Voith T211r
Horsepower (total):	165001-165028 - 700hp (520kW) 165029-165039 - 1,050hp (780kW)	165101-165117 - 1,050hp (780kW) 165118-165137 - 700hp (520kW)
Brake type:	Air EP	Air EP
Coupling type:	BSI	BSI
Multiple restriction:	Class 165, 166, 168, 172	Class 165, 166, 168, 172
Door type:	Bi-parting sliding plug	Bi-parting sliding plug
Body structure:	Welded aluminium	Welded aluminium
Operator:	Chiltern Railways	Great Western Railway
Owner:	Angel Trains	Angel Trains

Above & Below: *Class 165 front end layout (lower image also applicable to Class 166).*
1: Destination indicator, 2: Headlight, 3: Combined white marker and red tail light. 4: BSI coupling, 5: Coupling electrical box with roller cover. 6: Emergency air connection. On Chiltern sets (above) an air smooth cover has been placed around the light clusters. On GWR sets (below), the old style light block is retained. Both: **CJM**

Left Upper & Left Below:
The Class 165 'Networker Turbo' fleet were ordered by Network SouthEast to modernise Chiltern and Thames routes. Delivered in 1990-1993. The two and three car sets operate on both routes, 165/0s on Chiltern and 165/1s on GWR. In 2003-2005 Chiltern sets were upgraded with air conditioning and now have solid glazed windows. Two-car set No. 165021, is seen at South Ruislip. Three-car '165s' have an additional Motor Standard (MS), increasing seating by 106. Chiltern set No. 165037 is viewed in the latest Chiltern livery. As the Chiltern '165s' operate over tracks controlled by London Transport, they are fitted with 'trip-cocks', a bogie mounted device which strikes a track mounted activator when a signal is showing a red aspect, applying the trains brakes if the signal is passed at danger. The equipment is mounted on a 'shoe beam' on the leading bogie on the non-driving side. Both: **CJM**

Below: *Class 165 Chiltern Railways interior showing the 3+2 high-density layout. These 165/0s have fixed glazed windows as they are fitted with air conditioning. Interior of 165005 is shown.* **CJM**

Key Facts

- 'Networker Turbo' design introduced by Network SouthEast to replace first generation DMUs on Thames Valley and Chiltern routes.
- Fleet cascades now see the GW fleet working in the Bristol and Exeter area.
- Chiltern sets fitted with London Underground style 'trip-cock' equipment, mounted on leading bogie non-driving side.

'Networker Turbo' — Class 165

Right & Below: The Class 165/1s operate on GWR, formed as either two- or three-car sets. Until 2018, they were deployed on Thames Valley services. But after electrification and introduction of Class 387s, the '165s' cascaded and now operate branch Thames line services, the Didcot to Oxford shuttle, Reading to Basingstoke and Cardiff, Bristol and South West local services. Refurbished and fitted with a universal toilet, two-car set No. 165133 is shown right at Newport. Two-car Class 165/1s seat 157 in the high-density 2+3 style. Below is three-car set No. 165102 operating on the Reading-Basingstoke route. The three-car sets seat 257. Each vehicle of a '165' carries a 350hp (260kW) Perkins engine. Both: **CJM**

Left: The Class 165 intermediate Motor Standard (MS) seats 106. The seating on the two and three car vehicles being different (see table). Each Class 165 vehicle carries an underslung 350hp (260kW) Perkins engine. Great Western Class 165s can be identified from Class 166s, by having an opening hopper in each body side window. Car No. 55415 from set 165101 is illustrated at Newton Abbot. **Antony Christie**

Below Left: Class 165/1 Great Western Railway interior. When these sets were introduced they were for suburban operations in the Thames Valley, where patronage was high and thus a high-capacity 2+3 interior was fitted. It is possible in the future, when the fleet fully moves to the West Country, that a lower-density layout could be installed. Talk exists of strengthening the two-car sets to three and the three to either four or five cars by inserting former 'Networker' TS vehicles, this would follow re-engineering using a Hydrive to increase power output, if trials are successful. **CJM**

Below Right: Class 165/1 driving cab. Based on the Networker style, this layout uses a joint power and brake controller on the left side. **CJM**

2023 Fleet

165/0
165001	CRW
165002	CRW
165003	CRW
165004	CRW
165005	CRW
165006	CRW
165007	CRW
165008	CRW
165009	CRW
165010	CRW
165011	CRW
165012	CRW
165013	CRW
165014	CRW
165015	CRW
165016	CRW
165017	CRW
165018	CRW
165019	CRW
165020	CRW
165021	CRW
165022	CRW
165023	CRW
165024	CRW
165025	CRW
165026	CRW
165027	CRW
165028	CRW
165029	CRW
165030	CRW
165031	CRW
165032	CRW
165033	CRW
165034	CRW
165035	CRW
165036	CRW
165037	CRW
165038	CRW
165039	CRW

165/1
165101	GWR
165102	GWR
165103	GWR
165104	GWR
165105	GWR
165106	GWR
165107	GWR
165108	GWR
165109	GWR
165110	GWR
165111	GWR
165112	GWR
165113	GWR
165114	GWR
165116	GWR
165117	GWR
165118	GWR
165119	GWR
165120	GWR
165121	GWR
165122	GWR
165123	GWR
165124	GWR
165125	GWR
165126	GWR
165127	GWR
165128	GWR
165129	GWR
165130	GWR
165131	GWR
165132	GWR
165133	GWR
165134	GWR
165135	GWR
165136	GWR
165137	GWR

Rolling Stock Review : 2023-2024

Class 166   'Networker Turbo Express'

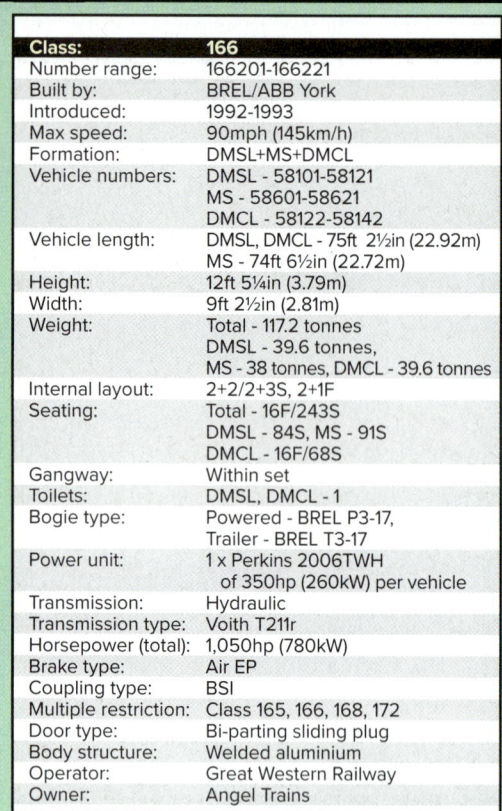

For history and pictures of this class, see Modern Locomotives Illustrated Issue 216

Class:	166
Number range:	166201-166221
Built by:	BREL/ABB York
Introduced:	1992-1993
Max speed:	90mph (145km/h)
Formation:	DMSL+MS+DMCL
Vehicle numbers:	DMSL - 58101-58121 MS - 58601-58621 DMCL - 58122-58142
Vehicle length:	DMSL, DMCL - 75ft 2½in (22.92m) MS - 74ft 6½in (22.72m)
Height:	12ft 5¼in (3.79m)
Width:	9ft 2½in (2.81m)
Weight:	Total - 117.2 tonnes DMSL - 39.6 tonnes, MS - 38 tonnes, DMCL - 39.6 tonnes
Internal layout:	2+2/2+3S, 2+1F
Seating:	Total - 16F/243S DMSL - 84S, MS - 91S DMCL - 16F/68S
Gangway:	Within set
Toilets:	DMSL, DMCL - 1
Bogie type:	Powered - BREL P3-17, Trailer - BREL T3-17
Power unit:	1 x Perkins 2006TWH of 350hp (260kW) per vehicle
Transmission:	Hydraulic
Transmission type:	Voith T211r
Horsepower (total):	1,050hp (780kW)
Brake type:	Air EP
Coupling type:	BSI
Multiple restriction:	Class 165, 166, 168, 172
Door type:	Bi-parting sliding plug
Body structure:	Welded aluminium
Operator:	Great Western Railway
Owner:	Angel Trains

Above & Right: *A fleet of 21 three-car Class 166s were ordered by Network SouthEast as the main line 'express' version of the 'Networker Turbo' titled 'Networker Turbo Express', with a top speed of 90mph (145km/h) and fitted with air conditioning. Originally based at Reading for Thames Valley use, the sets now operate from Bristol and Exeter. To cope with platform heights in the West of England, a 'Ride Height Modification' (RHM) to increase the height of the body was made, this has also been done on the Class 165s. To identify modified sets, a yellow warning triangle on the black front panel has been applied. The view top shows set No. 166218 working in Devon on the Exeter-Okehampton, Barnstaple, Exmouth and Paignton group of services, where up to four '166s' operate each day. The above view right shows the internal layout using the 2+3 seating style. The set shown No. 166202 sports an earlier FGW-based moquette, these are progressively being changed to the GWR grey and green style. Both:* **CJM**

2023 Fleet

166201	GWR
166202	GWR
166203	GWR
166204	GWR
166205	GWR
166206	GWR
166207	GWR
166208	GWR
166209	GWR
166210	GWR
166211	GWR
166212	GWR
166213	GWR
166214	GWR
166215	GWR
166216	GWR
166217	GWR
166218	GWR
166219	GWR
166220	GWR
166221	GWR

Above & Below: *The Class 166 sets can be easily identified from the Class 165 three-car sets, as the '166s' have only five non-opening 'hopper' windows in each coach, whereas the 165s have all windows with a top hopper. In 2023, First Great Western blue livery still applied to eight sets, set No. 166202 is illustrated at Severn Tunnel Junction. These sets have now had their first class seating area de-classified to standard. In the image below is an intermediate MS, (No. 58619 from set 166219) these seat 91 in the 2+3 style and have extra luggage space and a cycle stowage area at one end, identified by branding on the external doors (far end). Both:* **CJM**

Key Facts

- Introduced as 'Networker Turbo Express' stock to replace first generation DMU and loco-hauled stock on Thames routes from Paddington.

- Originally fitted with first class seating in both driving cars now declassified.

- Remained on Thames Valley lines until 2016-2018 when sets migrated to the Bristol area allocated to St Philips Marsh and given ride-height modifications.

- By the 2020s, sets can be found operating on routes to Cardiff, Portsmouth, Exeter, Paignton, Exmouth, Okehampton and Barnstaple.

'Clubman' — Class 168

Class:	168/0	168/1	168/2	168/3
Design:	'Networker' outline	'Turbostar' outline	'Turbostar' outline	'Turbostar' outline
Number range:	168001-168005	168106-168113	168214-168219	168321-168329
Built by:	Adtranz Derby	Adtranz/Bombardier Derby	Bombardier Derby	Adtranz Derby
Introduced:	1997-1998	2000-2002	2003-2006	2000-2001
Max speed:	100mph (160km/h)	100mph (160km/h)	100mph (160km/h)	100mph (161km/h)
Formation:	DMSL(A)+MSL+MS+DMSL(B)	168106 - 168107: DMSL(A)+MSL+MS+DMSL(B) 168108 - 168113: DMSL(A)+MS+DMSL(B)	168214, 168218-168219: DMSL(A)+MS+DMSL(B) 168215 - 168217: DMSL(A)+MS+MS+DMSL(B)	DMSL+DMSL
Vehicle numbers:	DMSL(A) - 58151-58155 MSL - 58651-58655 MS - 58451-58455 DMSL(B) - 58251-58255	DMSL(A) - 58156-58163 MSL - 58756-58757 MS - 58456-58463 DMSL(B) - 58256-58263	DMSL(A) - 58164 - 58169 MS - 58365-58367 MS - 58464-58469 DMSL(B) - 58264 - 58269	DMSL - 50301-50308/399 DMSL - 79301-79308/399
Vehicle length:	77ft 6in (23.62m)	77ft 6in (23.62m)	77ft 6in (23.62m)	77ft 6in (23.62m)
Height:	12ft 4½in (3.77m)	12ft 4½in (3.77m)	12ft 4½in (3.77m)	12ft 4½in (3.77m)
Width:	8ft 10in (2.69m)	8ft 10in (2.69m)	8ft 10in (2.69m)	8ft 10in (2.69m)
Weight:	168.8 tonnes DMSL(A) - 43.7 tonnes MSL - 41 tonnes MS - 40.5 tonnes DMSL(B) - 43.6 tonnes	168106-107 - 175.1 tonnes 168108-113 - 132.2 tonnes DMSL(A) - 45.2 tonnes MSL - 42.9 tonnes MS - 41.8 tonnes DMSL(B) - 45.2 tonnes	168214/218/219 - 134.9 tonnes 168215-168217 - 178.2 tonnes DMSL(A) - 45.4 tonnes MSL - 43.3 tonnes MS - 44 tonnes DMSL(B) - 45.5 tonnes	Total - 91.6 tonnes DMSL - 45.8 tonnes DMSL - 45.8 tonnes
Internal layout:	2+2	2+2	2+2	2+2
Seating:	Total - 275S DMSL (A) - 57S MSL - 73S MS - 77S DMSL(B) - 68S	168106-168107 - Total - 275S 168108-168113 - Total - 202S DMSL(A) - 57S MSL - 73S MS - 76S DMSL(B) - 69S	168214/218/219 - Total 202S 168215-168217 - Total 278S DMSL(A) - 57S MS - 76S MS - 76S DMSL(B) - 69S	Total - 128S DMSL - 59S DMSL - 69S
Gangway:	Within set	Within set	Within set	Within set
Toilets:	DMSL, MSL - 1	DMSL, MSL - 1	DMSL, MSL - 1	DMSL - 1
Bogie type:	Powered - P3-23 Trailer - T3-23	Powered - P3-23 Trailer - T3-23	Powered - P3-23 Trailer - T3-23	Powered - P3-23 Trailer - T3-23c
Power unit:	1 x MTU 6R183TD13H of 422hp (315kW) per car	1 x MTU 6R183TD13H of 422hp (315kW) per car	1 x MTU 6R183TD13H of 422hp (315kW) per car	1 x MTU 6R183TD13H of 422hp (315kW) per car
Transmission:	Hydraulic	Hydraulic	Hydraulic	Hydraulic
Transmission type:	Voith T211r	Voith T211r	Voith T211r	Voith T211r (168329 - ZF gearbox)
Horsepower (total):	1,688hp (1,260kW)	168106-107 - 1,688hp (1,260kW) 168108-113 - 1,266hp (945kW)	168214/18/19 - 1,266hp (945kW) 168215-217 - 1,688hp (1,260kW)	844hp (630kW)
Brake type:	Air EP	Air EP	Air EP	Air EP
Coupling type:	Outer - BSI, Inner - Bar	Outer - BSI, Inner - Bar	Outer - BSI, Inner - Bar	Outer - BSI, Inner - Bar
Multiple restriction:	Class 165, 166, 168, 172	Class 165, 166, 168, 172	Class 165, 166, 168, 172	Class 165, 166, 168, 172
Door type:	Bi-parting swing plug	Bi-parting swing plug	Bi-parting swing plug	Bi-parting swing plug
Body structure:	Welded aluminium, bolt-on steel ends	Welded aluminium, bolt-on steel ends	Welded aluminium, bolt-on steel ends	Welded aluminium bolt-on steel ends
Operator:	Chiltern Railways	Chiltern Railways	Chiltern Railways	Chiltern Railways
Owner:	Porterbrook Leasing	Porterbrook Leasing	Porterbrook Leasing	Porterbrook Leasing
Special features:	Chiltern ATP, trip-cocks	Chiltern ATP, trip-cocks	Chiltern ATP, trip-cocks	Chiltern ATP, trip-cocks
Notes:				Previously used on SWT and TPTE as 170301-170309

Below: *The original five Class 168s were the first new trains ordered post privatisation, when Chiltern Railways, via Porterbrook Leasing, placed an order with Adtranz for a fleet of DMUs, designated 'Clubman' stock by Adtranz. This DMU product line later became the Class 17x 'Turbostar' platform. The original five Class 168/0s were built with a 'Networker' style cab profile, this was changed on follow-on orders and 'Turbostar' builds. Class 168/0 set No. 168002 is seen near Banbury. The '168/0s' seat 275 in low-density 2+2 'Clubman' style. Toilets are provided in both driving cars and one intermediate vehicle. The fleet is painted in Chiltern Railways main line silver, grey and white livery and are based at Aylesbury depot.* **CJM**

2023 Fleet

168/0
168001 CRW
168002 CRW
168003 CRW
168004 CRW
168005 CRW

168/1
168106 CRW
168107 CRW
168108 CRW
168109 CRW
168110 CRW
168111 CRW
168112 CRW
168113 CRW

168/2
168214 CRW
168215 CRW
168216 CRW
168217 CRW
168218 CRW
168219 CRW

168/3
168321 CRW
168322 CRW
168323 CRW
168324 CRW
168325 CRW
168326 CRW
168327 CRW
168328 CRW
168329 CRW

Key Facts

- First of the 'Turbostar' platform originally known as 'Clubman'.
- First five sets sport early 'Networker' front end design.
- Class 168/1 and 168/2 built with 'Turbostar' cab end design, but incorporate two different light cluster styles.
- Mix of three and four car sets, some single MS cars were built to augment existing sets.
- Class 168/3s were cascaded from First TransPennine and previously operated with South West Trains as Class 170/3s, now modified as Class 168s.

For history and pictures of this class, see Modern Locomotives Illustrated Issue 216

Class 168   'Clubman'

Above: Additional Class 168s were delivered in 2000 to modernise the Chiltern route, with the first follow-on order for Class 168/1s, these were built to what become the standard 'Turbostar' profile. Additional intermediate vehicles were added in 2002. In 2023, Class 168/1s are formed in both three-car (six sets) and four-car (two sets) formations. Three-car set No. 168112 is shown with the three lamp clusters on each side of the front end. **CJM**

Below & Inset: A next follow-on order for '168s' was placed by Chiltern, funded by Porterbrook, for a fleet of six Class 168/2s, these are now formed as three, three-car and three, four-car sets. The front ends of the driving cars on this sub-class have large headlights and a smaller combined marker/tail light unit. Three-coach set No. 168218 is seen departing from Leamington Spa. Units are painted in standard Chiltern silver and white livery. The three-car Class 168/2s, introduced in 2003-2004 seat 202 passengers, while the four-car sets introduced from 2006, accommodate 278 passengers. All intermediate vehicles in Class 168/2s (inset image) are of the Motor Standard (MS) design, seating 76 in the 2+2 style, no toilets are provided in these coaches. Vehicle 58366 illustrated. All Class 168s are based at Aylesbury depot, but also receive attention at the Chiltern depot in Wembley, North London. **CJM**

Above: Interior of the 'Clubman' fleet uses the 2+2 group layout. Image shows the area directly behind the cab of Class 168/0 set No. 168005. **Nathan Williamson**

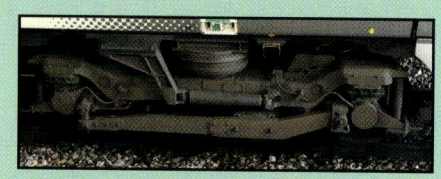

Above: To allow operation over London Transport lines, Class 168s are fitted with bogie mounted 'trip-cock' equipment. If a signal is red, a bar sticks up at the side of the track, if struck by the bogie mounted equipment, it will apply the trains brakes. When a signal is cleared to a proceed aspect, the lineside bar lowers and the train can proceed normally. The equipment is mounted only on the leading bogie on the non-driving side. **CJM**

Right: In 2015-2016 nine members of Class 168/3 were added to the Chiltern fleet. These sets have a complex history. They were originally introduced in 2000-2001 on South West Trains as Class 170/3s, based at Salisbury to work alongside the Class 159s, this did not work out well, and they were later transferred to TransPennine Express, working between Manchester and the East Coast via the Pennine route, remaining as Class 170/3. In 2015-2016 they were again moved, this time to Chiltern Railways, providing much needed extra capacity, as passenger growth had outstripped available stock. On transfer to Chiltern, the sets were overhauled, made compatible with other Class 168s and for uniformity, reclassified as Class 168/3. The sets now accommodate 128 standard class passengers in the 2+2 style. Set No. 168327 leads another set of the same sub-class at Leamington Spa. **CJM**

Hybrid-Flex 168329

Porterbrook Leasing and Rolls Royce/MTU developed a hybrid diesel-battery power pack for fitting to Class 168/3 No. 168329. The core is a Rolls Royce/MTU hybrid drive unit formed of an EU stage 5 compliant 6H 1800 engine coupled to an electric motor generator and a MTU 'energypack' battery system, operating as a self-charging hybrid. A new gearbox has been installed which allows re-generative braking. The system will allow the train to operate in battery mode around stations, during acceleration and deceleration, thus reducing emissions. The trial unit was originally tested on the Ecclesbourne Valley Railway in spring/summer 2021 before entering main line service with Chiltern in late 2021. In 2022-2023 the set was frequently deployed on the Aylesbury to Princes Risborough service.

Right: No. 168329 is seen at Honeybourne. **CJM**

'Turbostar' — Class 170

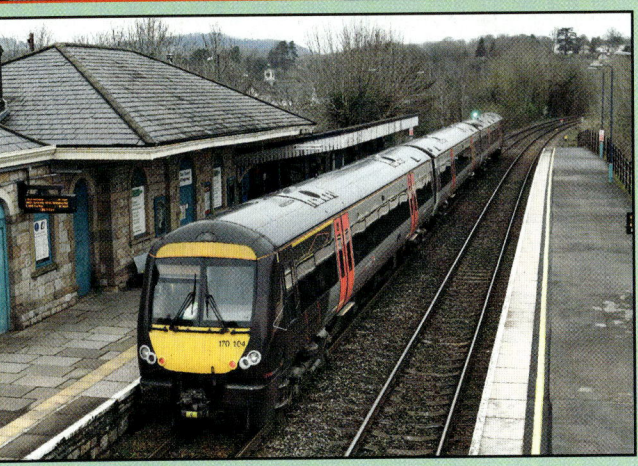

2023 Fleet

For history and pictures of this class, see Modern Locomotives Illustrated Issue 216

Left: The first of the Class 170 standard 'Turbostar' fleet emerged from Adtranz Derby in 1998, when two-car Class 170/1 sets were delivered to Midland Main Line. Ten sets were soon strengthened to three-car formation by adding a Motor Standard (MS) in 2001. Later refurbished, the sub-class now operate for CrossCountry Trains, on the east-west routes between Cardiff and Nottingham. The sets have nine first class seats in each DMCL vehicle. With its DMCL leading, three-set No. 170104 is seen at Chepstow. **CJM**

Below & Inset: The 10 three-car Class 170/1 sets incorporate a Motor Standard (MS) vehicle, these seat 80 in the 2+2 style (illustrated inset). These coaches do not have a toilet. The vehicles are fitted with a standard MTU 6R 183TD power unit. MS No. 55101 from set 170101 is shown at Gloucester. All CrossCountry sets carry XC silver and deep maroon livery with a black window band. **CJM**

Below: Six Class 170/3s operate, four for ScotRail and two for CrossCountry. The XC sets are internally the same as the Class 170/1 sets. No. 170398 is viewed from its DMSL end at Gloucester. The two XC 170/3s operate in a general pool with the three-car 170/1s. **CJM**

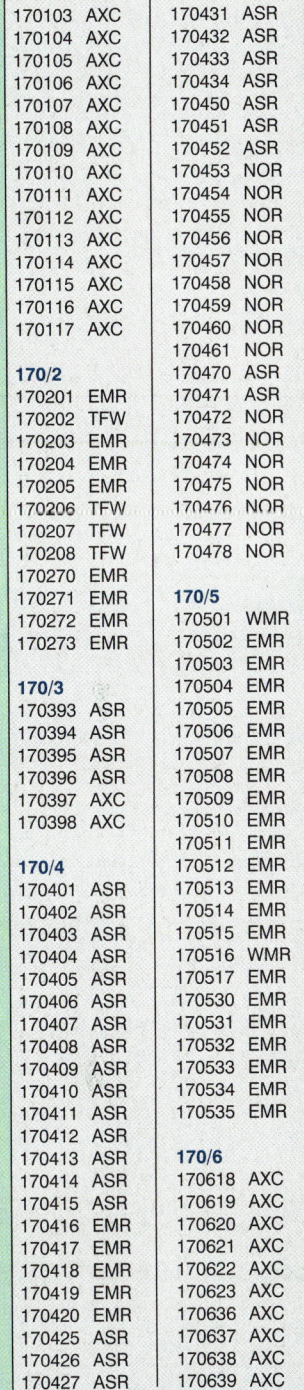

Below & Inset: By mid-2023, only three-car Transport for Wales (TfW) Class 170/2s remained in traffic on Welsh services and these are scheduled to move to East Midlands Trains in the near future. The front ends of sets differ slightly, the valance, either side of the BSI coupling, is much deeper than on earlier builds. After originally working for Greater Anglia, where limited first class seating was provided, the sets were modified to all standard class. Right, set No. 170201 displays full Transport for Wales white, grey and black livery. The inset image shows the intermediate Motor Standard Lavatory (MSL), these seat 68 in the 2+2 layout with a standard design toilet compartment at one end. The total seating on a TfW three-car set is 180. Both: **CJM**

170/1			
170101	AXC	170428	ASR
170102	AXC	170429	ASR
170103	AXC	170430	ASR
170104	AXC	170431	ASR
170105	AXC	170432	ASR
170106	AXC	170433	ASR
170107	AXC	170434	ASR
170108	AXC	170450	ASR
170109	AXC	170451	ASR
170110	AXC	170452	ASR
170111	AXC	170453	NOR
170112	AXC	170454	NOR
170113	AXC	170455	NOR
170114	AXC	170456	NOR
170115	AXC	170457	NOR
170116	AXC	170458	NOR
170117	AXC	170459	NOR
		170460	NOR
		170461	NOR
170/2		170470	ASR
170201	EMR	170471	ASR
170202	TFW	170472	NOR
170203	EMR	170473	NOR
170204	EMR	170474	NOR
170205	EMR	170475	NOR
170206	TFW	170476	NOR
170207	TFW	170477	NOR
170208	TFW	170478	NOR
170270	EMR		
170271	EMR	**170/5**	
170272	EMR	170501	WMR
170273	EMR	170502	EMR
		170503	EMR
170/3		170504	EMR
170393	ASR	170505	EMR
170394	ASR	170506	EMR
170395	ASR	170507	EMR
170396	ASR	170508	EMR
170397	AXC	170509	EMR
170398	AXC	170510	EMR
		170511	EMR
170/4		170512	EMR
170401	ASR	170513	EMR
170402	ASR	170514	EMR
170403	ASR	170515	EMR
170404	ASR	170516	WMR
170405	ASR	170517	EMR
170406	ASR	170530	EMR
170407	ASR	170531	EMR
170408	ASR	170532	EMR
170409	ASR	170533	EMR
170410	ASR	170534	EMR
170411	ASR	170535	EMR
170412	ASR		
170413	ASR	**170/6**	
170414	ASR	170618	AXC
170415	ASR	170619	AXC
170416	EMR	170620	AXC
170417	EMR	170621	AXC
170418	EMR	170622	AXC
170419	EMR	170623	AXC
170420	EMR	170636	AXC
170425	ASR	170637	AXC
170426	ASR	170638	AXC
170427	ASR	170639	AXC

Rolling Stock Review : 2023-2024

Class 170   'Turbostar'

Class:	170/1	170/2	170/3	170/4	170/5
Number range:	170101-170117	170201-170208 and 170270-170273	170393-170398	170401-170434 170450-170478 170921-170924*	170501-170535
Built by:	Adtranz Derby	170201-170208: Adtranz Derby 170270-170273: Bombardier Derby	Bombardier Derby	Adtranz Derby/ Bombardier Derby	Adtranz Derby
Introduced:	1998-1999 (2001 MS)	170201-170208 - 1999 170270-170273 - 2002	2002-2004	1999-2005	1999-2000
Max speed:	100mph (161km/h)	100mph (161km/h)	100mph (161km/h)	100mph (161km/h)	100mph (161km/h)
Formation:	170101-170110: DMSL+MS+DMCL 170111-170117: DMSL+DMCL	170201-170208: DMSL+MSL+DMSL 170270-170273: DMSL+DMSL	170393-170396: DMSL+MSL+DMSL 170397-170398: DMSL+MS+DMCL	170401-170420: DMCL(A)+MS+DMCL(B) 170425-170434/921-924: DMCL(A)+MS+DMSL(B) 170450-170461: DMSL(A)+MS+DMSL(B) 170470-170478: DMSL(A)+MS+DMSL(B)	170501-170517: DMSL(A)+DMSL(B) 170518-170523: DMSL+DMCL
Vehicle Numbers:	170101-170110: DMSL - 50101-110 MS - 55101-55110 DMCL - 79101-110 170111-170117: DMSL - 50111-50117 DMCL - 79111-79117	170201-170208: DMSL - 50201-50208 MSL - 56201-56208 DMSL - 79201-79208 170270-170273: DMSL - 50270-50273 DMSL - 79270-79273	170393 - 170396: DMCL - 50393-50396 MSL - 56393-56396 DMSL - 79393-79396 170397-170398: DMSL - 50397-50398 MS - 56397-56398 DMCL - 79397-79398	170401-170420: DMCL(A) - 50401-50424 MS - 56401-56424 DMCL(B) - 79401-79424 170425-170434: DMCL(A) - 50425-50434 MS - 56425-56434 DMCL(B) - 79425-79434 170450-170478: DMSL(A) - 50450-50478 MS - 56450-56478 DMSL(B) - 79450-79478	170501-170517: DMSL(A) - 50501-50517 DMSL(B) - 79501-79517 170518-170523: DMSL - 50518-50523 DMCL - 79518-79523
Vehicle length:	77ft 6in (23.62m)	77ft 6in (23.62m)	77ft 6in (23.62m)	77ft 6in (23.62m)	77ft 6in (23.62m)
Height:	12ft 4½in (3.77m)	12ft 4½in (3.77m)	12ft 4½in (3.77m)	12ft 4½in (3.77m)	12ft 4½in (3.77m)
Width:	8ft 10in (2.69m)	8ft 10in (2.69m)	8ft 10in (2.69m)	8ft 10in (2.69m)	8ft 10in (2.69m)
Weight:	170101-170110: Total - 132.8 tonnes 170111-170117: Total - 89.8 tonnes DMSL - 45 tonnes MS - 43.0 tonnes DMCL - 44.8 tonnes	170201-170208: Total - 133.7 tonnes 170270-170273: Total - 88.4 tonnes DMSL - 45 tonnes MSL - 45.3 tonnes DMSL - 43.4 tonnes	Total - 135.9-137.5 tonnes DMCL - 46.5 tonnes MSL - 44.7 tonnes MS - 43.1 tonnes DMSL - 46.3 tonnes	Total - 133.2 tonnes DMCL(A) - 45.8 tonnes DMSL - 46.3 tonnes MS - 41.4 tonnes	Total - 91.7 tonnes DMSL(A) - 45.8 tonnes DMSL(B) - 45.9 tonnes DMCL - 45.9 tonnes
Internal layout:	2+2F, 2+2S	2+1F, 2+2S	2+1F, 2+2S	2+1F, 2+2S	2+1F/2+2S
Seating:	170101-170110: Total - 9F/191S 170111-170117: Total - 9F/111S 170102-170110: DMSL - 59S MS - 80S DMCL - 9F/52S 170111-170117: DMSL - 59S DMCL - 9F/52S	170201-170208: Total - 180S 170270-170273: Total - 119S 170201-170208: DMSL - 46S MSL - 68S DMSL - 66S 170270-170273: DMSL - 57S DMSL - 62S	170393-170396: Total: 193S DMSL - 55S MSL - 71S DMSL - 67S 170397-170398: Total: 9F/191S DMSL - 59S MS - 80S DMCL - 9F/52S	170401-170434: 18F/168S 170450-170452: Total - 198S 170453-170478: Total - 198S 170401-170434: DMC(A) - 9F/43S MS - 76S DMC(B) - 9F/49S 170450-170452: DMSL(A) - 55S MS - 76S DMSL(B) - 67S 170453-170478: DMSL(A) - 55S MS - 76S DMSL(B) - 67S	170501-170517: Total - 122S DMSL(A) - 55S DMSL(B) - 67S 170518-170523: Total - 9F/111S DMSL - 59S DMCL - 9F/52S
Gangway:	Within set	Within set	Within set	Within set	Within set
Toilets:	DMSL - 1, DMCL - 1	One per vehicle	DMSL, DMCL, MS - 1	DMSL - 1, DMCL - 1	One per vehicle
Bogie type:	One P3-23c and one T3-23c per car	One P3-23c and one T3-23c per car	One P3-23c and one T3-23c per car	One P3-23c and one T3-23c per car	One P3-23c and one T3-23c per car
Power unit:	1 x MTU 6R 183TD 13H of 422hp (315kW) per car	1 x MTU 6R 183TD 13H of 422hp (315kW) per car	1 x MTU 6R 183TD 13H of 422hp (315kW) per car	1 x MTU 6R 183TD 13H of 422hp (315kW) per car	1 x MTU 6R 183TD 13H of 422hp (315kW) per car
Transmission:	Hydraulic Voith T211r	Hydraulic Voith T211r	Hydraulic Voith T211r	Hydraulic Voith T211r	Hydraulic Voith T211r
Horsepower:	3-car - 1,266hp (945kW) 2-car - 844hp (630kW)	3-car - 1,266hp (945kW) 2-car - 844hp (630kW)	1,266hp (945kW)	1,266hp (945kW)	844hp (630kW)
Brake type:	Air	Air	Air	Air	Air
Coupling type:	BSI	BSI	BSI	BSI	BSI
Multiple restriction:	Class 15x, 171-172	Class 15x, 171-172	Class 15x, 171-172	Class 15x, 171-172	Class 15x, 171-172
Door type:	Bi-parting slide plug	Bi-parting slide plug	Bi-parting slide plug	Bi-parting slide plug	Bi-parting slide plug
Special features:		RETB	Some RETB	Some RETB	Some RETB
Body structure:	Welded aluminium	Welded aluminium	Welded aluminium	Welded aluminium	Welded aluminium
Operator:	CrossCountry	Transport for Wales	ScotRail, CrossCountry	ScotRail, EMR, Northern	East Midlands, West Midlands
Owner:	Porterbrook Leasing	Porterbrook Leasing	Porterbrook Leasing	Porterbrook Leasing (46) Eversholt Leasing (5)	Porterbrook Leasing
Notes:				* Ex Class 171	

Key Facts

- Standard Adtranz/Bombardier 'Turbostar' second generation DMU.
- Original fleet ordered in two and three-car form for Midland Main Line.
- In 2023, sets operate with ScotRail, CrossCountry, West Midlands Trains, East Midlands Trains, Northern and Transport for Wales.
- The evolution of the 'Turbostar' platform saw different front end light cluster designs incorporated. Different style front valance panels are found of some sets.
- By 2023, with changes of use and operator, many different seating configurations and seat styles can be found. Further changes are expected.
- Hybrid diesel-battery technology is expected to be fitted to some sets.

'Turbostar' Class 170

170/6
170618-170639

Adtranz Derby

2000

100mph (161km/h)
170630-170635:
DMSL(A)+MS+DMSL(B)
170636-170639:
DMSL+MS+DMCL

170630-170635:
DMSL(A) - 50630-50635
MS - 56630-56635
DMSL(B) - 79630-79635
170636-170639:
DMSL - 50636-50639
MS - 56636-56639
DMCL - 79636-79639

77ft 6in (23.62m)
12ft 4½in (3.77m)
8ft 10in (2.69m)
Total - 134.1 tonnes
DMSL(A) - 45.8 tonnes
MS - 42.4 tonnes
DMSL(B) - 45.9 tonnes
DMCL - 45.9 tonnes

2+1F/2+2S
170630-170635:
Total - 196S
DMSL(A) - 55S
MS - 74S
DMSL(B) - 67S
170636 - 170639:
Total - 9F/191S
DMSL - 55S
MS - 80S
DMCL - 9F/52S

Within set
DMSL - 1, DMCL - 1
One P3-23c and
  one T3-23c per car
1 x MTU 6R 183TD 13H
  of 422hp (315kW) per car
Hydraulic Voith T211r
1,266hp (945kW)

Air
BSI
Class 15x, 171-172
Bi-parting slide plug
Some RETB
Welded aluminium
CrossCountry

Porterbrook Leasing

Above, Below Left & Below Right: Northern commenced using Class 170s in 2018-2019, by 2023 a fleet of 18 are based at Botanic Gardens depot in Hull. Northern '170s' were cascaded from ScotRail and refurbished, they were repainted in Northern white and blue livery. The above view, shows set No. 170472 at Hull. The Northern refurbishment provided a considerably improved interior ambiance, (shown below left) much in keeping with a new train, rather than a refurbish. The interiors are all standard class and use high-quality Chapman seating. A passenger information system is fitted and end of coach electronic displays providing news, weather and other information The view below right shows the intermediate Motor Standard Lavatory (MSL) from set No. 170455, these seat 76 in the 2+2 style with no toilet compartment. All: **CJM**

Above, Inset & Right: A major operator of Class 170s is Scottish Railways, with Class 170/3 and 170/4s allocated to Haymarket. Nos. 170401-170434 are deemed as 'Express' sets, while the remainder are general sets. First class seating is incorporated in Nos. 170401-170434/170450-170452. They sport Scottish Railways Saltire blue and white livery. Above, No. 170450 is seen at Edinburgh Haymarket with its DMC coach leading. The inset shown the Scottish Railways standard class 2+2 group seating in set No. 170412. Right is MSL No. 56410 from set 170410, with seating for 76 standard with two pairs of sliding plug doors. All: **CJM**

Class 170   'Turbostar'

Above, Left, & Below Left: Over recent years, much cascading of Class 170s has taken place as new trains have been commissioned. Originally five former Scottish sets, Nos. 170416-170420, were transferred to operate on EMR Regional services. Further units, previously with West Midlands Trains have followed and have by mid 2023 replaced EMR Class 156s. Extra sets have come from moving 171201-171202 and 171401-171402 from GTR Southern to EMR and reforming them back as Class 170s Nos. 170421-170424. At present these sets are classified 170/9 as they are still fitted with Dellner couplings and thus not compatible with other EMR '170s'. The EMR 170s are based at Derby, all have been refurbished and sport EMR aubergine and grey livery. The former GTR Southern sets sport EMR branded Southern green and white. Some Class 170/5s planned for transfer to EMR from West Midlands have been repainted in aubergine and grey livery with West Midlands branding. In the above view aubergine and grey liveried No. 170504 with WMR branding is seen at Hereford. On the left is No. 170509 in full EMR colours at Kidsgrove. Below left is ex-Southern No. 170924 seen at Sutton Parkway with EMR Regional branding. Below, the inset shows the 2+2 interior of a former West Midlands set. **CJM / Cliff Beeton / John Binch**

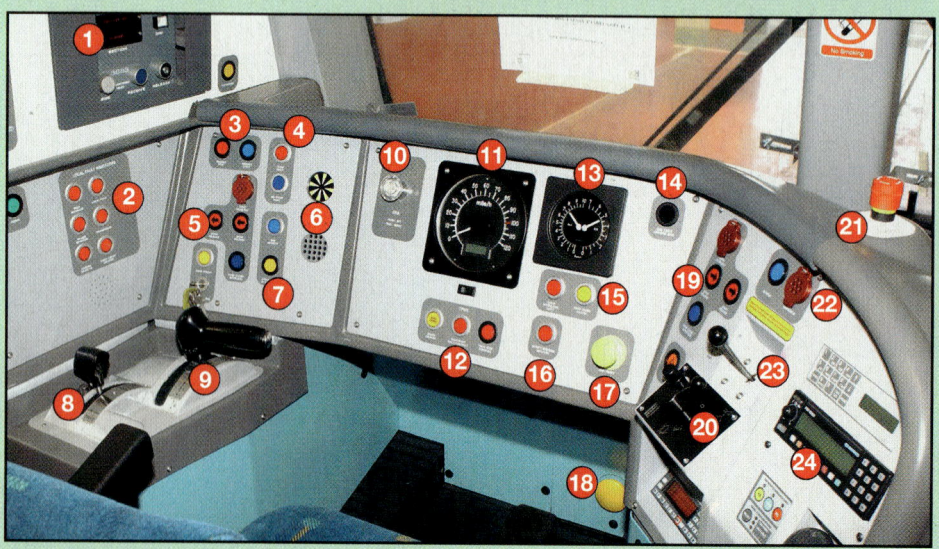

Left: Class 170 cab layout. 1: Radio Electronic Token Block (RETB) system, 2: Local fault indicators, 3: Engine start/stop buttons, 4: Fault indicators, 5: Left side door controls, 6: AWS system, 7: WSP/Sand indicators, 8: Master switch and key socket, 9: Power/brake controller, 10: Drivers reminder appliance, 11: Speedometer, 12: TPWS controls, 13: Main reservoir and brake cylinder pressure gauge, 14: Cab light switch, 15: Door interlock, passenger communication indicator, 16: Safety systems isolated indicator, 17: AWS reset button, 18: Pass comm hold over button, 19: Right side door controls, 20: Windscreen wiper controls, 21: Emergency brake plunger, 22: Couple/uncouple buttons, 23: Warning horn, 24: Communications panel. Cab illustrated from original Greater Anglia No. 170270 when delivered. **CJM**

'Turbostar' — Class 171

For history and pictures of this class, see *Modern Locomotives Illustrated* Issue 216

Class:	171/7	171/8
Number range:	171727-171730	171801-171812
Built by:	Bombardier Derby	Bombardier Derby
Introduced:	2003-2005	2004
Max speed:	100mph (161km/h)	100mph (161km/h)
Formation:	DMCL+DMSL	DMCL(A)+MS+DMCL(B)
Vehicle numbers:	DMCL - 50727-729/392	DMCL(A) - 50721-50723/50801 - 50806
	DMSL - 79727-729/392	MS - 54801 - 54806 / 56801-56806
		DMCL(B) - 79721-79726/79801 - 79806
Vehicle length:	77ft 6in (23.62m)	77ft 6in (23.62m)
Height:	12ft 4½in (3.77m)	12ft 4½in (3.77m)
Width:	8ft 10in (2.69m)	8ft 10in (2.69m)
Weight:	Total - 95.4 tonnes	Total - 180.4 tonnes
	DMCL - 47.6 tonnes	DMCL(A) - 46.5 tonnes
	DMSL - 47.8 tonnes	MS - 43.7 tonnes
		DMCL(B) - 46.5 tonnes
Internal layout:	2+1F, 2+2S	2+1F, 2+2S
Seating:	Total - 9F/107S	Total - 18F/167S
	DMCL - 9F/43S	DMCL(A) - 9F/43S
	DMSL - 64S	MS - 74S
		DMCL(B) - 9F/50S
Gangway:	Within set	Within set
Toilets:	DMCL, DMSL - 1	DMCL - 1
Bogie type:	One P3-23c and one T3-23c per car	One P3-23c and one T3-23c per car
Power unit:	1 x MTU 6R 183TD of 422hp (315kW) per car	1 x MTU 6R 183TD of 422hp (315kW) per car
Transmission:	Hydraulic	Hydraulic
Transmission type:	Voith T211r to ZF final drive	Voith T211r to ZF final drive
Horsepower (total):	844hp (630kW)	Total 1,266hp (944kW)
Brake type:	Air	Air
Coupling type:	Dellner 12	Dellner 12
Multiple restriction:	Class 171	Class 171
Door type:	Bi-parting slide plug	Bi-parting slide plug
Body structure:	Aluminium	Aluminium
Operator:	GTR	GTR
Owner:	Porterbrook Leasing	Porterbrook Leasing

2023 Fleet

171/7
171727 GTR
171728 GTR
171729 GTR
171730 GTR

171/8
171801 GTR
171802 GTR
171803 GTR
171804 GTR
171805 GTR
171806 GTR
171807 GTR
171808 GTR
171809 GTR
171810 GTR
171811 GTR
171812 GTR

Top, Above & Inset: *Govia Thameslink Railway / Southern non-electrified services are operated by a fleet of two and three-car Class 171 'Turbostar' units allocated to Selhurst. In 2023, Four two-car, and 12 three-car sets are in use. The Southern 'Turbostars' are classified 171 and not 170 as they are fitted with Dellner 12 couplings. All sets have first class seating in the DMCL vehicles for nine, located directly behind the driving cab. At top, two-car set No. 171728 is seen from its DMCL end. Above is three-car set No. 171804 viewed from its DMCL vehicle. Originally six of the 171/8s were four-car sets, but major reformations in 2022-2023 has resulted in the present fleet formations. The inset, shows the interior, set out in the 2+2 style using Chapman seating, mainly formed in the group style. In summer 2023 a Class 171/2 transferred to East Midlands was on hire to GTR Southern to cover a unit shortage.* **Antony Christie (2) / CJM**

Key Facts

- Derivative of '170' for GTR, Class 171 fitted with Dellner 12 coupling in place of BSI.
- Fleet reformed in 2022-2023 to form 4 two car and 12 3-car sets.
- Sets used on London-Uckfield and Ashford-Hastings services.

Left: *The Class 171 intermediate Motor Seconds (MSs) seat 74 in the 2+2 style, no toilet is provided, these are located one in each driving vehicle, one in each set in the DMC vehicle is of the universal type. No. 54608 from set 171806 is shown.* **CJM**

Rolling Stock Review : 2023-2024

Class 172 — 'Turbostar'

For history and pictures of this class, see *Modern Locomotives Illustrated* Issue 216

Class:	172/0	172/1	172/2	172/3
Number range:	172001-172008	172101-172104	172211-172222	172331-172345
Built by:	Bombardier Derby	Bombardier Derby	Bombardier Derby	Bombardier Derby
Introduced:	2010	2011	2011	2011
Max speed:	75mph (121km/h)	100mph (161km/h)	100mph (161km/h)	100mph (161km/h)
Formation:	DMS+DMS	DMSL+DMS	DMSL+DMS	DMSL+MS+DMS
Vehicle numbers:	DMSL - 59311-59318 DMS - 59411-59418	DMSL - 59111-59114 DMS - 59211-59214	DMSL - 50211-50222 DMS - 79211-79222	DMSL - 50331-50345 MS - 56331-56345 DMS - 79331-79345
Vehicle length:	DMSL - 76ft 3in (23.27m) DMS - 76ft 3in (23.27m)	DMSL - 76ft 3in (23.27m) DMS - 76ft 3in (23.27m)	DMSL - 76ft 3in (23.27m) DMS - 76ft 3in (23.27m)	DMSL - 76ft 3in (23.27m) MS - 76ft 6in (23.36m) DMS - 76ft 3in (23.27m)
Height:	12ft 4½in (3.77m)	12ft 4½in (3.77m)	12ft 4½in (3.77m)	12ft 4½in (3.77m)
Width:	8ft 8in (2.69m)	8ft 8in (2.69m)	8ft 8in (2.69m)	8ft 8in (2.69m)
Weight:	Total - 83.1 tonnes DMSL - 36.5 tonnes DMS - 41.6 tonnes	Total - 82.2 tonnes DMSL - 41.4 tonnes DMS - 40.8 tonnes	Total - 83.2 tonnes DMSL - 41.9 tonnes DMS - 41.3 tonnes	Total - 121.3 tonnes DMSL - 41.9 tonnes MS - 38.1 tonnes DMS - 41.3 tonnes
Internal layout:	2+2	2+2	2+2	2+2
Seating:	Total - 138S DMSL - 62S DMS - 76S	Total - 140S DMSL - 60S DMS - 80S	Total - 120S + 7 tip up DMSL - 52S + 4 tip up DMS - 68S + 3 tip up	Total - 192S + 7 tip up DMSL - 52S + 4 tip up MS - 72S + 8 tip up DMS - 68S + 3 tip up
Gangway:	Within set	Within set	Throughout	Throughout
Toilets:	DMSL-1	DMSL - 1	DMSL - 1	DMSL - 1
Bogie type:	B5006	B5006	B5006	B5006
Power unit:	1 x MTU 6H1800R83 of 484hp (360kW) per vehicle	1 x MTU 6H1800R83 of 484hp (360kW) per vehicle	1 x MTU 6H1800R83 of 484hp (360kW) per vehicle	1 x MTU 6H1800R83 of 484hp (360kW) per vehicle
Transmission:	Mechanical	Mechanical	Mechanical	Mechanical
Transmission type:	ZF	ZF	ZF	ZF
Horsepower:	968hp (720kW)	968hp (720kW)	968hp (720kW)	1,449hp (1,080kW)
Brake type:	Air EP	Air EP	Air EP	Air EP
Coupling type:	Outer - BSI, Inner - Bar	Outer - BSI, Inner - Bar	Outer - BSI, Inner - Bar	Outer - BSI, Inner - Bar
Multiple restriction:	Class 15x, 170, 172	Class 15x, 170, 172	Class 15x, 170, 172	Class 15x, 170, 172
Door type:	Twin-leaf sliding plug	Twin-leaf sliding plug	Twin-leaf sliding plug	Twin-leaf sliding plug
Body structure:	Aluminium body, steel cabs	Aluminium body, steel cabs	Aluminium body, steel cabs	Aluminium body, steel cabs
Operator:	West Midlands Railway	West Midlands Railway	West Midlands Railway	West Midlands Railway
Owner:	Angel Trains	Angel Trains	Porterbrook Leasing	Porterbrook Leasing

Key Facts

- Derivative of '170' fitted with a diesel-mechanical transmission and basic interior design.
- 172/0 sets originally worked for London Overground, with 172/1s on Chiltern Railways. 172/2 and 172/3 were built for Midland area use. All sets now with West Midlands Railway.
- Fleet is of mixed design, 172/0 and 172/1s do not have end gangways. 172/2 and 172/3 have end gangways and a revised cab layout.

Above: The Class 172 emerged in 2010, with a fleet of eight two-car sets for London Overground on the Gospel Oak to Barking line. After electrification, the low-density sets were cascaded to the West Midlands in 2019 and are now finished in WMR mauve and gold livery. Set No. 172001 is seen at Birmingham Moor Street. **CJM**

Below Left: Class 172/0 front end, also applicable to Class 172/1. 1: High level marker light, 2: Destination indicator, 3: Headlight, 4: White marker and red tail light, 5: Forward facing camera, 6: Emergency lamp bracket, 7: BSI coupling, 8: Electrical connection box, 9: Air horns. **CJM**

Below Right: The West Midland Railway upgrade for the 172/0 fleet was undertaken by Bombardier, Ilford, it included the addition of a universal toilet in the 593xx vehicle. The original style low-density seating was retained, now covered in WMR moquette, floors were also carpeted. The spacious 2+2 interior is shown from a door pocket position. In common with modern stock, the toilet and disabled area is marked by a blue panel on or above the near passenger doors. A high-quality passenger information system is installed. The interior of set No. 172002 is shown. **CJM**

2023 Fleet

172/0
172001 WMR	172217 WMR
172002 WMR	172218 WMR
172003 WMR	172219 WMR
172004 WMR	172220 WMR
172005 WMR	172221 WMR
172006 WMR	172222 WMR
172007 WMR	**172/3**
172008 WMR	172331 WMR
	172332 WMR
172/1	172333 WMR
172101 WMR	172334 WMR
172102 WMR	172335 WMR
172103 WMR	172336 WMR
172104 WMR	172337 WMR
	172338 WMR
172/2	172339 WMR
172211 WMR	172340 WMR
172212 WMR	172341 WMR
172213 WMR	172342 WMR
172214 WMR	172343 WMR
172215 WMR	172344 WMR
172216 WMR	172345 WMR

Rolling Stock Review : 2023-2024

'Turbostar' — Class 172

Right: The second batch of non-gangway Class 172s, consisted of four Class 172/1s, originally built for Chiltern Railways and funded by Angel Trains. These were formed DMSL+DMS, with accommodation for 140 standard class passengers. A universal toilet compartment was housed in the DMSL (or 591xx) vehicle. Originally, sets were allocated to Aylesbury and painted in Chiltern white and blue suburban livery. In spring 2021, the four sets were stored by Chiltern due to a downturn in passenger numbers during the Covid pandemic. In late 2021 the fleet was transferred to West Midlands Railway, joining the rest of the Class 172s based at Tyseley. Carrying West Midlands mauve and gold, set No. 172101 is seen at Smethwick Galton Bridge on 12 December 2022. **CJM**

Left: The two largest fleets of Class 172 were built for West Midlands use, consisting of 12 two-car Class 172/2s and 15 three-car Class 172/3s allocated to Tyseley. These have end gangways, seriously reducing the size of the cab and restricting forward vision. The all standard class sets have seating for 120 in a two-car and 192 in a three-car, all in the 2+2 style. Extra fold down seats are provided, with DMSL vehicles having 11, and DMS/MS coaches eight. The fleet is predominately used on local services on the Birmingham Snow Hill routes. Two-car set No. 172218 in mauve and gold livery arrives at Leamington Spa with a train from Coventry. **CJM**

Below Far Left: Seating on '172s' uses the 2+2 style, with good spacing between seats. An individual seat design is incorporated, with a mix of group and airline styles. Carpets are provided throughout, as is a good quality passenger information and public address system. Moquette is of standard West Midlands fleck style. **CJM**

Left & Below: Class 172/3s are formed with an intermediate Motor Standard (MS), seating 72 plus eight tip-up seats. The MS, No. 56341 from set 172341 is seen. The '172s' are mounted on lightweight B5006 bogies and have one powered and one trailer bogie per vehicle. The outer end bogies of driving cars being unpowered. Three-car set No. 172344 is seen at The Hawthorns. **CJM**

Rolling Stock Review : 2023-2024

Class 175 'Coradia'

For history and pictures of this class, see *Modern Locomotives Illustrated* Issue 216

Class:	175/0	175/1
Number range:	175001-175011	175101-175116
Built by:	Alstom, Birmingham	Alstom, Birmingham
Type:	Coradia 1000	Coradia 1000
Introduced:	1999-2001	1999-2001
Max speed:	100mph (161km/h)	100mph (161km/h)
Formation:	DMSL(A)+DMSL(B)	DMSL(A)+MSL+DMSL(B)
Vehicle numbers:	DMSL(A) - 50701-50711 DMSL(B) - 79701-79711	DMSL(A) - 50751-50766 MSL - 56751-56766 DMSL(B) - 79751-79766
Vehicle length:	75ft 7in (23.03m)	DMSL - 75ft 7in (23.03m) MSL - 75ft 5in (22.98m)
Height:	12ft 4in (3.75m)	12ft 4in (3.75m)
Width:	9ft 2in (2.79m)	9ft 2in (2.79m)
Weight:	Total - 99.5 tonnes DMSL(A) - 48.8 tonnes DMSL(B) - 50.7 tonnes	Total - 147.7 tonnes DMSL(A) - 50.7 tonnes MSL - 47.5 tonnes DMSL(B) - 49.5 tonnes
Internal layout:	2+2	2+2
Seating:	Total - 118S DMSL(A) - 54S DMSL(B) - 64S	Total - 186S DMSL(A) - 54S MSL - 68S, DMSL(B) - 64S
Gangway:	Within set	Within set
Toilets:	DMSL(A), DMSL(B) - 1	DMSL(A), MSL, DMSL(B) - 1
Bogie type:	ACR - Alstom MB1	ACR - Alstom MB1
Power unit:	One Cummins N14 of 450hp (335kW) per car	One Cummins N14 of 450hp (335kW) per car
Transmission:	Hydraulic	Hydraulic
Transmission type:	Voith T211rzze to ZF final drive	Voith T211rzze to ZF final drive
Horsepower (total):	900hp (670kW)	1,350hp (1,005kW)
Brake type:	Air	Air
Coupling type:	Outer - Scharfenberg, Inner - Bar	Outer - Scharfenberg, Inner - Bar
Multiple restriction:	Within type and Class 180	Within type and Class 180
Door type:	Single-leaf sliding plug	Single-leaf sliding plug
Body structure:	Steel	Steel
Operator:	Transport for Wales	Transport for Wales
Owner:	Angel Trains	Angel Trains

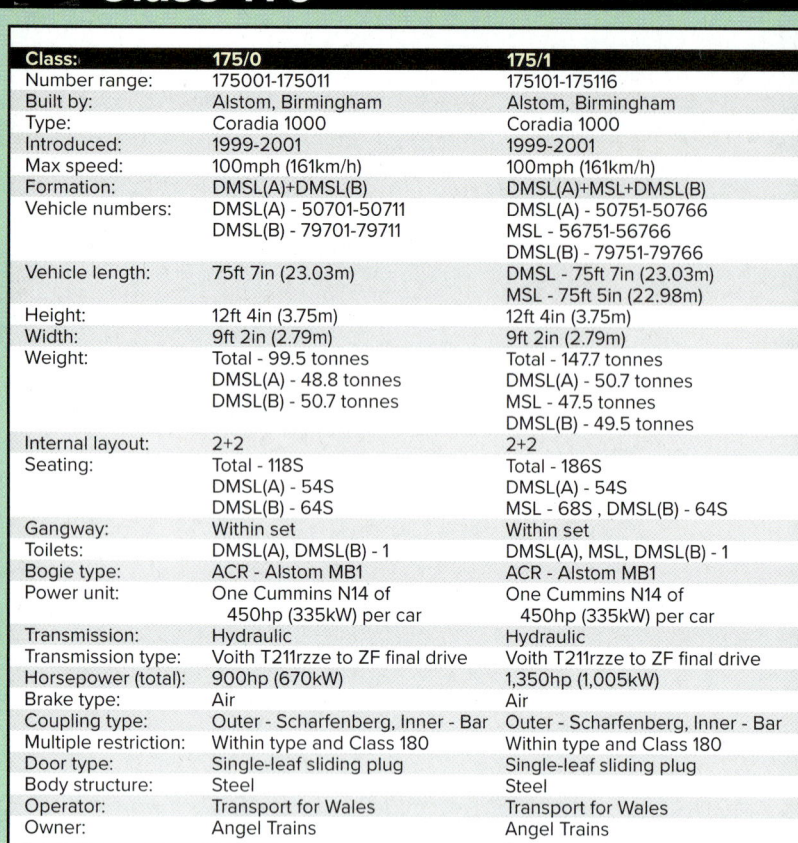

Key Facts

- Ordered by First Group for Welsh use, based at Chester. Part of a larger order made with Alstom for 'Coradia 1000' stock, including Class 180s.
- Built alongside 14 'high-speed' Class 180 sets for use by Great Western with streamlined cab ends.
- Sets now operated by Transport for Wales on longer distance services.
- Sets to be withdrawn after TfW Class 197s introduced in 2023-2024.

2023 Fleet

175/0
175001 TFW
175002 TFW
175003 TFW
175004 TFW
175005 TFW
175006 TFW
175007 TFW
175008 TFW
175009 TFW
175010 TFW
175011 TFW

175/1
175101 TFW
175102 TFW
175103 TFW
175104 TFW
175105 TFW
175106 TFW
175107 TFW
175108 TFW
175109 TFW
175110 TFW
175111 TFW
175112 TFW
175113 TFW
175114 TFW
175115 TFW
175116 TFW

Left: Class 175 driving cab. Based on the single power/brake controller style, located on the left side. **CJM**

Left: As part of a large First Group order for new stock, two and three-car Class 175s were ordered from Alstom and delivered between 1999-2001, to operate with the Welsh franchise, based at Chester, used on long-distance services. The units are part of the Alstom 'Coradia 1000' family. By 2023, the fleet is painted in the Transport for Wales white, grey and red livery. Class 175/0 covers 11 two-car sets, No. 175003 is seen at Cardiff. **CJM**

Below & Inset: The 16 three-car Class 175/1 sets, seat 186 standard class passengers in the 2+2 low-density style, mainly in the group layout. A toilet is provided in each vehicle, with one in the 507xx coach being of the universal access type. The 'Coradia' design uses single leaf sliding-plug doors, located at vehicle ends, feeding a transverse walkway from which access to the passenger saloons is made. The driving cab has its own swing plug door on each side, the cab can also be accessed from the passenger saloon. Set No. 175104 is seen at Ludlow. The inset image shows the intermediate MSL with seating for 69. These high-quality sets have only a limited period of use remaining on the Welsh network, as new stock is currently under production. A number of '175s' were taken out of service in early 2023 with serious engine defects which were the source of fires. Both: **CJM**

For history and pictures of this class, see Modern Locomotives Illustrated Issue 216.

'Coradia' — Class 180

Class:	180
Number range:	180101-180114
Built by:	Alstom, Birmingham
Type:	Coradia 1000
Introduced:	2000-2001
Max speed:	125mph (201km/h)
Formation:	DMSL(A)+MFL+MSL+MSLRB+DMSL(B)
Vehicle numbers:	DMSL(A) - 50901-50914 MFL - 54901-54914 MSL - 55901-55914 MSLRB - 56901-56914 DMSL(B) - 59901-59914
Vehicle length:	DMSL(A), DMSL(B) - 77ft 7in (23.71m) MFL, MSL, MSLRB - 75ft 5in (23.03m)
Height:	12ft 4in (3.75m)
Width:	9ft 2in (2.79m)
Weight:	Total - 252.5 tonnes DMSL(A) - 51.7 tonnes MFL - 49.6 tonnes MSL - 49.5 tonnes MSLRB - 50.3 tonnes DMSL(B) - 51.4 tonnes
Internal layout:	2+1F, 2+2S
Seating:	Total - 42F/226S DMSL(A) - 46S MFL - 42F MSL - 68S MSLRB - 56S DMSL(B) - 56S
Gangway:	Within set
Toilets:	DMSL(A), MFL, MSL, MSLRB, DMSL(B) - 1
Bogie type:	ACR - Alstom MB2
Power unit:	One Cummins QSK19 of 750hp (560kW) per car
Transmission:	Hydraulic
Transmission type:	Voith T312br to ZF final drive
Horsepower (total):	3,750hp (2,800kW)
Brake type:	Air
Coupling type:	Outer - Scharfenberg, Inner - Bar
Multiple restriction:	Within type and Class 175
Door type:	Single-leaf swing plug
Body structure:	Steel
Operator:	Grand Central Railway
Owner:	Angel Trains

Above & Below: A fleet of 14 Class 180s were built for First Great Western in 2000-2001, in summer 2023 just 11 remain in service with Grand Central, based at Heaton depot in Newcastle, working open access services from London King's Cross to Sunderland and Bradford. All have been refurbished, with each five-car now seating 42 first and 226 standard class passengers. Sets are finished in a striking Grand Central black and orange livery, with a number of units carrying cast nameplates. The image above shows set No. 180106 passing Otterington near Darlington. Below in an intermediate vehicle, a Motor Standard Lavatory Restaurant Buffet No. 56908 from set 180108. These vehicles seat 56 in the 2+2 style with a small catering outlet at one end. Standard class coaches have silver sliding plug door and the first class coach have gold coloured doors. **CJM / Antony Christie**

2023 Fleet

180101	GTL	180108	GTL
180102	GTL	180109	OLS
180103	GTL	180110	GTL
180104	GTL	180111	OLS
180105	GTL	180112	GTL
180106	GTL	180113	OLS
180107	GTL	180114	GTL

Above: In 2021, the DMSLB vehicle No. 59912 from set 180112, was given a green body band in place of orange, indicating it is equipped to operate from dual fuel Liquefied Natural Gas (LNG). It also carried suitable bodyside branding. **Antony Christie**

Above Left & Above Right: Grand Central Class 180 interiors. On the left is the luxury first class interior with leather reclining seats, while on the right is the 2+2 standard class interior, with good quality hinged arm-rest seats, mainly positioned in groups around tables. Both: **Antony Christie**

Left: Cast nameplate *Kirkgate Calling* as applied to MLSRB vehicle No. 56914 from set 180114 **Antony Christie**

Key Facts

- Originally ordered by First Group for Great Western use. Part of a larger order placed with Alstom for 'Coradia 1000' stock including Class 175s.

- Changes saw some sets move to the North West for a short time before moving to Grand Central - 10 sets and Hull Trains - 4 sets.

- Hull Trains sets were replaced by Class 802s in 2020. Then transferred to EMR.

- EMR sets withdrawn in May 2023 some to Grand Central or stored.

Class 185 'Desiro'

For history and pictures of this class, see Modern Locomotives Illustrated Issue 216

Class:	185
Number range:	185101-185151
Built by:	Siemens Transportation, Germany
Type:	Desiro UK
Introduced:	2005-2007
Max speed:	100mph (161km/h)
Formation:	DMCL+MSL+DMS
Vehicle numbers:	DMCL - 51101-51151
	MSL - 53101-53151
	DMS - 54101-54151
Vehicle length:	DMCL, DMS - 77ft 11in (23.76m)
	MSL - 77ft 10in (23.75m)
Height:	12ft 4in (3.75m)
Width:	9ft 3in (2.84m)
Weight:	Total - 163.1 tonnes
	DMCL - 55.4 tonnes
	MSL - 52.7 tonnes
	DMS - 55 tonnes
Internal layout:	2+2S, 2+1F
Seating:	Total - 15F/154S + 12 tip up
	DMCL - 15F/18S + 8 tip up
	MSL - 72S
	DMS - 64S + 4 tip up
Gangway:	Within set
Toilets:	MSL - 1, DMCL - 1
Bogie type:	Siemens
Power unit:	One Cummins QSK19 of 750hp (560kW) per car
Transmission:	Hydraulic
Transmission type:	Voith Turbopack T312 and SK-485 final drive
Horsepower (total):	2,250hp (1,680kW)
Brake type:	Air
Coupling type:	Outer: Dellner 12, Inner: Bar
Multiple restriction:	Within class only
Door type:	Bi-parting sliding plug
Body structure:	Aluminium
Operator:	TransPennine Trains
Owner:	Eversholt Leasing

185101 TPT	185112 TPT	185123 TPT	185134 TPT	185145 TPT
185102 TPT	185113 TPT	185124 TPT	185135 TPT	185146 TPT
185103 TPT	185114 TPT	185125 TPT	185136 TPT	185147 TPT
185104 TPT	185115 TPT	185126 TPT	185137 TPT	185148 TPT
185105 TPT	185116 TPT	185127 TPT	185138 TPT	185149 TPT
185106 TPT	185117 TPT	185128 TPT	185139 TPT	185150 TPT
185107 TPT	185118 TPT	185129 TPT	185140 TPT	185151 TPT
185108 TPT	185119 TPT	185130 TPT	185141 TPT	
185109 TPT	185120 TPT	185131 TPT	185142 TPT	
185110 TPT	185121 TPT	185132 TPT	185143 TPT	
185111 TPT	185122 TPT	185133 TPT	185144 TPT	

Above: *The TransPennine replacement stock introduced in 2005, consisted of a fleet of 51 Siemens 'Desiro UK' three-car DMUs. both driving vehicles have different body styles. Set No. 185150 seen at Brough illustrates its Driving Motor Standard (DMS) end, with seating for 64 and windows the full vehicle length. Passenger access is by two pairs of bi-parting plug doors on each side. A separate single leaf crew door is provided for cab access. All sets sport standard TransPennine Express colours.* **CJM**

Left: *Coupled between the two driving cars is a Motor Standard Lavatory (MSL), providing seating for 72 in the 2+2 style, with a mix of airline and group layout. The vehicle houses one standard type toilet. Vehicle No. 53150 is shown.* **CJM**

Key Facts

- Only Siemens Desiro diesel fleet operating in the UK.
- Built specifically for the TransPennine Trains operations.
- After Class 397 and 802 stock was introduced, Class 185 numbers were to be reduced, but by 2023 all the fleet remain.
- Fitted with low-density seating layout.

Above: *The Class 185 Driving motor Composite Lavatory (DMCL) has a totally different body on the drivers side to the DMS coach. The DMCL, shown above, has a solid side panel from inward of the cab to the first door pocket, where a universal toilet compartment is located. Seating in this vehicle is for 15 first and 18 standard. Set No. 185142 is seen south of York in June 2022.* **CJM**

Right: *Standard class interior, showing the low-density 2+2 seating style, with a mauve and grey colour moquette, each seat has fold down armrests, and an above seat glazed bottomed luggage shelf. The sets are carpeted throughout.* **Antony Christie**

'Civity' Class 195

For history and pictures of this class, see Modern Locomotives Illustrated Issue 243

Class:	195/0	195/1
Number range:	195001-195025	195101-195133
Built by:	CAF, Irun, Spain and Newport, UK Bodyshells from CAF Zaragoza	CAF, Irun, Spain and Newport UK Bodyshells from CAF Zaragoza
Type:	Civity	Civity
Introduced:	2018-2020	2018-2020
Max speed:	100mph (161km/h)	100mph (161km/h)
Formation:	DMSL+DMS	DMSL+MS+DMS
Vehicle numbers:	DMSL - 101001-101025 DMS - 103001-103025	DMSL - 101101-101133 MS - 102101-102133 DMS - 103101-103133
Train length:	157ft 8in (48.05m)	234ft 3in (71.4m)
Vehicle length:	DMS, DMSL - 78ft 8in (24.02m)	DMS, DMSL - 78ft 8in (24.02m) MS - 76ft 6in (23.35m)
Height:	12ft 7in (3.85m)	12ft 7in (3.85m)
Width:	8ft 9in (2.71m)	8ft 9in (2.71m)
Weight:	DMSL 43.9t, DMS 43.2t	DMSL 43.9t, MS 40.50t, DMS 43.2t
Internal layout:	2+2S	2+2S
Seating:	Fisa - Total - 108S DMSL: 45S, DMS: 63S	Fisa - Total -185S DMSL: 45S, MS: 76S, DMS: 63S
Gangway:	Within set only	Within set only
Toilets:	DMSL - 1	DMSL - 1
Bogie type:	CAF	CAF
Power unit:	One Rolls Royce/MTU 6H1800R85L of 523hp (390kW) per car	One Rolls Royce/MTU 6H1800R85L of 523hp (390kW) per car
Transmission:	Mechanical	Mechanical
Transmission type:	ZF Astronic 6-speed	ZF Astronic 6-speed
Horsepower (total):	1,046hp (780kW)	1,569hp (1,170kW)
Brake type:	Air	Air
Coupling type:	Dellner 12	Dellner 12
Multiple restriction:	Within class only	Within class only
Door type:	Bi-parting sliding plug	Bi-parting sliding plug
Body structure:	Aluminium	Aluminium
Operator:	Northern	Northern
Owner:	Eversholt Leasing	Eversholt Leasing

Above: *Class 195 front end layout. 1: High level marker light, 2: Destination indicator, 3: Forward facing camera, 4: Headlight, 5: Combined red tail and white marker light, 6: Running lights, 7: Lamp bracket, 8: Electrical connection box, 9: Dellner coupling, 10: Air horns in coupling pocket, 11: Obstacle deflector plate.* **CJM**

Above: *Spanish rolling stock builder CAF supplied 58 Class 195 DMU trains for the Northern franchise from their 'Civity' platform in 2019-2020. A fleet of 25 two-car Class 195/0 and 33 three-car Class 195/1 sets were built allocated to Newton Heath depot in Manchester. Nos. 195001-006, 195101-130 and the intermediate cars for Nos. 195131-195133 were fully built in Spain, while Nos. 195007-195025 and the driving cars for Nos. 195131-195133 had body shells fabricated in Spain, but were fitted out at the CAF plant in Llanwern, South Wales. The two-car sets seat 108 in the 2+2 layout, using a mix of group and airline styles. Set No. 195003 is shown passing Dore & Totley in June 2022. Each vehicle has two pairs of passenger-operated bi-parting plug doors.* **CJM**

Right: *The Class 195s are fitted with a European style driving cab, with the driver positioned in the middle of the front end. The driver's desk incorporates a number of UK features such as TPWS, AWS and radio systems. CCTV displays are provided on the left side, so the driver can have a clear image of the train/platform interface. A full train management system is also provided. A joint power/brake controller is located in the UK style on the left side.* **Antony Christie**

Class 195   'Civity'

2023 Fleet

195/0
195001	NOR
195002	NOR
195003	NOR
195004	NOR
195005	NOR
195006	NOR
195007	NOR
195008	NOR
195009	NOR
195010	NOR
195011	NOR
195012	NOR
195013	NOR
195014	NOR
195015	NOR
195016	NOR
195017	NOR
195018	NOR
195019	NOR
195020	NOR
195021	NOR
195022	NOR
195023	NOR
195024	NOR
195025	NOR

195/1
195101	NOR
195102	NOR
195103	NOR
195104	NOR
195105	NOR
195106	NOR
195107	NOR
195108	NOR
195109	NOR
195110	NOR
195111	NOR
195112	NOR
195113	NOR
195114	NOR
195115	NOR
195116	NOR
195117	NOR
195118	NOR
195119	NOR
195120	NOR
195121	NOR
195122	NOR
195123	NOR
195124	NOR
195125	NOR
195126	NOR
195127	NOR
195128	NOR
195129	NOR
195130	NOR
195131	NOR
195132	NOR
195133	NOR

Above: *Each Class 195 vehicle is powered by a Rolls Royce/MTU 6H1800R85L of 523hp (390kW), an under-slung unit driving a mechanical transmission. A three-car set develops 1,569hp (1,170kW). Three-car Class 195/1 No. 195103 is shown south of York in June 2022.* **CJM**

Left: *Class 195 intermediate Motor Standard (MS) has seating for 76 in the 2+2 style, plus four good quality fold down seats in door pockets. The '195s' are some of the few trains that carry a European Vehicle Number (EVN), applied below the UK number. In the case shown 95 70 00 103129 4 (95 = type of train DMU, 70 = registration country UK, 00 are space digits, 103129 = vehicle number, 4 = check digit. All vehicles are air conditioned and have no opening windows.* **CJM**

Above Left: *Seating in Class 195s is provided in the 2+2 low-density style, with a welcoming open vehicle ambience. Seating is a mix of airline and group layouts. Seats without fixed tables have a good-quality fold down table, attached to the seat in front. Power sockets are provided throughout and a free wi-fi service is available (this was tested in spring 2022 and was of poor and slow quality). The main passenger saloon in shown, looking towards the inner end of a driving coach from the cab end door. Above seat luggage racks are provided for light items. A good-quality passenger information system is fitted on the panels by the door pockets.* **CJM**

Above Right: *Adjacent to some door pockets is a stand back area where luggage can be stowed. A fold down seat is also provided. Unlike on some builds of new train, on the 195s passengers are able to both open and close doors once released by the train crew. Door positions have an emergency door release handle and an emergency call point.* **CJM**

Key Facts

- First CAF built diesel-units to operate in mainland UK.
- Some vehicles/sets built in Spain, others fitted out at CAF facility at Llanwern, Newport, South Wales.
- Vehicle structure identical to Class 331 electric fleet.
- All sets sport Northern livery.

For history and pictures of this class, see *Modern Locomotives Illustrated* Issue 243

'Civity' — Class 196

Key Facts

- A fleet of 26 CAF built 'Civity' two and four-car sets are currently being delivered to West Midlands Railway.
- Six 2-car sets will operate on the new East West Rail between Oxford-Bedford from 2024.
- Vehicle shells assembled in Spain, fitted out at the CAF Llanwern plant in Wales.
- All sets sport mauve and gold WMR colours.

2023 Fleet

196/0		
196001 WMR	196010 WMR	196106 WMR
196002 WMR	196011 WMR	196107 WMR
196003 WMR	196012 WMR	196108 WMR
196004 WMR		196109 WMR
196005 WMR	**196/1**	196110 WMR
196006 WMR	196101 WMR	196111 WMR
196007 WMR	196102 WMR	196112 WMR
196008 WMR	196103 WMR	196113 WMR
196009 WMR	196104 WMR	196114 WMR
	196105 WMR	

Class:	196/0	196/1
Number range:	196001-196012	196101-195114
Built by:	CAF, Irun, Spain and Newport, UK Bodyshells from CAF Zaragoza	CAF, Irun, Spain and Newport UK Bodyshells from CAF Zaragoza
Type:	Civity	Civity
Introduced:	2020-2021	2020-2021
Max speed:	100mph (161km/h)	100mph (161km/h)
Formation:	DMSL+DMS	DMSL+MS+MS+DMS
Vehicle numbers:	DMSL - 121001-121012 DMS - 124001-124012	DMSL - 121101-121114 MS - 122101-122114, MS - 123101-123114 DMS - 124101-124114
Train length:	157ft 8in (48.05m)	315ft 4in (96.1m)
Vehicle length:	DMS, DMSL - 78ft 8in (24.02m)	DMS, DMSL - 78ft 8in (24.02m) MS - 76ft 6in (23.35m)
Height:	12ft 7in (3.85m)	12ft 7in (3.85m)
Width:	8ft 9in (2.71m)	8ft 9in (2.71m)
Weight:	Total 84.1 tonnes DMSL - 41.4T, DMS - 42.7T	164.2 tonnes DMSL - 41.4T, MS - 39.8T, MS - 40.3T, DMS - 42.7T
Internal layout:	2+2	2+2
Seating:	Total - 134S (+4TU) (Fisa) DMSL - 58S (+4TU) DMS - 76S	Total - 304S (+4TU) (Fisa) DMSL - 58S (+4TU) MS - 88S, MS - 82S DMS - 76S
Gangway:	Throughout	Throughout
Toilets:	DMSL - 1	DMSL - 1
Bogie type:	CAF	CAF
Power unit:	One Rolls Royce/MTU 6H1800R85L of 523hp (390kW) per car	One Rolls Royce/MTU 6H1800R85L of 523hp (390kW) per car
Transmission:	Mechanical	Mechanical
Transmission type:	ZF Astronic 6-speed	ZF Astronic 6-speed
Horsepower (total):	1,046hp (780kW)	1,569hp (1,170kW)
Brake type:	Air	Air
Coupling type:	Dellner 12	Dellner 12
Multiple restriction:	Within class only	Within class only
Door type:	Bi-parting sliding plug	Bi-parting sliding plug
Body structure:	Aluminium	Aluminium
Operator:	West Midlands Railway	West Midlands Railway
Owner:	Infracapital and Deutsche Asset Management	Infracapital and Deutsche Asset Management

Top, Above, Below Left & Below Right: After Abellio was awarded a new West Midlands Railway franchise in late-2017, new train orders followed. This included 26 'Civity' platform DMUs from CAF, with two sub-classes, 12 two-car sets of Class 196/0 and 14 four-car units of Class 196/1. They have replaced the Class 170 fleet. Most Class 196s have been built at the CAF plant in Llanwern, Newport, South Wales, with some vehicles coming from Spain. Sets are similar to the Class 197s operated by TfW, but have a more basic, suburban, interior. Above, is four-car set No. 196107 at Wellington in May 2023. These sets, of Class 196/1, have two Motor Standard (MS) coaches formed between the driving motor cars. This gives a traction output to 2,092hp (1,560kW) per set. The above inset image shows the inner end of the DMSL (121xxx) coach, where the universal toilet compartment and wheelchair seating is located. All '196' vehicles carry their full European Vehicle Number or EVN. The image below left shows a two-car Class 196/0 set, No. 196005, these sets have one universal type toilet in the DMCL coach. Right below, is the interior, this is set out for standard class occupancy using the 2+2 low density style with seats arranged in groups and airline format, airline seats have a fold down table on the rear of the seat ahead, all seats have power and USB charging sockets. Very little luggage space is provided and can cause serious issues on the Shrewsbury to Birmingham corridor. All: **CJM**

Class 197 'Civity'

Class:	197/0	197/1
Number range:	197001-197051	197101-197126
Built by:	CAF Llanwern, UK Bodyshells from CAF Zaragoza	CAF Llanwern, UK Bodyshells from CAF Zaragoza
Type:	Civity	Civity
Introduced:	2021-2022 - on delivery	2021-2022 - on delivery
Max speed:	100mph (161km/h)	100mph (161km/h)
Formation:	DMSL+DMS	DMSL+MS+DMS
Vehicle numbers:	DMSL - 131001-131051 DMS - 133001-133051	DMSL - 131101-131126 MS - 132101-132126 DMS - 133191-133126
Train length:	157ft 8in (48.05m)	234ft 3in (71.4m)
Vehicle length:	DMS, DMSL - 78ft 8in (24.02m)	DMS, DMSL - 78ft 8in (24.02m) MS - 76ft 6in (23.35m)
Height:	12ft 7in (3.85m)	12ft 7in (3.85m)
Width:	8ft 9in (2.71m)	8ft 9in (2.71m)
Weight:	Total - 84.3tonnes	Total - 124.4 tonnes
Internal layout:	2+2	2+2
Seating:	Total - 116S (+6TU) Fainsa Spohia DMSL - 42S (+6TU), Catering area DMS - 74S	Total - 188S (+6TU) Fainsa Spohia DMSL - 42S (+6TU), catering area MS - 72S, DMS - 74S
Gangway:	Throughout	Throughout
Toilets:	DMSL - 1	DMSL, MS - 1
Bogie type:	CAF	CAF
Power unit:	One Rolls Royce/MTU 6H1800R85L of 523hp (390kW) per car	One Rolls Royce/MTU 6H1800R85L of 523hp (390kW) per car
Transmission:	Mechanical	Mechanical
Transmission type:	ZF Astronic 6-speed	ZF Astronic 6-speed
Horsepower (total):	1,046hp (780kW)	1,569hp (1,170kW)
Brake type:	Air	Air
Coupling type:	Dellner 12	Dellner 12
Multiple restriction:	Within class only	Within class only
Door type:	Bi-parting sliding plug	Bi-parting sliding plug
Body structure:	Aluminium	Aluminium
Owner/Operator:	Transport for Wales	Transport for Wales
Notes:	21 sets (197003/022-041) fitted with ETCS for Cambrian use	Plan for 14 3-cars to have one driving car arranged with higher grade seating

Above: *Class 197 front end equipment. 1: High level marker light, 2: Destination display, 3: Forward facing camera, 4: Gangway doors, 5: Headlight, 6: Combined red tail and white marker light, 7: Lamp bracket, 8: Warning horns, 9: Dellner coupling with electrical connection box below. Unit 197002 is shown.* **Cliff Beeton**

Left: *Transport for Wales (TfW) have now introduced two- and three-car Class 197 DMUs, built by CAF at their plant in Wales. The fleet is part of the 'Civity' platform and are similar to the Class 196s. When fully introduced in 2024, they will replace Class 158 and 175s. 21 two-cars (197003/022-041) will be fitted with ETCS for operation over the Cambrian routes. Two-car set No. 197011 is seen at Bangor in March 2023 with a Holyhead to Manchester Airport service. The sets carry full TfW grey and red livery with no yellow end, but the gangway door is finished in red. Sets will be based at the CAF operated Chester depot, with light maintenance carried out at Machynlleth.* **CJM**

Right: *A fleet of 26 three car Class 197s are currently under commissioning, these seat 188 passengers and will be used on longer distance services, such as the Holyhead/Manchester to Cardiff and West Wales routes. Some sets are scheduled to have a limited number of upgraded seats for an advanced level of travel comfort. A light catering provision is available on both two and three-car sets. Three-car set No. 197102 is illustrated.* **CJM**

Left: *Class 197 interior, showing the high-quality 2+2 layout, with a mix of group and airline seating. Airline seats have fold down tables on the rear of the seat in front. Luggage stacks are provided near door pockets and an above seat light luggage shelf is provided. Each seat has a plug and USB charging point, coat hooks are provided and an electronic seat reservation system in installed (but not working when RSR did a sample ride). One of the better interiors of a modern train in recent years, but still with quite hard seats.* **CJM**

2023 Fleet

197/0
197001 TFW
197002 TFW
197003 TFW
197004 TFW
197005 TFW
197006 TFW
197007 TFW
197008 TFW
197009 TFW
197010 TFW
197011 TFW
197012 TFW
197013 TFW
197014 TFW
197015 TFW
197016 TFW
197017 TFW
197018 TFW
197019 TFW
197020 TFW
197021 TFW
197022 TFW
197023 TFW
197024 TFW
197025 TFW
197026 TFW
197027 TFW
197028 TFW
197029 TFW
197030 TFW
197031 TFW
197032 TFW
197033 TFW
197034 TFW
197035 TFW
197036 TFW
197037 TFW
197038 TFW
197039 TFW
197040 TFW
197041 TFW
197042 TFW
197043 TFW
197044 TFW
197045 TFW
197046 TFW
197047 TFW
197048 TFW
197049 TFW
197050 TFW
197051 TFW

197/1
197101 TFW
197102 TFW
197103 TFW
197104 TFW
197105 TFW
197106 TFW
197107 TFW
197108 TFW
197109 TFW
197110 TFW
197111 TFW
197112 TFW
197113 TFW
197114 TFW
197115 TFW
197116 TFW
197117 TFW
197118 TFW
197119 TFW
197120 TFW
197121 TFW
197122 TFW
197123 TFW
197124 TFW
197125 TFW
197126 TFW

Key Facts

- A fleet of 77 two and three-car CAF 'Civity' sets on delivery for TfW longer distance services.
- All sets allocated to Chester.
- Sets have limited catering provision, with an area set aside in the DMSL coach. An at seat services is offered on longer distance trains.
- 21 sets are fitted with ETCS for operation over the Cambrian routes.
- ETCS fitted sets will be out-based at Machynlleth for an extended period.
- Sets are owned and operated by Transport for Wales.

'Voyager' — Class 220

220001	AXC	220018	AXC
220002	AXC	220019	AXC
220003	AXC	220020	AXC
220004	AXC	220021	AXC
220005	AXC	220022	AXC
220006	AXC	220023	AXC
220007	AXC	220024	AXC
220008	AXC	220025	AXC
220009	AXC	220026	AXC
220010	AXC	220027	AXC
220011	AXC	220028	AXC
220012	AXC	220029	AXC
220013	AXC	220030	AXC
220014	AXC	220031	AXC
220015	AXC	220032	AXC
220016	AXC	220033	AXC
220017	AXC	220034	AXC

For history and pictures of this class, see Modern Locomotives Illustrated Issue 221

Below: In 1997 at the start of privatisation after the CrossCountry franchise was awarded to Virgin Rail Group, an order was placed for a fleet of 34 Bombardier 'Voyager' (non-tilt) trains and 44 'Super Voyager' (tilting) sets, formed into four and five-car trains. Accommodation was inferior to the vehicles they replaced, with the shorter trains causing serious overcrowding, with a lack of luggage space and problems with toilets. The Class 220s are now owned by Beacon Rail and operated by Arriva CrossCountry, based at Central Rivers depot, Burton. Following internal changes, each set now has 26 first and 174 standard class seats. Coaches have two single leaf plug doors. Originally an on-board 'shop' was provided, but these were removed and catering is now provided by a trolley. The Class 220s are formed with one driving car designated for first class travel, with 26 seats in the 2+1 style, this coach also houses the catering galley behind the driving cab. The vehicle is identified by a yellow cant rail band and a yellow panel on the electric connection box of the coupling. Set No. 220018 leads No. 220002 through Dore & Totley on 13 June 2022 with its DMFL (first class) car leading. **CJM**

Class:	220
Number range:	220001-220034
Built by:	Bombardier Transportation*
Type:	Voyager
Introduced:	2000-2001
Max speed:	125mph (201km/h)
Formation:	DMSL+MS+MSL+DMFL
Vehicle numbers:	DMSL - 60301-60334
	MS - 60701-60734
	MSL - 60201-60234
	DMFL - 60401-60434
Vehicle length:	77ft 6in (23.62m)
Height:	12ft 4in (3.76m)
Width:	8ft 11in (2.72m)
Weight:	Total - 194.7 tonnes
	DMSL - 51.2 tonnes
	MS - 45.9 tonnes
	MSL - 46.7 tonnes
	DMFL - 50.9 tonnes
Internal layout:	2+1F, 2+2S
Seating:	Total - 26F/174S
	DMSL - 42S
	MS - 66S
	MSL - 66S
	DMFL - 26F
Gangway:	Within set
Toilets:	DMSL, MSL, DMFL - 1
Bogie type:	Bombardier B5005
Power unit:	1 x Cummins QSK19 of 700hp (520kW) per car
Transmission:	Electric
Transmission package:	8 x Alstom Onix 800 per train
Horsepower:	2,800hp (2,059kW)
Operating range:	1,350 miles (2,173km)
Brake type:	Air, EP rheostatic
Route availability:	2
Coupling type:	Outer: Dellner 12, Inner: Bar
Multiple restriction:	Class 220, 221 and Class 57/3
Door type:	Single-leaf swing plug
Body construction:	Steel
Operator:	Arriva CrossCountry
Owner:	Beacon Rail

* Body shells assembled in Belgium, fitted out at Bombardier plants in Wakefield, UK, and Brugge, Belgium.

Below: On Voyager stock, toilets are provided in three vehicles. The area in the MS vehicle originally housed the shop, but is now modified to provide much needed extra luggage space. With its DMSL leading, set No. 220029 leads 'Super Voyager' No. 221124 south of York on 15 June 2022. All XC sets carry silver and deep maroon colours with pink doors. **CJM**

Left: The two intermediate vehicles of the Class 220 sets are classified Motor Standard (MS) and Motor Standard Lavatory (MSL). Both have a like body structure. Entrance is by a single leaf sliding plug door at either end, feeding a transverse vestibule with automatic bi-parting doors feeding the passenger saloon. Both vehicles seat 66 in the 2+2 style in a mix of group and airline seating styles. Originally, the non-toilet fitted coach housed a shop, but these have now been removed. Extra luggage space is now provided. Above seat racks are fitted for light weight items. The standard class interior is shown, with seats in both group and airline style, in a mix of red and blue moquette. **Antony Christie**

Key Facts

- Ordered by Virgin Rail as part of a total fleet replacement project for both CrossCountry and West Coast routes.
- All '220s' are now operated by Arriva CrossCountry.
- Interior changes have been made to improve the passenger environment and increase luggage space.
- Sets are based at Central Rivers depot, Burton and maintained by Bombardier / Alstom.

Class 221 'Super Voyager'

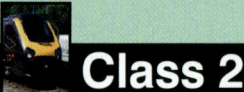

For history and pictures of this class, see *Modern Locomotives Illustrated* Issue 221

Class:	221 (5-car)	221 (5-car)	(4-car)
Number range:	221101-221118, 221142-221143	221119-221135, 221137-221139	221136/140/141/144
Built by:	Bombardier Transportation*	Bombardier Transportation*	
Type:	Super Voyager	Super Voyager	
Introduced:	2001-2002	2001-2002	
Max speed:	125mph (201km/h)	125mph (201km/h)	
Formation:	DMSL+MSL(A)+MSL(B)+MSRMB+DMFL	DMSL+MS+MSL(A)+MSL(B)+DMFL	
Vehicle numbers:	DMSL - 60351-60368, 60392-60393	DMSL - 60369-60390, 60391	
	MSL(A) - 60951-60968, 60992-60993	MS - 60769-60790, 60791	
	MSL(B) - 60851-60868, 60994, 60794	MSL(A) - 60969-60990, 60991	
	MSRMB - 60751-60768, 60792-60793	MSL(B) - 60869-60890	
	DMFL - 60451-60468, 60492-60493	DMFL - 60469-60490, 60491	
Vehicle length:	77ft 6in (23.67m)	77ft 6in (23.67m)	
Height:	12ft 4in (3.75m)	12ft 4in (3.75m)	
Width:	8ft 11in (2.73m)	8ft 11in (2.73m)	
Weight:	Total - 278.3 tonnes	Total - 5-car - 280.7 tonnes, 4-car - 226.3 tonnes	
	DMSL - 58.9 tonnes	DMSL - 58.5 tonnes	
	MSRMB - 53.1 tonnes	MS - 54.1 tonnes	
	MSL(A) - 56.6 tonnes	MSL(A) - 54.8 tonnes	
	MSL(B) - 53.1 tonnes	MSL(B) - 54.4 tonnes	
	DMFL - 56.6 tonnes	DMFL - 58.9 tonnes	
Internal layout:	2+1F, 2+2S	2+1F, 2+2S	
Seating:	5-car Total - 26F/230S	5-car Total - 26F/236S	
		4-car Total - 26F/176S	
	DMSL - 42S	DMSL - 42S	
	MSL(A) - 68S	MS - 66S	
	MSL(B) - 68S	MSL(A) - 66S	
	MSRMB - 52S	MSL(B) - 62S	
	DMFL - 26F	DMFL - 26F	
Gangway:	Within set	Within set	
Toilets:	DMSL, MSL, DMFL - 1	DMSL, MSL, DMFL - 1	
Bogie type:	Bombardier HVP	Bombardier HVP	
Power unit:	1 x Cummins QS19 of 520kW (700hp) at 1800rpm per car	1 x Cummins QS19 of 520kW (700hp) at 1800rpm per car	
Transmission:	Electric	Electric	
Traction motor:	10 Alstom Onix 800 per train	10 or 8 x Alstom Onix 800 per train	
Horsepower:	3,500hp (2,574kW)	5-car - 3,500hp (2,574kW)	
		4-car - 2,800hp (2,059kW)	
Operating range:	1,200 miles (1,931km)	1,200 miles (1,931km)	
Brake type:	Air, EP rheostatic	Air, EP rheostatic	
Route availability:	4	4	
Coupling type:	Outer: Dellner 12, Inner: Bar	Outer: Dellner 12, Inner: Bar	
Multiple restriction:	Class 220/221 only and Class 57/3	Class 220/221 only and Class 57/3	
Door type:	Single-leaf swing plug	Single-leaf swing plug	
Body construction:	Steel	Steel	
Operator:	Avanti West Coast / Grand Central	Arriva CrossCountry	
Owner:	Beacon Rail	Beacon Rail	

* Body shells assembled in Belgium, fitted out at Bombardier plants in Wakefield, UK, and Brugge, Belgium.

Key Facts

- Tilt version of 'Voyager' design ordered by Virgin for CrossCountry routes to allow higher speeds on curves.
- Franchise changes led to 20 sets operating for Avanti West Coast and 24 sets with CrossCountry.
- Tilt system isolated on CrossCountry sets.
- Four 4-car sets are in use with CrossCountry.

2023 Fleet

221101	AWC
221102	AWC
221103	AWC
221104	AWC
221105	AWC
221106	AWC
221107	AWC
221108	AWC
221109	AWC
221110	AWC
221111	AWC
221112	AWC
221113	AWC
221114	AWC
221115	AWC
221116	AWC
221117	AWC
221118	AWC
221119	AXC
221120	AXC
221121	AXC
221122	AXC
221123	AXC
221124	AXC
221125	AXC
221126	AXC
221127	AXC
221128	AXC
221129	AXC
221130	AXC
221131	AXC
221132	AXC
221133	AXC
221134	AXC
221135	AXC
221136	AXC
221137	AXC
221138	AXC
221139	AXC
221140	AXC
221141	AXC
221142	GTL
221143	GTL
221144	AXC

Below: In 2023, Avanti West Coast operate a fleet of 18 five-car Class 221 'Super Voyager' sets Nos. 221101-221118, these retain an operational tilt system. Based at Central Rivers depot, Burton, they operate London to North Wales coast services, as well as a few Birmingham and Anglo-Scottish duties. Sets have been 'refreshed' and sport a latest Avanti West Coast livery, retaining the original silver body, offset by dark ends and Avanti branding. Sets 221117 and 221101 pass near Kilsby on 14 February 2023 with the 15.02 Euston to Holyhead. The standard class driving car is leading. **CJM**

Right: One driving car of each set is set out for first class occupancy, with 2+1 seating for 26. The coach also houses a small galley area inwards of the cab. The first class vehicle is branded with a yellow cant rail band and a yellow cover to the coupling electrical box. Large '1' signs are applied to the bodyside. The DMF of set No. 221114 is shown. **CJM**

'Super Voyager' — Class 221

Right: The interior of the Voyager fleet is not favoured by many passengers, who consider it cramped, of poor design with a lack of space for anything but a handbag. Standard class seating is in the 2+2 style, On Avanti West Coast seats dark blue/grey fleck, seats are a mix of group and aircraft style. Due to the 'Voyager' body profile, allowing for tilt, very limited luggage rack space is provided at cant rail height. The interior of 221107 is shown. **CJM**

Right Middle: Originally the fleet of 44 tilting 'Super Voyager' trains were built as part of the Virgin Group fleet replacement order for the CrossCountry and West Coast franchises. These sets were mainly in five-car formation and designed to be able to take curves at increased speeds, by allowing the vehicle body to tilt, but avoiding passenger discomfort. The tilt system required a more robust bogie design and huge expenditure on vehicle and infrastructure systems. The tilting system on todays CrossCountry fleet is now isolated. In 2023, 24 '221' sets are operated by Arriva CrossCountry, operating in a common pool with the Class 220s, found throughout the AXC main line network. Four sets, Nos. 221136/140/141/144 operate in a four-car formation. Set No. 221144 is seen near Dawlish Warren on 8 July 2022 with the 16.27 Plymouth to Leeds. **CJM**

Right: The three intermediate coaches of the Class 221 are of the same body profile. Two vehicles have a universal toilet compartment and one has seating and luggage space where the original shop was located, seating is between 62-68. Externally the coaches are mounted on a more robust tilting style bogie. 62 seat MS No. 60869 from set 221119 is shown. **CJM**

Right Below: The introduction of the 'Voyager' Class 220 and 221 fleets, saw new standards of driver comfort, with a well designed cab layout, produced with significant input from driver's and the ASLE&F trade union. It is based on the European style with a single power/brake controller operated by the driver's left hand and has all controls within easy reach and vision. Space on the desk is provided for fitting of ERTMS in the future. On 'Voyager' stock the driver is in overall control of the passenger doors. The cab is the same on Class 220 and 221 stock, except on the '221s' extra controls are provided for the tilt system. The cab of No. 221142 when brand new is shown. **CJM**

■ In summer 2023 two off-lease, previously Avanti West Coast operated Class 221s Nos. 221142 and 221142 were leased for around a year to Grand Central to cover for a prolonged shortage and planned extended maintenance of Class 180s. The two sets will be based at Crofton and are planned for use on the Kings Cross-Bradford route, but will be gauge cleared for the Sunderland route. A possibility exists that further off-lease 'spare' Class 221s might follow.

Rolling Stock Review : 2023-2024

Class 222 'Meridian'

For history and pictures of this class, see Modern Locomotives Illustrated Issue 221

Class:	222/0 (7-car)	222/4 (5-car)	222/1 (5-car)
Number range:	222001-222004	222005-222023	222101-222104
Built by:	Bombardier, Brugge	Bombardier, Brugge	Bombardier, Brugge
Type:	Meridian	Meridian	Meridian
Introduced:	2004-2005	2004-2005	2005
Max speed:	125mph (201km/h)	125mph (201km/h)	125mph (201km/h)
Formation:	DMF+MF+MF+MSRMB MS+MS+DMS	DMF+MC+MSRMB+ MS+DMS	DMF+MC+MSRMB+MS+DMS or DMF+MF+MC+MSRMB+DMS
Vehicle numbers:	DMF - 60241-60244 MF - 60345-46/60445-46 MF - 60341-44 MSRMB - 60621-60624 MS - 60561-60564 MS - 60544-60554 DMS - 60161-60164	DMF - 60245-60263 MC - 60347/442-447, 60918-60933 MSRMB - 60627-60643 MS - 60531-60567 DMS - 60167-60183	DMF - 60271-60274 MC - 60571-60574 MSRMB - 60681-60684 MF/MS - 60555-556/60441-60443 DMS - 60191-60194
Vehicle length:	78ft 2in (23.85m)	78ft 2in (23.85m)	78ft 2in (23.85m)
Height:	12ft 4in (3.75m)	12ft 4in (3.75m)	12ft 4in (3.75m)
Width:	8ft 11in (2.73m)	8ft 11in (2.73m)	8ft 11in (2.73m)
Weight:	Total - 337.8 tonnes DMS - 52.8 tonnes MF - 46.8 tonnes MF - 46.8 tonnes MSRMB - 48 tonnes MS - 47 tonnes DMSO - 49.4 tonnes	Total - 249 tonnes DMF - 52.8 tonnes MC - 48.6 tonnes MSRMB - 49.6 tonnes MS - 47 tonnes DMS - 51 tonnes	Total - 197.3 tonnes DMF - 52.8 tonnes MC - 47.1 tonnes MSRMB - 48 tonnes DMSO - 49.4 tonnes
Internal layout:	2+1F, 2+2S	2+1F, 2+2S	2+1F, 2+2S
Seating:	Total - 106F/236S DMF - 22F MF - 42F MF - 42F MSRMB - 62S MS - 68S DMS - 38S	Total - 50F/192S DMF - 22F MC - 28F/22S MSRMB - 62S MS - 68S DMS - 40S	Total - 33F/216S DMF - 22F MC - 11F/46S MSRMB - 62S MS - 68 DMS - 40S
Gangway:	Within set	Within set	Within set
Toilets:	DMF, MF, MS, DMS - 1	DMF, MC, MS, DMS - 1	DMF, MC, DMS - 1
Bogie type:	Bombardier B5005	Bombardier B5005	Bombardier B5005
Power unit:	1 x Cummins QSK19R of 560kW (750hp) per car	1 x Cummins QSK19R of 560kW (750hp) per car	1 x Cummins QSK19R of 560kW (750hp) per car
Transmission:	Electric: 14 x Alstom Onix 800 per train	Electric: 10 x Alstom Onix 800 per train	Electric: 10 x Alstom Onix 800 per train
Horsepower:	5,250hp (3,920kW)	3,750hp (2,800kW)	3,750hp (2,800kW)
Operating range:	1,350 miles (2,173km)	1,350 miles (2,173km)	1,350 miles (2,173km)
Brake type:	Air, EP rheostatic	Air, EP rheostatic	Air, EP rheostatic
Route availability:	4	2	2
Coupling type:	Outer: Dellner 12, Inner: Bar	Outer: Dellner 12, Inner: Bar	Outer: Dellner 12, Inner: Bar
Multiple restriction:	Class 222	Class 222	Class 222
Door type:	Single-leaf swing plug	Single-leaf swing plug	Single-leaf swing plug
Body construction:	Steel	Steel	Steel
Operator:	East Midlands Railway	East Midlands Railway	East Midlands Railway
Owner:	Eversholt Leasing	Eversholt Leasing	Eversholt Leasing

Key Facts

- Development of 'Voyager' non-tilt design for Midland Main Line.
- Originally formed as four, five and nine car sets.
- A fleet of four four-car sets originally used by Hull Trains as Class 222/1 were later taken over by the East Midlands franchise and have now been reformed as five-car sets.
- Design changes can be found between the Class 220 and 222 product lines.
- Class 222s will go off lease when IET stock introduced.

2023 Fleet

222/0 (7-car)
222001 EMR
222002 EMR
222003 EMR
222004 EMR
222005 EMR

222/0 (5-car)
222006 EMR
222007 EMR
222008 EMR
222009 EMR
222010 EMR
222011 EMR
222012 EMR
222013 EMR
222014 EMR
222015 EMR
222016 EMR
222017 EMR
222018 EMR
222019 EMR
222020 EMR
222021 EMR
222022 EMR
222023 EMR

222/1
222101 EMR
222102 EMR
222103 EMR
222104 EMR

Below: Built for the Midland Mainline between London - Nottingham/Sheffield, the '222s' are a modified 'Voyager'. Originally, the order was for nine and four-car sets, but reformations resulted in six seven-car and 17 five-car sets. In 2022 two seven-car sets were disbanded giving vehicles to Class 222/1s to form extra five-car sets. Sets have minor structural differences to the '220/221' design, including the front end shape. Sets carry EMR InterCity aubergine, white and grey. Seven-car set No. 222001 is seen from its first class end. These sets seats 106 first and 236 standard class passengers. **CJM**

Left: One of the five-car sets No. 222009 is viewed from its standard class driving car which has a slightly different window arrangement to the first class vehicles. Also the train door inward from the cab is for passenger use, while on first class vehicles this is a staff door feeding the catering area. These differences are noticeable if these two images are compared. Construction of the Class 810 Hitachi stock is now well advanced with the first sets are due for testing this year. The Class 222s are expected to be phased out of service in 2024-2025. **CJM**

'Meridian' — Class 222

Right: When the Midland Main Line Class 222/0 order was placed, Hull Trains ordered four, four-car sets of like design, classified as 222/1. These were later absorbed into the Midland fleet, following introduction of new stock. Until spring 2022 they operated in their 'as built' four-car form, but in early 2022 two of the seven-car Class 222/0 sets donated two intermediate vehicles, which were then formed one in each Class 222/1 to make five-car sets. Ex-Hull Trains five-car, set No. 222101 is illustrated in June 2022 from its first class end.
CJM

Right: The Class 222 Motor Standard Motor Standard Buffet (MS/MSRMB) vehicles share a standard body design. MSRMB No. 60636 is illustrated, these seat 62 in the 2+2 style and have a catering area at on end (near end in picture). The majority of seats are in the group layout. Two passenger operated single-leaf swing-plug doors are provided on each side, these feed a vestibule area from which access to the passenger compartment is made. No. 60555 is shown carrying standard EMR Intercity colours. As these are non-tilt trains, they are mounted on the lighter weight bogie design.
Antony Christie

Right: The Class 222 interior uses the 2+2 layout in standard class and the 2+1 style in first. Standard class has a mix of group and airline layouts, while the first class seating is all arranged in groups around large tables. The seating on the original East Midlands sets and the original Hull Trains stock is slightly different. The standard class interior of No. 222015 is shown, although above seat luggage shelves are provided they are very small due to the shape of the vehicle. **Antony Christie**

Class 230

For history and pictures of this class, see Modern Locomotives Illustrated Issue 243

Class:	230/0 (VIV)	230/0 (VIV)	230/0 (WMR)	230/0 (TfW)
Number range:	230001	230002/011 (In USA)	230003-230005	230006-230010
Built by:	Vivarail, Long Marston	Vivarail, Long Marston	Vivarail, Long Marston	Vivarail, Long Marston
Original vehicles:	Met-Cam, Birmingham	Met-Cam, Birmingham	Met-Cam, Birmingham	Met-Cam, Birmingham
Introduced:	2016, as battery 2021	2018	2019	2020-2023
Original vehicles:	1979-1983	1979-1983	1979-1983	1979-1983
Max speed:	60mph (97km/h)	60mph (97km/h)	60mph (97km/h)	60mph (97km/h)
Formation:	DMS(A)+TSL+DMS(B)	BDM+BDM	DMS(A)+DMS(B)	DMS(A)+TS+DMSL(B)
Vehicle numbers:	DMS(A) - 300001 / TSL - 300201 / DMS(B) - 300101	BDM(A) - 300002 [TS - 300202] / BDM(B) - 300102	DMS(A) - 300003-300005 / DMS(B) - 300103-300105 / DMSL(B) - 300106-300110	DMS(A) - 300006-300010 / TS - 300206-300210
Vehicle length:	DMS - 60ft 3in (18.37m) / TSL - 59ft 5in (18.12m)	BDM - 60ft 3in (18.37m)	DMS - 60ft 3in (18.37m) / TS - 59ft 5in (18.12m)	DMS - 60ft 3in (18.37m)
Height:	11ft 11in (3.62m)	11ft 11in (3.62m)	11ft 11in (3.62m)	11ft 11in (3.62m)
Width:	9ft 4in (2.85m)	9ft 4in (2.85m)	9ft 4in (2.85m)	9ft 4in (2.85m)
Weight:	Total - 76 tonnes / DMS(A) - 28 tonnes / TSL - 20 tonnes / DMS(B) - 28 tonnes	Total - 50 tonnes / BDM(A) - 27 tonnes / BDM(B) - 23 tonnes	Total - 65.46 tonnes / DMS(A) - 32.73 tonnes / DMS(B) - 32.73 tonnes / DMS(B) - 29 tonnes	Total - 92 tonnes / DMS(A) - 29 tonnes / TS - 34 tonnes
Internal layout:	2+2, Longitudinal	2+2	2+2, Longitudinal	2+2, Longitudinal
Seating:	Total - 160S	Total - 79s / DMS(A)-60S, DMS(B)-44 / DMS(B)-37S	Total - 104S / DMS(A)-46S, TS-50S	Total - 133S
Gangway:	Within set	Within set	Within set	Within set
Toilets:	TSL - 1	- DMS(B) - 1	DMS(B) - 1	
Bogie type:	Bombardier Flexx 1000	Bombardier Flexx 1000	Bombardier Flexx 1000	Bombardier Flexx 1000
Power unit:	Battery - 6 x Hoppecke (3 below each DM)	Battery 60x 24V 200hp (150kW) per car	DMS - Two Ford Duratorq for charging, batteries in DMSs	TS - 4 x gensets for battery
Transmission:	Battery Electric	Electric	Electric	Electric
Transmission package:	Strukton Rail IGBT	Strukton Rail IGBT	Strukton Rail IGBT	Strukton Rail IGBT
Traction motor type:	LT118	LT118	LT118	LT118
Horsepower:	800hp (597kW)	800hp (597kW)	1,200hp (895kW)	tba
Brake type:	Air	Air	Air	Air
Route availability:	2	2	2 2	
Coupling type:	Wedgelock	Wedgelock	Wedgelock	Wedgelock
Multiple restriction:	Not fitted	Not fitted	Not fitted	Not fitted
Door type:	Single-leaf sliding	Single-leaf sliding	Single-leaf sliding	Single-leaf sliding
Body construction:	Aluminium	Aluminium	Aluminium	Aluminium
Operator:	GWR	Vivarail in USA as 2-car	GWR	Transport for Wales

Key Facts

- Re-use of withdrawn LUL D-78 surface line stock, rebuilt too main line standards.
- 230001 used for design development and trials, now owned by Great Western Railway for battery 'fast-charge' trials on West Ealing-Greenford line.
- First production trains introduced on West Midlands Railway Bedford-Bletchley route in 2019 withdrawn in 2022 after collapse of Vivarail. Later sold to Great Western for possible further use.
- Five Transport for Wales sets, delayed during conversion and Covid pandemic. Entered service on Bidston-Wrexham line in 2023.
- 67 un-converted D78 cars owned by Great Western for possible use.

2023 Fleet

230001	GWR	230007	TFW
230002	VIV(USA)	230008	TFW
230003	GWR	230009	TFW
230004	GWR	230010	TFW
230005	GWR	230011	VIV(USA)
230006	TFW		

Below, Bottom & Bottom Right: *Vivarail purchased over 200 withdrawn London Underground 'D'-stock vehicles when they were replaced by 'S' stock. The plan was to produce two, three or four-car passenger trains, fitted with high-quality interiors. The original 'D' stock bodies were not life expired and the chance was taken to rebuild as a low-cost option for fleet replacement. A prototype three-car set, No. 230001 was converted in 2016, powered by underslung automotive diesel engines. In 2019 three two-car sets (230003-230005) entered commercial service on the Marston Vale line between Bedford and Bletchley, illustrated upper left. These sets remained in service until December 2022 when Vivarail went into receivership and closed down. The sets were then stored and in mid 2023 were sold to Great Western Railway for possible future use. In 2020-2022, five, three-car sets were purchased by Transport for Wales for use on the Bidston-Wrexham line, massive delays were experienced due to the Covid Pandemic and major technical issues with the stock, later followed some overheating and fires. Further delays were experienced after Vivarail collapsed, but the units were introduced in spring 2023. These units in TfW grey and orange livery are based at Birkenhead North depot, operated by Merseyrail. The TFW sets seat 133 standard class passengers, have a universal toilet compartment in one driving car. In the illustration below left, set No. 230007 is seen at Bidston from its DMS(A) coach. The image was recorded during the testing period and two sets are coupled together, usually only a single set is used. In the below right view, an intermediate coach is seen, these carry the diesel gen-sets, which power the traction batteries mounted below both driving cars.* All: **CJM**

Class 230

Above: For Class 230 use, the original LUL driving cab has been totally rebuilt to main line standards, including the fitting of AWS and TPWS equipment. A combined power/brake controller is located on the left side. The cab of set No. 230001 is shown. **CJM**

Middle & Far Right: Class 230 front end equipment, for TfW and GWR sets. 1: High lever marker light, 2: Ventilation panels, 3: Destination indicator, 4: Headlight, 5: Combined white marker and red tail light, 6: Anti-climber plate, 7: Air horns, 8: Wedgelock coupling, 9: Deflector plate. Both: **CJM**

Right, Middle Right, Bottom Right & Below: Before the demise of Vivarail in December 2022, the company entered a partnership with Great Western Railway to operate a year long Government supported battery traction trial on the West Ealing to Greenford route, using three-car set No. 230001. The set, fitted with the Vivarail 'fast-charge' system was under development when Vivarail ceased trading. Great Western purchased the rights to the Vivarail 'fast-charge' system, set 230001 and 67 stored ex LUL D stock vehicles. They also took on the Vivarail team responsible for the project. Although originally planned for introduction in 2023, the trial will now commence in early 2024, with set 230001 based at Reading and out-based at West Ealing. The set was at Long Marston in mid 2023 being set up for operation. The set, carrying its Vivarail Venturer name is seen (right) at Long Marston on 21 June 2023. The intermediate TSL is seen middle right, this coach has two passenger doors, the disabled seating area and retention toilet, no traction equipment is carried on this vehicle. The two driving cars each carry three battery packs which provide the only power. These are charged by the 'fast-charge' system on each vehicle, consisting of a retractable slipper or shoe, which through lineside beacons lowers onto the power rails which are located in the bay platform at West Ealing. Charging takes around eight minutes. Below the slipper or shoe is shown 'A' and the track mounted power rails 'B'. The right bottom image shows the bodyside branding. All: **CJM**

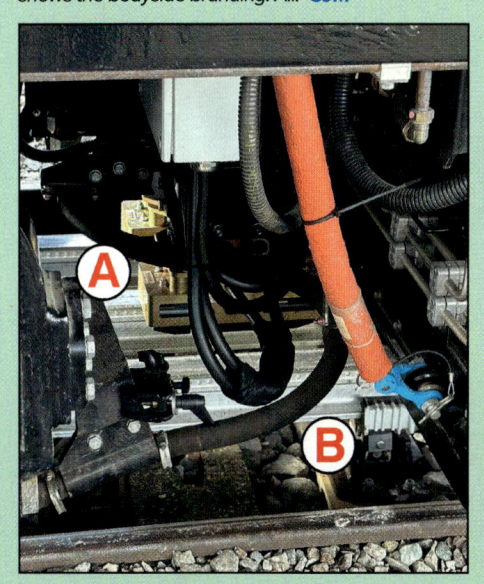

Rolling Stock Review : 2023-2024

Class 231 — 'Flirt'

Class:	231
Number range:	231001-231011
Built by:	Stadler, Bussnang, Switzerland
Body shells:	Stadler, Bussnang, Switzerland & Szolnok, Hungary
Type:	Flirt
Years introduced:	2022-2023
Max speed:	90mph (145km/h)
Formation:	DMS+TSL+PP+TS+DMS
Vehicle numbers:	DMS - 381001-381011
	TS- 381201-381211
	PP- 381401-381411
	TSL- 381301-381311
	DMS - 381101-381111
Train length:	264ft 8in (80.7m)
Vehicle length:	DMS - 74ft 3in (20.81m), TS/TSL - 50ft (15.22m), Power-Pack - 21ft 9in (6.69m)
Height:	13ft 0in (3.95m)
Width:	DMS, TS/TSL - 8ft 9in (2.72m), PP - 9ft 1in (2.82m)
Floor height:	3ft 1.7in (960mm)
Internal layout:	2+2S
Seating:	Total: 170S (+ 39 Tip up)
	DMS - 40S (+12TU), TS - 52S (+10), PP - 0, TSL 38S (+5TU), DMS 40S (+12TU)
Gangway:	Within set
Toilets:	1 universal
Weight:	Total -157.7T, DMS 39.5, TS 24.4, PP 28.5, TSL 25.4, DMS 39.9
Brake type:	Air (regenerative)
Bogie type:	Stadler, Jacob
Transmission:	Diesel-electric
Power:	4 x Deutz V8/16 of 645hp (480kW)
Traction motor type:	4 x TSA
Power at rail:	1500kW (2,012hp)
Fuel tank capacity:	506gal (2300lit)
Wheel diameter (New):	Motor bogie - 34.25in (870mm), Trailer bogie - 29.92in (760mm)
Coupling type:	Outer - Dellner 10, Inner - Jacob
Multiple restriction:	Within Class (up to three sets)
Door type:	Bi-parting sliding plug
Fittings:	Sanding and scrubber block
Construction:	Aluminium
Operator:	Transport for Wales
Owner:	SMB/Equitix Leasing

Above: *Class 231 front end layout. 1: High-level marker light, 2: Destination display, 3: Forward facing camera (in window top), 4: White frontal lights, 5: Red rear lights, 6: Horns (in coupling pocket), 7: Electrical connection box (with roller door), 10: Dellner 10 coupling.* **CJM**

2023 Fleet

231001	TFW
231002	TFW
231003	TFW
231004	TFW
231005	TFW
231006	TFW
231007	TFW
231008	TFW
231009	TFW
231010	TFW
231011	TFW

Left: *In January 2023, the first of 11 Stadler four-car Class 231 'Flirt' sets for Transport for Wales went into service, initially on the Rhymney line. Eventually the sets will be deployed on the South Wales to Cheltenham service. In traditional Flirt style, the sets are formed of four passenger vehicles and a central 'Power-Pack' coach, housing four Deutz diesel-alternator packs. The '231s' have two pairs of sliding plug doors on either side of driving cars and a single pair of doors on intermediate coaches. The fleet is based at Cardiff Canton and carries TfW grey and red livery without a yellow warning end. Set No. 231010 is seen arriving at Cardiff on 30 January 2023 with a service from Rhymney.* **CJM**

Below: *The four between vehicle bogies on the '231s' are of the Jacob articulated type. The Power-Pack vehicle of set No. 231010 is shown. Electric power generated from the four engines, two mounted either side of a central gangway pass to the driving cars, where the traction motors are located.* **CJM**

Below: *In keeping with the general 'Flirt' product platform, the interior is of a very high standard, with 2+2 seating with a mix of well spaced airline and group layouts. Phone/lap-top charging points are provided, as well as a clear ceiling mounted passenger information system. Set 231008 is shown.* **CJM**

For history and pictures of this class, see Modern Locomotives Illustrated Issue 225

Class 318

2023 Fleet

318250	ASR
318251	ASR
318252	ASR
318253	ASR
318254	ASR
318255	ASR
318256	ASR
318257	ASR
318258	ASR
318259	ASR
318260	ASR
318261	ASR
318262	ASR
318263	ASR
318264	ASR
318265	ASR
318266	ASR
318267	ASR
318268	ASR
318269	ASR
318270	ASR

Class:	318
Number range:	318250-318270
Built by:	BREL York
Years introduced:	1985-1986
Max speed:	90mph (145km/h)
Formation:	DTSOL+PMSO+DTSO
Vehicle numbers:	DTSOL - 77240-77259/77288
	PMSO - 62866-62885/62890
	DTSO - 77260-77279/77289
Vehicle length:	DTSOL/DTSO - 65ft 0¾in (19.83m)
	PMSO - 65ft 4¼in (19.92m)
Height:	12ft 1½in (3.70m)
Width:	9ft 3in (2.82m)
Weight:	Total - 107.5 tonnes
	DTSOL - 30 tonnes
	PMSO - 50.9 tonnes
	DTSO - 26.6 tonnes
Internal layout:	2+2/2+3 high density
Seating:	Total: 208S
	DTSOL - 55S, PMSO - 79S,
	DTSO - 74S
Gangway:	Within set
Toilets:	DTSOL - 1
Brake type:	Air (Westcode)
Bogie type:	DTSOL/DTSO - BREL BT13
	PMSO - BREL BP20
Power collection:	25kV ac overhead
Traction motor type:	4 x Brush TM2141
Horsepower:	1,438hp (1,072kW)
Coupling type:	Outer - Tightlock, Inner - Bar
Multiple restriction:	Within type only
Door type:	Bi-parting sliding
Construction:	Steel
Operator:	ScotRail
Owner:	Eversholt Leasing

Right: *Class 318 front end layout. Originally these sets incorporated a front end gangway, but this was removed on refurbishment to improve the cab area.*
1: Destination indicator, 2: Position of original gangway door, 3: Forward facing camera, 4: Lamp bracket, 5: Headlight, 6: Combined white marker and red tail light, 7: Tightlock coupling, 8: Warning horns, 9: Electrical connection box, 10: Manual uncoupling lever, 11: 'Drum switch' controlling electrical connections. **CJM**

Below & Right: *A fleet of 21 BREL York Works built Class 318s were launched in 1985-1986 for the Ayrshire Coast route between Glasgow and Ayr/Ardrossan. They were a follow-on order to the now withdrawn Class 317 design. During refurbishment, the end gangway doors were removed and an improved driving cab fitted. All sets now have a universal access toilet in the DTSOL vehicle. Set No. 318261 is seen arriving at Glasgow Central. The inset image right, shows the interior, in common with other Scottish EMU classes, these have been refreshed in recent years with Scottish Railways moquette, yellow grab handles and poles. '318' seating is in the 3+2 style except at coach ends where 2+2 is used. Although under overall control of the train crew, doors have both open and close button.* Both: **CJM**

Above: *Coupled between the two driving cars is a Pantograph Motor Standard Open (PMSO), seating 79 is a mix of a 2+2 and 2+3 seating layout. This coach houses all traction equipment and is mounted on heavy BP20 bogies. Car No. 62875 from set No. 318259 is shown from its pantograph end.* **CJM**

Left: *The revised, much deeper, cab side window on the non-driving side, installed at the same time as the end gangways were removed as part of a cab improvement project. Set 318268 shown.* **CJM**

Key Facts

- Fleet of 21 Class 318s built as three-car units for Scottish use.
- Originally sets were fitted with cab-end gangway doors, these were removed when refurbished to provide an improved cab. Cab side windows also enlarged.
- All sets are painted in Scottish Railways 'Saltire' blue livery.
- Fitted with Brush traction equipment.

Rolling Stock Review : 2023-2024

Class 319

Class:	319/0	319/2	319/3	319/4
Number range:	319011-319012	319214-319220	319361-319386	319429-319457
Former number range:	-	319014-319020	319161-319186	319029-319057
Built by:	BREL York	-	-	-
Years introduced:	1987	1987-1988	1990	1988-1989
Rebuilt:	-	1996-1997	1997-1999	1997-1999
Rebuilt by:	-	Railcare, Wolverton	Alstom, Eastleigh	Railcare, Wolverton
Max speed:	100 mph (161km/h)	100 mph (161km/h)	100 mph (161km/h)	100 mph (161km/h)
Formation:	DTSO(A)+PMSO+TSOL+DTSO(B)	DTSO+PMSO+TSOL+DTCO	DTSO(A)+PMSO+TSOL+DTSO(B)	DTCO+PMSO+TSOL+DTSO
Vehicle numbers:	DTSO(A) - 77311-77313 (odd) PMSO - 62901-62902 TSOL - 71782-71783 DTSO(B) - 77310-77312 (even)	DTSO - 77317-77329 (odd) PMSOL - 62904-62910 TSOL - 71785-71791 DTCO - 77316-77328 (even)	DTSO - 77459-77497, 77973-77983 (odd) PMSO - 63043-63062, 63094-63098 TSOL - 71929-71948, 71979-71984 DTSO(B) - 77458-77496, 77974-77984 (even)	DTCO - 77347-77355, 77445-77451 (odd) PMSO - 62919-62971 PTSOL - 71800-71876 DTSO - 77346-773354, 77444-77450 (even)
Vehicle length:	DTSO(A) - 65ft ¾in (19.83m) DTSO(B) - 65ft ¾in (19.83m) PMSO - 65ft 4¼in (19.93m) TSOL - 65ft 4¼in (19.93m)	DTSO - 65ft ¾in (19.83m) DTCO - 65ft ¾in (19.83m) PMSOL - 65ft 4¼in (19.93m) TSOL - 65ft 4¼in (19.93m)	DTSO(A) - 65ft ¾in (19.83m) DTSO(B) - 65ft ¾in (19.83m) PMSO - 65ft 4¼in (19.93m) PTSOL - 65ft 4¼in (19.93m)	DTCO - 65ft ¾in (19.83m) DTSO - 65ft ¾in (19.83m) PMSO - 65ft 4¼in (19.93m) TSOL - 65ft 4¼in (19.93m)
Height:	11ft 9in (3.58m)	11ft 9in (3.58m)	11ft 9in (3.58m)	11ft 9in (3.58m)
Width:	9ft 3in (2.82m)	9ft 3in (2.82m)	9ft 3in (2.82m)	9ft 3in (2.82m)
Weight:	Total - 136.5 tonnes DTSO(A) - 28.2 tonnes PMSO - 49.2 tonnes TSOL - 31 tonnes DTSO(B) - 28.1 tonnes	Total - 136.5 tonnes DTSO - 28.2 tonnes PMSOL - 49.2 tonnes TSOL - 31 tonnes DTCO - 28.1 tonnes	Total - 140.3 tonnes DTSO(A) - 29 tonnes PMSO - 50.6 tonnes TSOL - 31 tonnes DTSO(B) - 29.7 tonnes	Total - 136.5 tonnes DTSO(A) - 28.2 tonnes PMSO - 49.2 tonnes TSOL - 31 tonnes DTSO(B) - 28.1 tonnes
Internal layout:	2+2, 2+3	2+1F, 2+2S	2+2, 2+3	2+1F, 2+2, 2+3S
Seating:	Total - 319S DTSO(A) - 82S PMSO - 82S TSOL - 77S DTSO(B) - 78S	Total - 18F/212S DTSO - 64S PMSOL - 60S TSOL - 52S DTCO - 18F/36S	Total - 302S DTSO(A) - 72S PMSO - 79S TSOL - 74S DTSO(B) - 77S	Total - 12F/263S DTCO - 12F/51S PMSO - 74S TSOL - 67S DTSO - 71S
Gangway:	Within set, emergency end doors	Within set, emergency end doors	Within set, emergency end doors	Within set, emergency end doors
Toilets:	TSOL - 2	MSOL - 2, TSOL - 1	TSOL - 2	TSOL - 2
Brake type:	Air (Westcode)	Air (Westcode)	Air (Westcode)	Air (Westcode)
Bogie type:	DTSO, TSOL - BREL T3-7 PMSO - BREL P7-4	DTSO, DTCO, TSOL - T3-7 PMSO - BREL P7-4	DTSO, TSOL - BREL T3-7 PMSO - BREL P7-4	DTSO, DTCO, TSOL - BREL T3-7 PMSO - BREL P7-4
Power collection:	25kV ac overhead and 750V dc third rail§	25kV ac overhead and 750V dc third rail	25kV ac overhead and 750V dc third rail§	25kV ac overhead and 750V dc third rail§
Traction motor type:	4 x GEC G315BZ	4 x GEC G315BZ	4 x GEC G315BZ	4 x GEC G315BZ
Horsepower:	1,438hp (1,072kW)	1,438hp (1,072kW)	1,438hp (1,072kW)	1,438hp (1,072kW)
Coupling type:	Outer - Tightlock, Inner - Bar	Outer - Tightlock, Inner - Bar	Outer - Tightlock, Inner - Bar	Outer - Tightlock, Inner - Bar
Multiple restriction:	Within Class 319 series	Within Class 319 series	Within Class 319 series	Within Class 319 series
Door type:	Bi-parting sliding	Bi-parting sliding	Bi-parting sliding	Bi-parting sliding
Construction:	Steel	Steel	Steel	Steel
Operator:	West Midlands, ROG	West Midlands	Off lease, Northern	Off lease, West Midlands
Owner:	Porterbrook Leasing	Porterbrook Leasing	Porterbrook Leasing	Porterbrook Leasing
Sub-class differences:	Original phase 1 units		Original phase 2 units	

Notes: § - sets operated by Northern have 750V third rail equipment removed

Left: Class 319 front end layout. 1: Original route display, 2: Route indicator, 3: Emergency front end door for tunnel working, 4: White marker light, 5: Headlight, 6: Red tail light, 7: Tightlock coupling, 8: Electrical connection box, 9: Emergency uncoupling lever, 10: Emergency air connection, 11: Air vent for cab. **CJM**

■ Porterbrook Leasing and the Birmingham Centre for Rail Research and Education have developed a hydrogen powered Class 319, now classified as 799/2, able to operate on 25kV ac overhead, 750V dc third rail plus a hydrogen fuel cell. The project will be known as 'Hydro Flex'.

Key Facts

- Built in two batches by BREL for Network SouthEast 'Thameslink' services.
- Dual voltage AC 25kV /DC 750V trains.
- In 2023 sets were in use with Northern, West Midlands Railway and ROG.
- Many sets are off-lease, and stored. A large number have been rebuilt as bi- or tri-mode BMUs for Northern, Great Western and Transport for Wales as Class 769 'Flex' sets.

319/0		319371	OLS
319011	ROG	319372	NOR
		319375	NOR
319/2		319378	NOR
319214	WMR	319379	NOR
319215	WMR	319381	NOR
319217	WMR	319383	NOR
319219	WMR	319384	OLS
319220	WMR	319385	NOR
		319386	NOR
319/3			
319361	NOR	**319/4**	
319366	NOR	319433	WMR
319367	NOR	319454	OLS
319368	NOR	319457	WMR
319369	NOR		
319370	NOR		

Most of the off-lease sets marked are stored at Long Marston and available to take part in the Porterbrook Class 769 'Flex' project. Units identified for conversion are removed from this list.

Class 319

Above, Right & Below: *A number of refurbished Class 319s operate with Northern, based at Allerton (Liverpool). These have seen a change to the layout of the TSO vehicle, which now incorporates a universal access toilet compartment at the MSO end. This saw one window plated over in the TSO. Many sets have been taken off lease following introduction of Class 331s. The Class 319s only have a limited life with Northern, as when extra Class 323s are cascaded from West Midlands, they will be withdrawn. Set No. 319369 is seen at Liverpool Lime Street with its DTSO(A) leading. The image right shows a PMSO vehicle. These seat 79 in a mix of 2+2 and 2+3 seating. The pantograph is mounted at the far end and the main traction equipment is housed between the two BREL T7-4 power bogies. This vehicle 63047 carries the older Northern branding with the name in lower case letters. Refurbishment of the Class 319 fleet, was undertaken at Wolverton Works, this saw the high-density 2+2, 2+3 seating retained, but using improved individual style seats. A Passenger Information System (PIS) was also installed. The decor originally used Northern fleck and later two-tone blue moquette and yellow seat furniture and handrails.* **Antony Christie (2) / CJM**

Right & Below: *In mid 2023, nine Class 319s were still on the books of West Midlands Train (London Northwestern) operating peak hour services on the Euston to Northampton line, covering for the late delivery and commissioning of Class 730 stock. The sets in various liveries are scheduled to be withdrawn by the end of 2023. Right, is London Northwestern grey, green and black liveried No. 319457, while below is unbranded ex Thameslink-liveried No. 319005 (since withdrawn). Both:* **CJM**

Right: *London Northwestern Railway Class 319 interior, showing the old First Group moquette. This 319/2 has a 2+2 seating layout, with seats in a mix of group and airline.* **Adrian Paul**

Rolling Stock Review : 2023-2024 101

Class 320

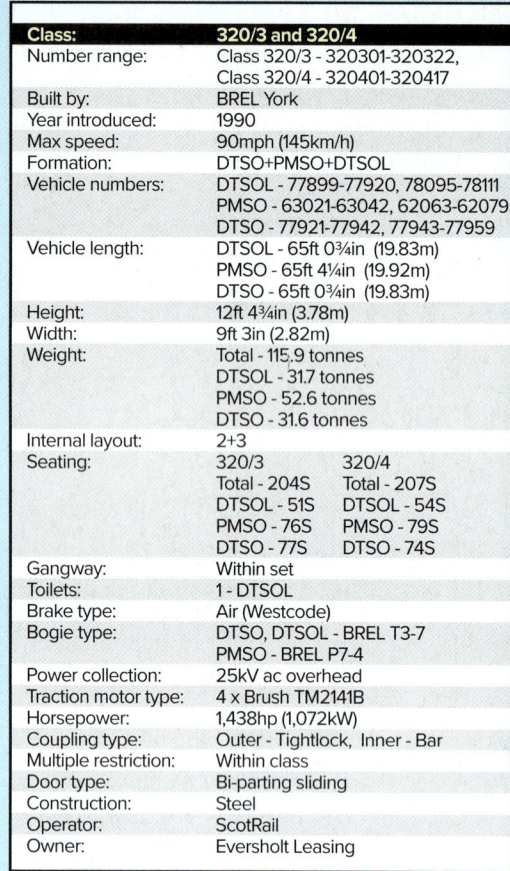

For history and pictures of this class, see Modern Locomotives Illustrated Issue 225

Class:	320/3 and 320/4
Number range:	Class 320/3 - 320301-320322, Class 320/4 - 320401-320417
Built by:	BREL York
Year introduced:	1990
Max speed:	90mph (145km/h)
Formation:	DTSO+PMSO+DTSOL
Vehicle numbers:	DTSOL - 77899-77920, 78095-78111 PMSO - 63021-63042, 62063-62079 DTSO - 77921-77942, 77943-77959
Vehicle length:	DTSOL - 65ft 0¾in (19.83m) PMSO - 65ft 4¼in (19.92m) DTSO - 65ft 0¾in (19.83m)
Height:	12ft 4¾in (3.78m)
Width:	9ft 3in (2.82m)
Weight:	Total - 115.9 tonnes DTSOL - 31.7 tonnes PMSO - 52.6 tonnes DTSO - 31.6 tonnes
Internal layout:	2+3
Seating:	320/3 320/4 Total - 204S Total - 207S DTSOL - 51S DTSOL - 54S PMSO - 76S PMSO - 79S DTSO - 77S DTSO - 74S
Gangway:	Within set
Toilets:	1 - DTSOL
Brake type:	Air (Westcode)
Bogie type:	DTSO, DTSOL - BREL T3-7 PMSO - BREL P7-4
Power collection:	25kV ac overhead
Traction motor type:	4 x Brush TM2141B
Horsepower:	1,438hp (1,072kW)
Coupling type:	Outer - Tightlock, Inner - Bar
Multiple restriction:	Within class
Door type:	Bi-parting sliding
Construction:	Steel
Operator:	ScotRail
Owner:	Eversholt Leasing

Above & Middle Right: *In 1990 22 three-car Class 320/3s were built by BREL York for the Strathclyde network, based on the Class 321. The sets had a middle motor coach with a pantograph flanking two driving trailers. In 2023, the sets accommodate 204 in the 2+3 style. Above right, Set No. 320312 is seen at Glasgow Central. To provide extra stock, 12 Class 321/4s were cascaded to Scotland in 2016-2019. They were reduced to three-car formation, by the removal of the TSO coach. These sets were renumbered into the Class 320 series, keeping the last three digits of their original '321' number. All '320s' are finished in Scottish Railways blue and white livery, operating in a common pool. Seating is slightly different in the two sub-classes, with 207 seats in the 320/4s. Middle Right, No. 320416 is seen at Glasgow Central. Both:* **CJM**

Below: *The interior of the Class 320s uses the high-density 2+3 style, with a mix of airline and group layout. All vehicles have a good quality passenger information system, and some pleasing, locally themed murals can be found on the end bulkhead walls. Small side mounted tables are provided below windows.* **CJM**

Above: *The intermediate Pantograph Motor Standard Open (PMSO), No. 63037 from set 320317 taken from the pantograph end.* **CJM**

2023 Fleet

320/3
320301	ASR
320302	ASR
320303	ASR
320304	ASR
320305	ASR
320306	ASR
320307	ASR
320308	ASR
320309	ASR
320310	ASR
320311	ASR
320312	ASR
320313	ASR
320314	ASR
320315	ASR
320316	ASR
320317	ASR
320318	ASR
320319	ASR
320320	ASR
320321	ASR
320322	ASR

320/4
320401	ASR
320402	ASR
320404	ASR
320411	ASR
320412	ASR
320413	ASR
320414	ASR
320415	ASR
320416	ASR
320417	ASR
320418	ASR
320420	ASR

Key Facts

- Follow-on order to Class 321 design, built by BREL York for BR Scottish Region.
- Twenty-two original sets, built as standard class only, based at Glasgow Shields Road.
- To enlarge the fleet, a batch of 12 redundant Class 321s were transferred to Scotland in 2016-2019, reduced to three-car length, modified and reclassified as 320/4.
- All units carry Scottish Railways 'Saltire' blue/white livery.

Class 321

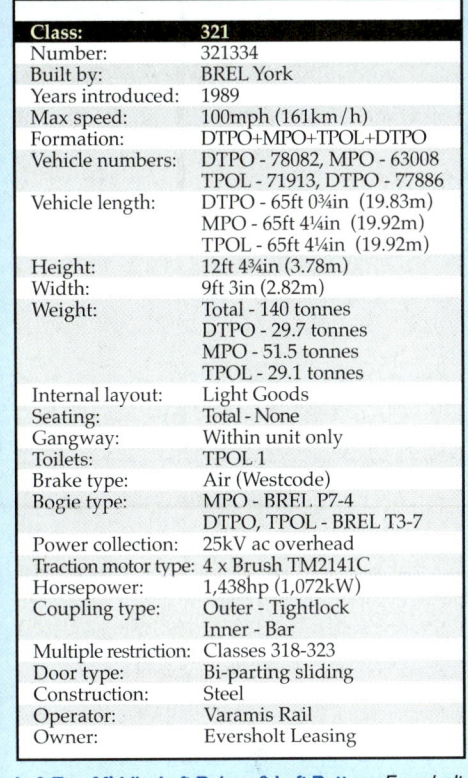

Class:	321
Number:	321334
Built by:	BREL York
Years introduced:	1989
Max speed:	100mph (161km/h)
Formation:	DTPO+MPO+TPOL+DTPO
Vehicle numbers:	DTPO - 78082, MPO - 63008 TPOL - 71913, DTPO - 77886
Vehicle length:	DTPO - 65ft 0¾in (19.83m) MPO - 65ft 4¼in (19.92m) TPOL - 65ft 4¼in (19.92m)
Height:	12ft 4⅜in (3.78m)
Width:	9ft 3in (2.82m)
Weight:	Total - 140 tonnes DTPO - 29.7 tonnes MPO - 51.5 tonnes TPOL - 29.1 tonnes
Internal layout:	Light Goods
Seating:	Total - None
Gangway:	Within unit only
Toilets:	TPOL 1
Brake type:	Air (Westcode)
Bogie type:	MPO - BREL P7-4 DTPO, TPOL - BREL T3-7
Power collection:	25kV ac overhead
Traction motor type:	4 x Brush TM2141C
Horsepower:	1,438hp (1,072kW)
Coupling type:	Outer - Tightlock Inner - Bar
Multiple restriction:	Classes 318-323
Door type:	Bi-parting sliding
Construction:	Steel
Operator:	Varamis Rail
Owner:	Eversholt Leasing

Left Top, Middle, Left Below, & Left Bottom: *Eversholt Leasing in partnership with Wabtec and Ricardo Rail have launched the Swift project, to convert redundant Class 321 passenger EMUs and re-purpose them into light freight carrying multiple units. Seating has been removed, checker plate flooring installed and side 'trolley' tethering points added to all vehicles, enabling small containerised or trolley mounted freight to be transported, with the ability to load or off-load from a standard passenger platform. Set No. 321334 has been adapted as a prototype and in late 2022 was leased to new light rail freight haulier Varamis Rail to operate a weekday service between Mossend in Scotland and Birmingham International. Several other off-lease Class 321s are available for conversion if a extra sets are sought by Varamis or another operator comes along. Set 321334 carries Swift Express Freight blue livery with bodyside graphics. If further sets are converted the interior is likely to be modified in a slightly different way. The complete set is seen in the top image and the intermediate trailer in the middle picture. Left Below and Left Bottom are the interior of vehicles, showing the strengthened floors and tethering points for trollies. On this conversion some isolating cocks which were originally under passenger seats have been retained in their original position and ringed in steel frames, further converts will have this equipment re-positioned. Four further sets have been converted by Gemini Rail, Wolverton Nos. 321407/419/428/429. All:* **CJM**

321334 VAR

Right: *In April 2023 the final Class 321s were taken out of service by Greater Anglia, these were the 'Renatus' sets (321301-321330) refurbished with an AC traction package, air conditioning and with universal toilet facilities. Rather than immediate scrap, owner Eversholt has been trying to find a new operator. It is possible some could be rebuilt as freight EMUs. Set No. 321328 is seen at Shenfield.* **CJM**

Rolling Stock Review : 2023-2024   103

VISIT OUR ONLINE SHOP
TO VIEW OUR FULL RANGE OF SPECIAL MAGAZINES ABOUT **THE RAILWAY INDUSTRY**

Key Shop

shop.keypublishing.com/specials

Locomotive Directory — ~~£9.99~~ **£5.00**

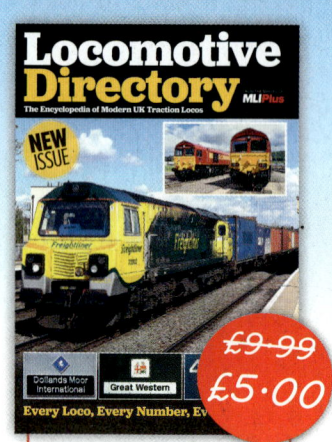

This 148-page special publication is the author's lifelong work to document each & every modern traction (diesel, electric, gas-turbine and bi-mode) locomotive to have operated on the UK main railway systems from the early 1920s to the present day.

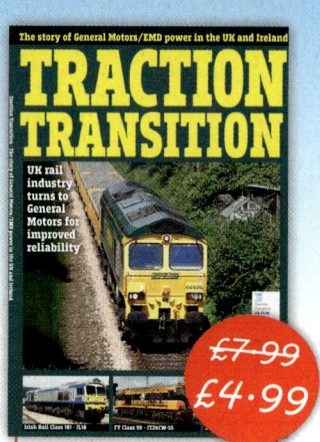

Traction Transition — ~~£7.99~~ **£4.99**

It was not until 1961 that locos from the main US loco builder General Motors (GM), started to operate in Ireland. This 132-page special magazine tells the Story of General Motors / EMD power in the UK and Ireland.

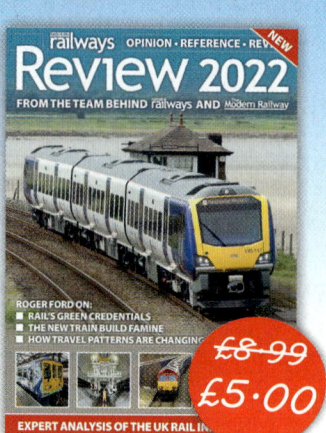

Review 2022 — ~~£8.99~~ **£5.00**

With the imperative to rebuild revenue post-Covid and the push for net zero, the expert editorial team assesses the prospects for the forthcoming year on the railway, describing its key challenges and opportunities.

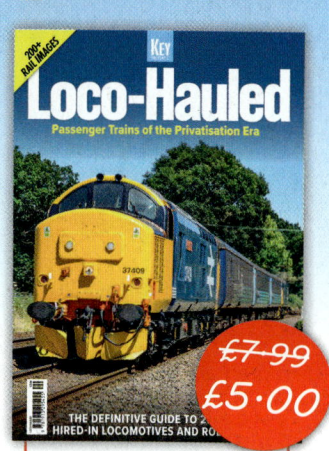

Loco-Hauled — ~~£7.99~~ **£5.00**

This lavishly illustrated publication will take a photographic look back at the past 25 years of such workings, beginning with the Welsh Valleys and running right through to the last hurrah of the recently ended Fife Circle operation.

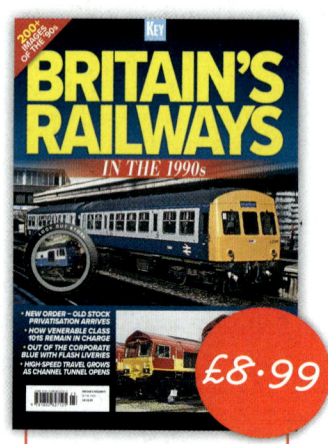

Britain's Railways in the 1990s — **£8.99**

The '90s arguably saw the greatest change of all in our rail system with privatisation, booming passenger numbers and the arrival of locomotives not built in the UK. This special publication traces those happenings in detail, year by year.

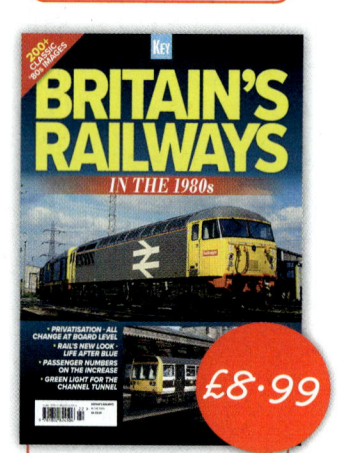

Britain's Railways in the 1980s — **£8.99**

The 1980s was a decade of change for Britain's railways. All of these changes are charted year by year in this eagerly awaited fourth title in this lavishly illustrated series.

Review 2023 — **£8.99**

As well as reflecting of the successes of 2022, such as the opening of Crossrail and progress with HS2, the expert editorial team attempts to chart the likely way forward for the year to come.

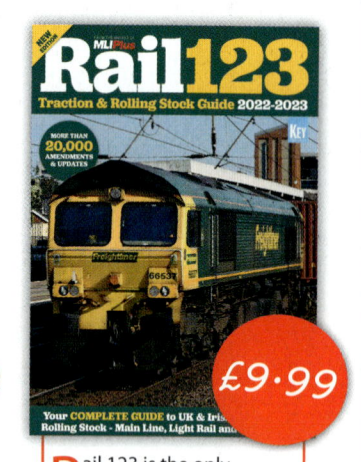

Rail 123 — **£9.99**

Rail 123 is the only publication to list ALL vehicles in one easy to follow, colour coded list. The new edition of Rail 123 includes more than 20,000 changes and updates from the 2021-2022 edition.

FREE P&P* when you order online at...
shop.keypublishing.com/specials
Call +44 (0)1780 480404 *(Monday to Friday 9am-5.30pm GMT)*

Also available from **W.H Smith** and all leading newsagents. Or download from **Pocketmags.com** or your native app store - search *Railway Specials*

SUBSCRIBERS don't forget to use your **£2 OFF DISCOUNT CODE!**

Discounted products on this page not valid in conjunction with any other promotions on the shop.

*Free 2nd class P&P on all UK & BFPO orders. Overseas charges apply.

Class 323

For history and pictures of this class, see Modern Locomotives Illustrated Issue 228

Class:	323
Number range:	323201-323243
Built by:	Hunslet TPL, Leeds
Years introduced:	1992-1993
Max speed:	90mph (145km/h)
Formation:	DMSO(A)+PTSOL+DMSO(B)
Vehicle numbers:	DMSO(A) - 64001-64043
	PTSOL - 72201-72243
	DMSO(B) - 65001-65043
Vehicle length:	DMSO - 76ft 8¾in (23.37m)
	PTSOL - 76ft 10¾in (23.44m)
Height:	12ft 4¾in (3.78m)
Width:	9ft 2¼in (2.80m)
Weight:	Total - 121.4 tonnes
	DMSO(A) - 41.0 tonnes
	PTSOL - 39.4 tonnes
	DMSO(B) - 41.0 tonnes
Internal layout:	2+3
	323223-323225 - 2+2
Seating:	Total - 275S
	DMSO(A) - 97S
	PTSOL - 81S + 3 tip up
	DMSO(B) - 97S
Gangway:	Within unit only
Toilets:	PTSOL - 1
Brake type:	Air (Westcode)
Bogie type:	Powered - RFS BP62
	Trailer - RFS BT52
Power collection:	25kV ac overhead
Traction motor type:	4 x Holec DMKT 52/24
Horsepower:	1,566hp (1,168kW)
Coupling type:	Outer - Tightlock, Inner - Bar
Multiple restriction:	Class 317-323
Door type:	Bi-parting sliding plug
Construction:	Welded aluminium alloy
Operator:	West Midlands Railway, Northern
Owner:	Porterbrook Leasing

Key Facts

- Only Hunslet TPL built passenger fleet for the UK.
- Built for Birmingham CrossCity electrification and Provincial services in the Manchester area.
- Three sets (323223-323225) built with reduced seating and extra luggage space for Manchester Airport services (now removed).
- West Midlands fleet to be replaced by Class 730s, 17 '323s' to transfer to Northern.

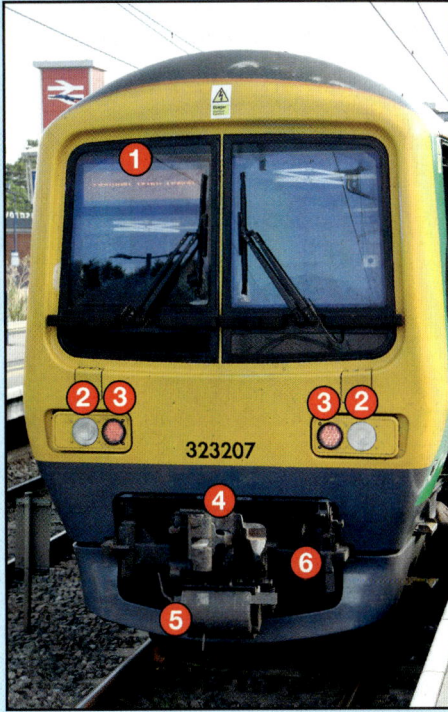

Right: *Class 323 front end layout. 1: Destination indicator, 2: Headlight, 3: Combined white marker and red tail light, 4: Tightlock coupling, 5: Coupling electric control box. 6: Air horns in rear of coupling pocket. Set No. 323207, a West Midlands set, is illustrated.* **CJM**

Below: *Currently, the Class 323s, based at Birmingham Soho depot, operate the Birmingham CrossCity line. The 23m long vehicles, formed into three-vehicle sets, seat 275 standard class passengers in high-density style. West Midlands sets carry white, black and gold livery, shown on set No. 323220 at Aston. When replaced by Class 730s, 17 WM Class 323s will move to Northern at Allerton.* **John Binch**

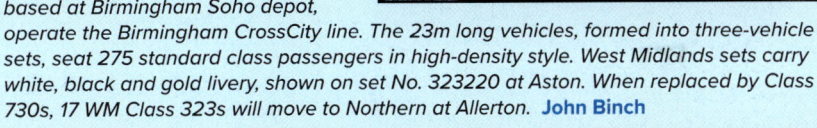

Right: *In recent years, the intermediate PTSOL, has been rebuilt to house a universal access toilet compartment, with an adjacent wheelchair position. This area is identified by blue branding on the external doors. In 2021-2022 the West Midlands sets were all named after towns they serve, using 'stick-on' names on the side of the PTSOL. No. 323240 Erdington is illustrated. The Soho depot badge is also applied.* **CJM**

Left: *Class 323 interior, showing the 2+3 high-density seating, with the latest West Midlands grey and orange fleck moquette and orange square style grab handles.* **Adrian Paul**

Right: *In early 2023 set No. 323221 was repainted in original Regional Railways grey and green livery. It is recorded near Bromsgrove on 14 February 2023.* **CJM**

Rolling Stock Review : 2023-2024

Class 323

323201	WMR
323202	WMR
323203	WMR
323204	WMR
323205	WMR
323206	WMR
323207	WMR
323208	WMR
323209	WMR
323210	WMR
323211	WMR
323212	WMR
323213	WMR
323214	WMR
323215	WMR
323216	WMR
323217	WMR
323218	WMR
323219	WMR
323220	WMR
323221	WMR
323222	WMR
323223	NOR
323224	NOR
323225	NOR
323226	NOR
323227	NOR
323228	NOR
323229	NOR
323230	NOR
323231	NOR
323232	NOR
323233	NOR
323234	NOR
323235	NOR
323236	NOR
323237	NOR
323238	NOR
323239	NOR
323240	WMR
323241	WMR
323242	WMR
323243	WMR

2023 Fleet

Above: *In 2023, a fleet of 17 Class 323 three-car sets operate for Northern in the Manchester/Liverpool-Crewe area. A batch of 17 are scheduled to transfer from West Midlands after Class 730s are introduced. The fleet of 34 will then form the backbone of electric services in the area. The Allerton based sets seat 284 and meet the latest PRM requirements. Set No. 323233 is seen at Crewe.* **CJM**

Left: *Coupled between the driving cars is a Pantograph Trailer Standard Open with Lavatory (PTSOL), following refurbishment this now houses a universal style toilet compartment, located at the end opposite to the pantograph. This area is identified by the plating over of two original windows and replacing one large window with a small unit. Adjacent to the universal toilet is the disabled area, with wheelchair parking, the area is identified by a blue band above the passenger doors, (close end in image). Car No. 72233 is shows from set No. 323233.* **Antony Christie**

Left: *The refurbished Northern Class 323 interior looks very neat, it retains the original 3+2 seating layout, but now with Northern two-tone blue fleck moquette. The vehicles retain hopper windows as no air conditioning is provided. This view shows the area looking twards the end of a coach with the door at the end leading to te adjacent coach. It clearly illustrates the very limited walking space between seats.* **CJM**

■ In summer 2023, it was confirmed that West Midlands sets 323202/03/05-13/17-19/41-43 will transfer to Northern as soon as a suficent number of Class 730s are commissioned. The ex-WMR sets will be fitted with charging points and TV screens over the next few momnths before transfer.

'Railnet' — Class 325

For history and pictures of this class, see Modern Locomotives Illustrated Issue 223

Class:	325
Number range:	325001-325016
Built by:	Adtranz Derby
Year introduced:	1995-1996
Max speed:	100mph (161km/h)
Formation:	DTPMV(A)+MPMV+TPMV+DTPMV(B)
Vehicle numbers:	DTPMV(A) - 68300-68330 (even numbers)
	MPMV - 68340-68355
	TPMV - 68360-68375
	DTPMV(B) - 68301-68331 (odd numbers)
Vehicle length:	DTPMV - 65ft 0¾in (19.83m)
	MPMV/TPMV - 65ft 4¼in (19.92m)
Height:	12ft 4¼in (3.77m)
Width:	9ft 3in (2.82m)
Weight:	Total - 138.4 tonnes
	DTPMV(A) - 29.1 tonnes
	MPMV - 49.5 tonnes
	TPMV - 30.7 tonnes
	DTPMV(B) - 29.1 tonnes
Internal layout:	Open
Seating:	None - Parcel/Mail space
Gangway:	Not fitted
Toilets:	Not fitted
Brake type:	Air (EP/auto)
Bogie type:	Powered - Adtranz P7-4
	Trailer - Adtranz T3-7
Power collection:	25kV ac overhead,
	750V dc third rail (isolated/removed)
Traction motor type:	4 x GEC G315BZ
Horsepower:	1,438hp (1,072kW)
Coupling type:	Drop-head buck-eye/screw
Multiple restriction:	Within type, TDM wired
Door type:	Roller shutter
Construction:	Steel
Operator:	DB-C
Owner:	Royal Mail

Key Facts

- Built to operate 'Mail by Rail' services.
- Mail bays isolated from cab area for security.
- One set withdrawn and broken up, many other sets stored out of use at Crewe.
- Originally fitted for dual AC/DC operation, third rail equipment now isolated or removed.
- TDM system fitted when built to operate with TDM fitted locos. (now isolated).

Above: *Class 325 driving cab, based on the 'Networker' style with a single power/brake controller on the left side. These sets could operate from AC or DC power systems as well as in push-pull mode using the TDM jumper system. In recent years the DC third rail equipment and the TDM facility has been isolated.* **CJM**

Right Above & Right Middle: *Introduced in 1995 to update the movement of 'Mail by Rail', the Class 325 dual-voltage 'Railnet' trains saw little use as planned. Soon after delivery, Royal Mail ceased using the railway for mail transport. These Royal Mail owned trains, are based on the dual-voltage Class 319, using a 'Networker' style cab end. They are open vehicles with straps and lugs to hold standardised Royal Mail trolleys, access is by two roller-shutter doors on either side of each vehicle, with the train crew having no access to the mail area. Sets are not gangway fitted, increasing security. After a long period of storage, a reduced 'Mail by Rail' operation now operates on weekdays between London, the North West and Scotland, usually using eight or 12-car trains. One '325' was broken up to provide spares several years ago. In 2023 the operational sets are worked by DB-Cargo. In the upper view, set No. 325009 heads south near Daventry. The right lower image TPMV coach No. 68349, these vehicles weigh 30.7 tonnes. All:* **CJM**

2023 Fleet

325001	DBC
325002	DBC
325003	DBC
325004	DBC
325005	DBC
325006	DBC
325007	DBC
325008	DBC
325009	DBC
325011	DBC
325012	DBC
325013	DBC
325014	DBC
325015	DBC
325016	DBC

Below & Below Right: *Trailer mail coach No. 68372 from set No. 325013, shows the exterior design of the coach, with two roller shutter doors on each side, adjacent to each door at eye height in a tamper or warning light, which illuminates blue if a fault is detected. On the right, set No. 325002 is seen heading north on the approach to Stafford on 14 September 2022 with a Princess Royal Distribution Centre to Glasgow main service. All sets are based at Crewe electric depot and maintained by DB-Cargo.* **Antony Christie / CJM**

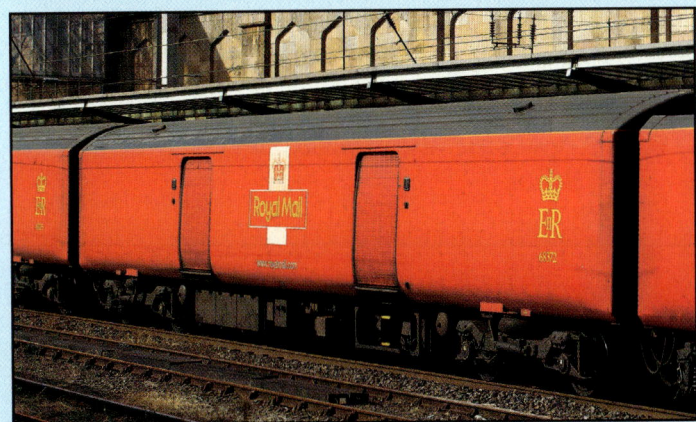

Rolling Stock Review : 2023-2024

Class 331 'Civity'

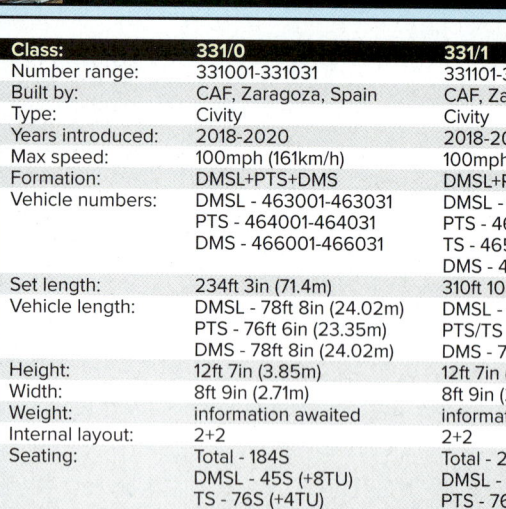

For history and pictures of this class, see Modern Locomotives Illustrated Issue 243

Class:	331/0	331/1
Number range:	331001-331031	331101-331112
Built by:	CAF, Zaragoza, Spain	CAF, Zaragoza, Spain
Type:	Civity	Civity
Years introduced:	2018-2020	2018-2020
Max speed:	100mph (161km/h)	100mph (161km/h)
Formation:	DMSL+PTS+DMS	DMSL+PTS+TS+DMS
Vehicle numbers:	DMSL - 463001-463031	DMSL - 463101-463112
	PTS - 464001-464031	PTS - 464101-464112
	DMS - 466001-466031	TS - 465101-465112
		DMS - 466101-466112
Set length:	234ft 3in (71.4m)	310ft 10in (94.75m)
Vehicle length:	DMSL - 78ft 8in (24.02m)	DMSL - 78ft 8in (24.02m)
	PTS - 76ft 6in (23.35m)	PTS/TS - 76ft 6in (23.35m)
	DMS - 78ft 8in (24.02m)	DMS - 78ft 8in (24.02m)
Height:	12ft 7in (3.85m)	12ft 7in (3.85m)
Width:	8ft 9in (2.71m)	8ft 9in (2.71m)
Weight:	information awaited	information awaited
Internal layout:	2+2	2+2
Seating:	Total - 184S	Total - 260S
	DMSL - 45S (+8TU)	DMSL - 45S (+8TU)
	TS - 76S (+4TU)	PTS - 76S (+4TU)
	DMS - 63S (+7TU)	TS - 76S (4TU)
		DMS - 63S (+7TU)
Gangway:	Within set	Within set
Toilets:	DMSL - 1	DMSL - 1
Brake type:	Air, Regeneration	Air, Regeneration
Bogie type:	CAF	CAF
Power collection:	25kV ac overhead	25kV ac overhead
Traction motor type:	Traction Systems Austria	Traction Systems Austria
Horsepower:	1,475hp (1,100kW)	1,475hp (1,100kW)
Coupling type:	Outer - Dellner, Inner - Bar	Outer - Dellner, Inner - Bar
Multiple restriction:	Within class only	Within class only
Door type:	Bi-parting sliding plug	Bi-parting sliding plug
Construction:	Aluminium	Aluminium
Operator:	Northern Rail	Northern Rail
Owner:	Eversholt Leasing	Eversholt Leasing

Above: Class 331 driving cab, is of a modern futuristic state-of-the-art design, it occupies the full vehicle width, with the driving position in the middle. A combined power brake controller is fitted, operated by the driver's left hand. A full Train Management System (TMS) is installed. Note the UK required TPWS equipment located on the raised display on the right side, above the radio. **Tony Miles**

Above: The CAF-built Class 331s, were built in Spain and shipped to the UK via the docks at Portbury, Bristol, they come in both three- (Class 331/0) and four-car (Class 331/1) formations. The fleet is owned by Eversholt, and operated by Northern, allocated to Liverpool Allerton (331/0) or Leeds Neville Hill (331/1) depots. The two sub-classes are the same, except for an extra Trailer Standard (TS) in four-car units. Sets carry Northern white and blue livery. The design incorporates one large cab front window, with the driver positioned centrally. In the above view, three-car set No. 331021 is seen at Layton station on the Blackpool North branch. **Antony Christie**

Left: The intermediate vehicle on '331/0' sets is a Pantograph Trailer Standard Open (PTSO), with its pantograph located at the Driving Motor Standard end. The intermediate vehicle, No. 464018 from set 331018 is captured from its pantograph end at Liverpool Lime Street. The '331s' carry full European Vehicle Numbering (EVN). **Antony Christie**

Rolling Stock Review : 2023-2024

'Civity' — Class 331

331/0	
331001	NOR
331002	NOR
331003	NOR
331004	NOR
331005	NOR
331006	NOR
331007	NOR
331008	NOR
331009	NOR
331010	NOR
331011	NOR
331012	NOR
331013	NOR
331014	NOR
331015	NOR
331016	NOR
331017	NOR
331018	NOR
331019	NOR
331020	NOR
331021	NOR
331022	NOR
331023	NOR
331024	NOR
331025	NOR
331026	NOR
331027	NOR
331028	NOR
331029	NOR
331030	NOR
331031	NOR

331/1	
331101	NOR
331102	NOR
331103	NOR
331104	NOR
331105	NOR
331106	NOR
331107	NOR
331108	NOR
331109	NOR
331110	NOR
331111	NOR
331112	NOR

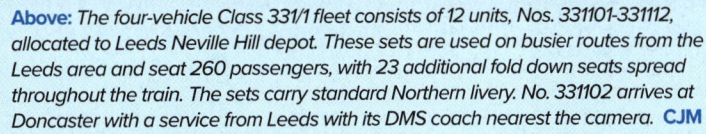

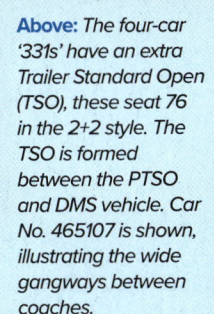

2023 Fleet

Above: *The four-vehicle Class 331/1 fleet consists of 12 units, Nos. 331101-331112, allocated to Leeds Neville Hill depot. These sets are used on busier routes from the Leeds area and seat 260 passengers, with 23 additional fold down seats spread throughout the train. The sets carry standard Northern livery. No. 331102 arrives at Doncaster with a service from Leeds with its DMS coach nearest the camera.* **CJM**

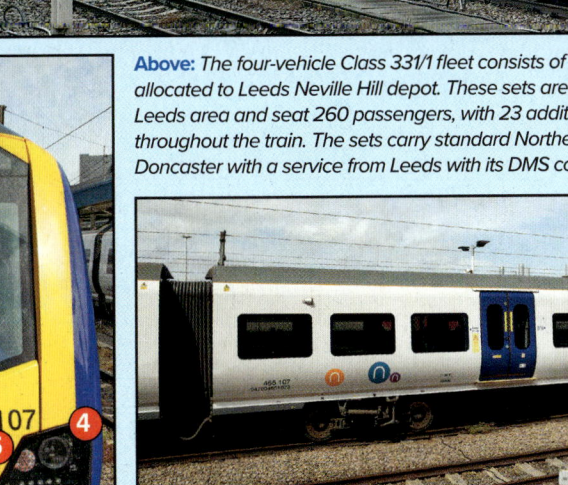

Above: *Class 331 front end layout and equipment. 1: High level marker light, 2: Destination indicator, 3: Forward facing camera, 4: Headlight, 5: Combined red tail and white marker light, 6: Running lights, 7: Lamp bracket, 8: Electrical coupling connection box, 9: Dellner coupling, 10: Air horns in coupling pocket, 11: Obstacle deflector plate.* **Antony Christie**

Above: *The four-car '331s' have an extra Trailer Standard Open (TSO), these seat 76 in the 2+2 style. The TSO is formed between the PTSO and DMS vehicle. Car No. 465107 is shown, illustrating the wide gangways between coaches.* **Antony Christie**

Above: *The interior of the Class 331 is the same layout-style as found on the CAF-built Class 195 diesel units. It includes a good quality 2+2 layout, with a mix of group and airline seating styles, the group seats have a good table. phone/lap-top charging points are by seats and above seat racks are provided for light luggage. The view above shows a DMSL coach, looking towards the universal toilet compartment at the inner end. Left is a close-up of a facing seating group, showing the quality table and folding arm-rests in the middle and corridor side. Both:* **CJM**

Key Facts

- Two and three car versions operating with Northern.
- Sets can be found throughout the Northern area allocated to Allerton (Liverpool) and Neville Hill (Leeds) depots.
- Body structure is the same as used on the Class 195 DMUs built by CAF.

Rolling Stock Review : 2023-2024

Class 333

Class:	333
Number range:	333001-333016
Built by:	CAF Spain, Siemens Germany
Years introduced:	2000, TSO added 2002-2003
Max speed:	100mph (161km/h)
Formation:	DMSO(A)+PTSOL+TSO+DMSO(B)
Vehicle numbers:	DMSO(A) - 78451-78481 (odd numbers)
	PTSOL - 74461-74476
	TSO - 74477-74492
	DMSO(B) - 78452-78482 (even numbers)
Vehicle length:	DMSO - 77ft 10¾in (23.74m)
	PTSOL/TSO - 75ft 11in (23.14m)
Height:	12ft 1½in (3.70m)
Width:	9ft 0¼in (2.75m)
Weight:	Total - 185.1 tonnes
	DMSO(A) - 50.6 tonnes
	PTSOL - 46.0 tonnes
	TSO - 38.5 tonnes
	DMSO(B) - 50.0 tonnes
Internal layout:	2+3
Seating:	Total - 353S + 6 tip up
	DMSO(A) - 90S
	PTSOL - 73S (+7TU)
	TSO - 100S
	DMSO(B) - 90S
Gangway:	Within set
Toilets:	PTSOL - 1
Brake type:	Air
Bogie type:	CAF
Power collection:	25kV ac overhead
Traction motor type:	4 x Siemens
Horsepower:	1,877hp (1,400kW)
Coupling type:	Outer - Dellner 10, Inner - Bar
Multiple restriction:	Within class
Door type:	Bi-parting sliding plug
Construction:	Steel
Operator:	Northern
Owner:	Angel Trains

For history and pictures of this class, see Modern Locomotives Illustrated Issue 233

Left: Class 333 front end layout and equipment. 1: High level marker light, 2: Forward facing camera, 3: Destination indicator, 4: White marker light, 5: Headlight, 6: Red tail light, 7: Electrical connection box above coupling, 8: Dellner 10 coupling, 9: Windscreen washer filler port, 10: Warning horns in coupling pocket. **Antony Christie**

Below: Class 333s contain two 23m long intermediate vehicles, the TSO seats 100 in the 2+3 style using a mix of group and airline layout. No. 74481 is seen, showin two pairs of passenger operated bi-parting plug doors on each side. **Adrian Paul**

2023 Fleet

333001	NOR
333002	NOR
333003	NOR
333004	NOR
333005	NOR
333006	NOR
333007	NOR
333008	NOR
333009	NOR
333010	NOR
333011	NOR
333012	NOR
333013	NOR
333014	NOR
333015	NOR
333016	NOR

Above: The Class 333s are operated on the Aire Valley lines from Leeds to Skipton, Bradford and Ilkley, they are based at Leeds Neville Hill depot. All are refurbished and carry the latest Northern colours. The '333s' were the first Siemens/CAF sets to operate in the UK and were of the same external design as the now withdrawn Heathrow Express Class 332 stock. Set No. 333005 is seen at Shipley. **Adrian Paul**

Key Facts

- Originally delivered as three-car sets, augmented to four vehicles in 2003.
- Based at Neville Hill depot, Leeds and used on Aire Valley routes from Leeds to Skipton, Bradford and Ilkley.
- Major fleet refurbishment in 2019-2020.
- Fleet to be retained by Northern.

Above & Right: Class 333 interiors were fully refurbished in 2019-2020, this project included installing a universal access toilet compartment in the Pantograph Trailer Standard coach. Seating throughout the train is in the 2+3 high-density style, using Northern blue moquette and yellow fittings. Seats do not have armrests and are quite narrow. A passenger information system (PIS) is fitted and above seat light luggage racks are provided. Above, is a view from the first door pocket looking towards the cab. On the left is general view looking along the length of a PTSOL vehicle with the toilet compartment at the far end. A full interior CCTV system is fitted. Both: **CJM**

'Juniper' — Class 334

Class:	334
Number range:	334001-334040
Built by:	Alstom, Washwood Heath
Type:	Juniper
Years introduced:	1999-2002
Max speed:	90mph (145km/h)
Formation:	DMSO(A)+PTSO+DMSO(B)
Vehicle numbers:	DMSO(A) - 64101-64140
	PTSO - 74301-74340
	DMSO(B) - 65101-65140
Vehicle length:	DMSO(A) - 69ft 0¾in (21.05m)
	PTSO - 65ft 4¼in (19.92m)
	DMSO(B) - 69ft 0¾in (21.05m)
Height:	12ft 3in (3.73m)
Width:	9ft 2¾in (2.81m)
Weight:	Total - 124.6 tonnes
	DMSO(A) - 42.6 tonnes
	PTSO - 39.4 tonnes
	DMSO(B) - 42.6 tonnes
Internal layout:	DMSO 2+2, PTSO 2+3
Seating:	Total - 178S
	DMSO(A) - 64S, PTSO - 55S
	DMSO(B) - 59S
Gangway:	Within set
Toilets:	PTSO - 1
Brake type:	Air (regenerative)
Bogie type:	Alstom DMSO - LTB3, PTSO - TBP3
Power collection:	25kV ac overhead
Traction motor type:	4 x Alstom Onix 800
Horsepower:	1,448hp (1,080kW)
Coupling type:	Outer - Dellner, Inner - Bar
Multiple restriction:	Within class only
Door type:	Bi-parting sliding plug
Construction:	Steel
Operator:	ScotRail
Owner:	Eversholt Leasing

For history and pictures of this class, see Modern Locomotives Illustrated Issue 225

Left: *Class 334 front end layout. All three of the 'Juniper' platform UK fleets have different front ends and different cab layouts. Front end equipment positions are 1: Warning horns behind grille, 2: High level marker light, 3: Destination indicator, 4: Red tail light, 5: White marker light, 6: Headlight, 7: Lamp bracket, 8: Dellner coupling with gathering horn above, 9: Electrical connection box with roller door. When built Tightlock couplings were fitted, Dellner couplings were fitted in 2015-2016.* **Ian Lothian**

2023 Fleet

334001	ASR
334002	ASR
334003	ASR
334004	ASR
334005	ASR
334006	ASR
334007	ASR
334008	ASR
334009	ASR
334010	ASR
334011	ASR
334012	ASR
334013	ASR
334014	ASR
334015	ASR
334016	ASR
334017	ASR
334018	ASR
334019	ASR
334020	ASR
334021	ASR
334022	ASR
334023	ASR
334024	ASR
334025	ASR
334026	ASR
334027	ASR
334028	ASR
334029	ASR
334030	ASR
334031	ASR
334032	ASR
334033	ASR
334034	ASR
334035	ASR
334036	ASR
334037	ASR
334038	ASR
334039	ASR
334040	ASR

Key Facts

- Only AC version of Alstom 'Juniper' product platform.
- Designed for use in Glasgow/Edinburgh area.
- Slow delivery and test period due to issues with 'Juniper' design.
- Delivered in Carmine and Cream SPT livery, repainted in blue/white 'Saltire' colours.
- Originally fitted with Tightlock couplings, changed to Dellner couplings in 2015-2016.

Above, Inset & Right: *Between 1999-2002 a fleet of 40 Alstom 'Juniper' three-car EMUs were built for outer suburban services in Scotland, around the Edinburgh and Glasgow areas. These non-gangwayed sets are standard class only, seating 183. The two driving cars are motor vehicles (DMSO), with an intermediate Pantograph Trailer Standard Open (PTSO) carrying the pantograph and housing the transformer. Sets are currently finished in Scottish Railways blue and white livery. Above, set No. 334027 arrives at Haymarket, bound for Edinburgh Waverley. The inset image above shows the spacious interior, using the 2+3 style with most seats in groups, covered in ScotRail moquette. Two power sockets for phone and laptop charging are located by each seat bay. The view right, shows the exterior of the intermediate PTSO, viewed from the pantograph toilet end, where two windows are blanked out to house the toilet. The retention of the window positions, allows changes of the vehicles use in the future if needed. In 2023 it was announced a major refresh of the Class 334s would commence in 2024. All:* **CJM**

Class 345 'Aventra'

For history and pictures of this class, see *Modern Locomotives Illustrated* Issue 243

Class:	345
Number range:	345001-345070
Built by:	Bombardier Derby
Type:	Aventra
Years introduced:	2017-2020
Max speed:	90mph (145km/h)
Formation:	DMS(A)+PMS(A)+MS(A)+MS(B)+TS+ MS(C)+MS(D)+PMS(B)+DMS(B)
Vehicle numbers:	DMS(A) - 340101 - 340170 PMS(A) - 340201 - 340270, MS(A) - 340301 - 340370 MS(B) - 340401 - 340470, TS - 340501 - 340570 MS(C) - 340601 - 340670, MS(D) - 340701 - 340770 PMS(B) - 340801 - 340870, DMS(B) - 340901 - 340970
Train length:	673ft (205m)
Vehicle length:	73ft 9in (22.5m)
Height:	To be advised
Width:	9ft 2in (2.78m)
Weight:	Total - 318.4t DMS(A)-39t, PMS(A)-37.1t, MS(A)-36.5t, MS(B)-31.4t, TS-29.7t, MS(C)-37.2t, MS(D)-37.2t, PMS(B)-37.1t, DMS(B)-39t
Internal layout:	Longitudunal and 4 bays of 2+2 in MS & PTS
Seating:	Total - 412S, Capacity for 1,500 DMS(A) - 46S, PMS(A) - 46S, MS(A) - 46S, MS(B) - 49S, TS - 38S, MS(C) - 49S, MS(D) - 46S, PMS(B) - 46S, DMS(B) - 46S
Gangway:	Within set
Toilets:	Not fitted
Brake type:	Air (regenerative)
Bogie type:	Bombardier Flexx Eco
Power collection:	25kV ac overhead
Traction motor type:	20 x Bombardier 265kW
Horsepower:	7,107hp (5,300kW)
Coupling type:	Outer - Dellner, Inner - Bar
Multiple restriction:	Emergency only
Door type:	Bi-parting sliding plug (3 pairs per coach side)
Construction:	Aluminium
Operator:	CrossRail (Elizabeth Line)
Owner:	345 Rail Leasing (Equitix Invest, Nat West & SMBC)
Fittings:	TPWS - Liverpool Street-Shenfield and Paddington to Reading ECTS - Heathrow Airport section CBTC - Paddington-Abbey Wood/Pudding Mill

345001	CRO	345036	CRO
345002	CRO	345037	CRO
345003	CRO	345038	CRO
345004	CRO	345039	CRO
345005	CRO	345040	CRO
345006	CRO	345041	CRO
345007	CRO	345042	CRO
345008	CRO	345043	CRO
345009	CRO	345044	CRO
345010	CRO	345045	CRO
345011	CRO	345046	CRO
345012	CRO	345047	CRO
345013	CRO	345048	CRO
345014	CRO	345049	CRO
345015	CRO	345050	CRO
345016	CRO	345051	CRO
345017	CRO	345052	CRO
345018	CRO	345053	CRO
345019	CRO	345054	CRO
345020	CRO	345055	CRO
345021	CRO	345056	CRO
345022	CRO	345057	CRO
345023	CRO	345058	CRO
345024	CRO	345059	CRO
345025	CRO	345060	CRO
345026	CRO	345061	CRO
345027	CRO	345062	CRO
345028	CRO	345063	CRO
345029	CRO	345064	CRO
345030	CRO	345065	CRO
345031	CRO	345066	CRO
345032	CRO	345067	CRO
345033	CRO	345068	CRO
345034	CRO	345069	CRO
345035	CRO	345070	CRO

Above: Class 345 front end layout. The '345's were the first units to take advantage of changes in legislation to allow the removal of the yellow warning front end. 1: High level marker light, 2: Destination indicator, 3: Combined white marker and red tail light, 4: Headlight, 5: Warning horns behind grille, 6: Dellner coupling (not for multiple working), 7: Emergency air connection. **CJM**

Below: Designed and built for use on the London east-west 'Elizabeth Line', 70 nine-car Bombardier 'Aventra' sets are based at Old Oak Common. These open plan trains, with mainly longitudinal seating, are high-density Metro stock operating between Reading / Heathrow Airport in the west and Shenfield / Abbey Wood in the east. Set No. 345033 calls at Brentwood with a service bound for Shenfield. **CJM**

Key Facts

- First units to operate without yellow warning ends since 1960s.
- TfL would like another four Class 345s, in 2023 subject to financial review.
- Some sets operated as 7 car formations until mid-2022.
- Only 'Aventra' sets with three door groups per car side.

Right: Until 2022, some short seven-vehicle sets were used and short running was operated between Reading/Heathrow and Paddington, through the central core and Liverpool Street to Shenfield. Today the full route is open with all nine car trains working 24 trains per hour in each direction under central London. Trains terminate at both Heathrow Terminal 4 and 5. Set No. 345047 is seen at Heathrow Terminal 5 with a service bound for Abbey Wood. **CJM**

'Aventra' — Class 345

Right & Inset: A Pantograph Motor Standard (PMS) coach is coupled to each driving car, with a pantograph mounted at the outer end. Each PMS seats 46 in the longitudinal seating style. To reduce station dwell times and loading/unloading, three pairs of passenger doors are fitted on each side of each coach. At some stations, the middle door set is isolated due to curvature. Below is the wide between vehicle gangway, fully air sealed and give an excellent view throughout the train increasing security. Both: **CJM**

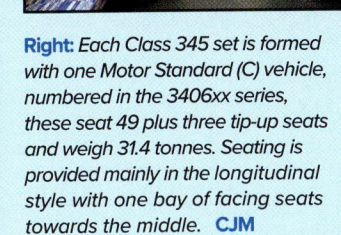

Right: Each Class 345 set is formed with one Motor Standard (C) vehicle, numbered in the 3406xx series, these seat 49 plus three tip-up seats and weigh 31.4 tonnes. Seating is provided mainly in the longitudinal style with one bay of facing seats towards the middle. **CJM**

Below: All intermediate coaches with the exception of the Trailer Standard (TS) formed in the middle of the formation, have one bay of facing seats, illustrated here from a door pocket position. Usually these seats fill up first, especially with longer distance travellers. **CJM**

Right: When the Class 345s were first introduced the Transport for London (TFL) badge was carried From 2022 the Elizabeth Line branding was applied. **CJM**

Right: Class 345 driving cab. This was a new design, introduced for the Bombardier, later Alstom 'Aventra' product platform, it is based on the left side driving position, and incorporates a wrap-around desk with the combined power/brake controller on the left side. A Train Management System (TMS) is incorporated, as are screens able to provide good quality views of the exterior of the train, to enable the driver to have a full understanding of platform/train interface. Trains operate under the Driver Only Operation Passenger (DOOP) system, with the driver fully responsible for the doors and safe operation of the train. On board staff travel on each train. **CJM**

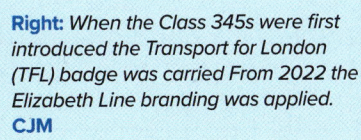

Rolling Stock Review : 2023-2024

Class 350   'Desiro'

For history and pictures of this class, see *Modern Locomotives Illustrated* Issue 178

Class	350/1	350/2	350/3	350/4
Number range:	350101-350130	350231-350267	350368-350377	350401-350410
Built by:	Siemens - Vienna, Austria, and Duewag, Germany	Siemens - Duewag, Germany	Siemens - Duewag, Germany	Siemens - Duewag, Germany
Type:	Desiro	Desiro	Desiro	Desiro
Years introduced:	2004-2005	2008-2009	2014	2014
Max speed:	110mph (177km/h)	100mph (161km/h)	110mph (177km/h)	110mph (177km/h)
Formation:	DMSO(A)+TSO+PTSO+DMSO(B)	DMSO(A)+TSO+PTSO+DMSO(B)	DMSO(A)+TSO+PTSO+DMSO(B)	DMSO(A)+TSO+PTSO+DMSO(B)
Vehicle numbers:	DMSO(A) - 63761-63790 TSO - 66811-66840 PTSO - 66861-66890 DMSO(B) - 63711-63740	DMSO(A) - 61431-61467 TSO - 65231-65267 PTSO - 67531-67567 DMSO(B) - 61531-61567	DMSO(A) - 60141-60150 TSO - 60511-60520 PTSO - 60651-60660 DMSO(B) - 60151-60160	DMSO(A) - 60691-60700 TSO - 60901-60910 PTSO - 60941-60950 DMSO(B) - 60671-60680
Vehicle length:	66ft 9in (20.34m)	66ft 9in (20.34m)	66ft 9in (20.34m)	66ft 9in (20.34m)
Height:	12ft 1½in (3.69m)	12ft 1½in (3.69m)	12ft 1½in (3.69m)	12ft 1½in (3.69m)
Width:	9ft 2in (2.79m)	9ft 2in (2.79m)	9ft 2in (2.79m)	9ft 2in (2.79m)
Weight:	Total - 179.3 tonnes DMSO(A) - 48.7 tonnes TSO - 36.2 tonnes PTSO - 45.2 tonnes DMSO(B) - 49.2 tonnes	Total - 166.1 tonnes DMSO(A) - 43.7 tonnes TSO - 35.3 tonnes PTSO - 42.9 tonnes DMSO(B) - 44.2 tonnes	Total - 170 tonnes DMSO(A) - 44.2 tonnes TSO - 36.2 tonnes PTSO - 44.6 tonnes DMSO(B) - 45 tonnes	Total - 170 tonnes DMSO(A) - 44.2 tonnes TSO - 36.2 tonnes PTSO - 44.6 tonnes DMSO(B) - 45 tonnes
Internal layout:	2+2	2+2, 2+3	2+1F / 2+2S	2+1F / 2+2S
Seating:	Total - 226S (+9TU) DMSO(A) - 60S TSO - 56S PTSO - 50S (+9TU) DMSO(B) - 60S	Total - 267S (+9TU) DMSO(A) - 70S TSO - 56S PTSO - 61S + 9 tip up DMSO(B) - 70S	Total - 230S DMSO(A) - 60S TSO - 56S PTSO - 50S + 9 tip up DMSO(B) - 60s	Total - 197S DMSO(A) - 56S TSO - 56S PTSO - 42S DMSO(B) - 56S
Gangway:	Throughout	Throughout	Throughout	Throughout
Toilets:	TSO, PTSO - 1	TSO, PTSO - 1	TSO, PTSO - 1	TSO, PTSO - 1
Brake type:	Air (regenerative)	Air (regenerative)	Air (regenerative)	Air (regenerative)
Bogie type:	SGP SF5000	SGB SF5000	SGB SF5000	SGB SF5000
Power collection:	25kV ac overhead	25kV ac overhead	25kV ac overhead	25kV ac overhead
Traction motor type:	4 x Siemens 1TB2016-0GB02	4 x Siemens 1TB2016-0GB02	4 x Siemens 1TB2016-0GB02	4 x Siemens 1TB2016-0GB02
Horsepower:	2,680hp (2,000kW)	2,680hp (2,000kW)	2,680hp (2,000kW)	2,680hp (2,000kW)
Coupling type:	Outer - Dellner 12, Inner - Bar	Outer - Dellner 12, Inner - Bar	Outer - Dellner 12, Inner - Bar	Outer - Dellner 12, Inner - Bar
Multiple restriction:	Within class	Within class	Within class	Within class
Door type:	Bi-parting sliding plug	Bi-parting sliding plug	Bi-parting sliding plug	Bi-parting sliding plug
Construction:	Aluminium	Aluminium	Aluminium	Aluminium
Operator:	West Midlands	West Midlands	West Midlands	West Midlands
Owner:	Angel Trains	Porterbrook Leasing	Angel Trains	Angel Trains
Notes:	Wired for 750V dc operation	Due off lease 2024		

Above & Inset: Four sub-classes of Siemens 'Desiro' are operated by West Midland Railways, all are based at the Siemens Northampton depot. Sets are virtually identical to the SWR Class 450s. The '350s' are designed for AC (overhead) and DC (third rail) operation, with the Class 350/1 sub-class fitted with shoe gear. No. 350121 is seen from its DMSO(B) vehicle, painted in London North Western Railway livery at Daventry. The inset image shows Trailer Composite Open No. 66832. These vehicles are now set out for standard class occupation seating 56. The coach also has one standard toilet compartment. **CJM / Antony Christie**

Left: Class 350 front end layout, applicable to all sub-classes. 1: White marker light, 2: Destination indicator, 3: Headlight, 4: Red tail light, 5: Gangway door, 6: Dellner 12 coupling, 7: Electric coupling connection box, 8: Air warning horns, 9: Adjustable obstacle deflector plate. The ridged 'blocks' on either side of the lower front are anti-climber plates, if two trains impacted end-on the teeth would engage and reduce the level of over-riding. **CJM**

Key Facts

- Part of Siemens 'Desiro' family.
- Each sub-classes has slightly different technical details and interior layout.
- Class 350/2 sets restricted to 100mph (161km/h) running.
- Class 350/2 sets, owned by Porterbrook, scheduled to be returned to lease owner by 2024.
- Final 10 sets (350/4), Originally leased to TransPennine, transferred to West Midlands in early 2020.
- First class seating was removed on all LNW trains from May 2023, first class area declassified.

'Desiro' Class 350

Above, Below & Inset: In 2008-2009, 37 Class 350/2 four-car sets entered service, funded by Porterbrook Leasing, these had a top speed of 100mph (161km/h). The fleet has a higher seating capacity of 243 standard against 202 on the earlier Class 350/1. In 2023, sets carry a mix of London Northwestern Railway liveries, some branded in full, others just with a name banner. Above, set No. 350239 is seen near South Kenton from its DMSO(B) end. Below, the intermediate TSO a vehicle now seating 56 standard class seats. The coach also houses a standard toilet compartment (close end in picture). Vehicle 65259 is shown from set 350259. The inset picture shows the rather cramped 3+2 seating installed on this sub-class. Above seat luggage racks are fitted and a quality passenger information system installed. Seating shown is in is in London Midland green.
CJM / Nathan Williamson / Antony Christie

Above: In 2014, a fleet of 10 Class 350/3s were delivered to London Midland, financed by Angel Trains. These sets were set out with low-density seating, having 24 first and 206 standard class seats per unit, using a mix of airline and facing group layouts. Although operated by West Midlands Railway, the Class 350s are branded London Northwestern Railway, a trading title of West Midlands Railway, they are deployed on longer-distance services. Most sets now carry standard grey and green LNW livery, as shown on set No. 350376 heading North at Daventry, with its DMSO(B) nearest the camera. **CJM**

Class 350   'Desiro'

Left: Until May 2023, the Class 350s were formed with an intermediate Trailer Composite Open (TCO), which had a small first class seating area in the middle. After WMRs decision to remove all first class seating from the May 2023 timetable change, the first class area was declassified to standard seating. The area is shown offering a more luxurious 2+2 seating style. **CJM**

350/1		350245	WMR
350101	WMR	350246	WMR
350102	WMR	350247	WMR
350103	WMR	350248	WMR
350104	WMR	350249	WMR
350105	WMR	350250	WMR
350106	WMR	350251	WMR
350107	WMR	350252	WMR
350108	WMR	350253	WMR
350109	WMR	350254	WMR
350110	WMR	350255	WMR
350111	WMR	350256	WMR
350112	WMR	350257	WMR
350113	WMR	350258	WMR
350114	WMR	350259	WMR
350115	WMR	350260	WMR
350116	WMR	350261	WMR
350117	WMR	350262	WMR
350118	WMR	350263	WMR
350119	WMR	350264	WMR
350120	WMR	350265	WMR
350121	WMR	350266	WMR
350122	WMR	350267	WMR
350123	WMR		
350124	WMR	**350/3**	
350125	WMR	350368	WMR
350126	WMR	350369	WMR
350127	WMR	350370	WMR
350128	WMR	350371	WMR
350129	WMR	350372	WMR
350130	WMR	350373	WMR
		350374	WMR
350/2		350375	WMR
350231	WMR	350376	WMR
350232	WMR	350377	WMR
350233	WMR		
350234	WMR	**350/4**	
350235	WMR	350401	WMR
350236	WMR	350402	WMR
350237	WMR	350403	WMR
350238	WMR	350404	WMR
350239	WMR	350405	WMR
350240	WMR	350406	WMR
350241	WMR	350407	WMR
350242	WMR	350408	WMR
350243	WMR	350409	WMR
350244	WMR	350410	WMR

Left Above & Left: The final batch of 10 Class 350s, classified 350/4 emerged in 2014, ordered for First TransPennine, to work between Manchester and Edinburgh/Glasgow via the West Coast. Sets were transferred to West Midlands Railway from late 2019, after '397s' were introduced. On these sets, seating is provided for 19 first and 178 standard in the low-density 2+2 style. In the above view, set No. 350403 is shown from its DMSO(A) at Daventry. In the left image, the Trailer Composite open coach (now trailer standard) is seen from set No. 350404.
CJM / Antony Christie

Left: Class 350 driving cab. The cab area of the Class 350s is very cramped and offers limited visibility due to the presence of the end gangway. All 'Desiro' sets, except the Class 360s, share a similar cab design. The power and brake controller is on the left side using a 'hammer head' handle. **CJM**

■ With the Porterbrook Leasing Class 350/2 sets due off lease by 2024, the owners have announced a plan to convert some sets to battery-electric under the 'Battery-Flex' title.

'Electrostar' Class 357

For history and pictures of this class, see Modern Locomotives Illustrated Issue 194

Class:	357/0	357/2 & 357/3	
Number range:	357001-357046	357/2 - 357201-357211 357/3 - 357312-357328	
Built by:	Adtranz Derby	Adtranz/Bombardier Derby	
Years introduced:	1999-2001	Delivered - 2001-2002 As 357/3 - 2015	
Max speed:	100mph (161km/h)	100mph (161km/h)	
Formation:	DMSO(A)+MSO+PTSOL+DMSO(B)	DMSO(A)+MSO+PTSOL+DMSO(B)	
Vehicle numbers:	DMSO(A) - 67651-67696 MSO - 74151-74196 PTSOL - 74051-74096 DMSO(B) - 67751-67796	DMSO(A) - 68601-68628 MSO - 74701-74728 PTSOL - 74601-74628 DMSO(B) - 68701-68728	
Vehicle length:	DMSO(A), DMSO(B) - 68ft 1in (20.75m) PTSOL, MSO - 65ft 11½in (20.10m)	DMSO(A), DMSO(B) - 68ft 1in (20.75m) PTSOL, MSO - 65ft 11½in (20.10m)	
Height:	12ft 4½in (3.77m)	12ft 4½in (3.78m)	
Width:	9ft 2¼in (2.80m)	9ft 2¼in (2.80m)	
Weight:	Total - 157.6 tonnes DMSO(A) - 40.7 tonnes MSO - 36.7 tonnes PTSOL - 39.5 tonnes DMSO(B) - 40.7 tonnes	Total - 157.6 tonnes DMSO(A) - 40.7 tonnes MSO - 36.7 tonnes PTSOL - 39.5 tonnes DMSO(B) - 40.7 tonnes	
Internal layout:	2+2, 2+3	2+2, 2+3	
Seating:	Total - 278S + 4 tip up DMSO(A) - 71S MSO - 78S PTSOL - 58S + 4 tip up DMSO(B) - 71S	357/2 357201-211 Total - 278S + 4 tip up DMSO(A) - 71S MSO - 78S PTSOL - 58S + 4 tip up DMSO(B) - 71S	357/3 357312-328 Total - 220S DMSO(A) - 56S MSO - 50S PTSOL - 56S DMSO(B) - 58S
Gangway:	Within set	Within set	
Toilets:	PTSOL - 1	PTSOL - 1	
Brake type:	Air (rheostatic/regen)	Air (rheostatic/regen)	
Bogie type:	Power - Adtranz P3-25 Trailer - Adtranz T3-25	Power - Adtranz P3-25 Trailer - Adtranz T3-25	
Power collection:	25kV ac overhead (750v dc equipped)	25kV ac overhead (750v dc equipped)	
Traction motor type:	6 x Adtranz (250kW)	6 x Adtranz (250kW)	
Horsepower:	2,010hp (1,500kW)	2,010hp (1,500kW)	
Coupling type:	Outer - Tightlock, Inner - Bar	Outer - Tightlock, Inner - Bar	
Multiple restriction:	Within class only	Within class only	
Door type:	Bi-parting sliding plug	Bi-parting sliding plug	
Operator:	c2c	c2c	
Owner:	Porterbrook Leasing	Angel Trains	
Sub class variation:	Porterbrook fleet	Angel Trains fleet	Low density with extra standing room
Note:	To be fitted with Bombardier ETCS and EBICab 2000 ATP		

Above: *Class 357 'Electrostar' front end layout and equipment positions. 1: White marker light, 2: Destination indicator, 3: Forward facing camera, 4: Headlight, 5: Red tail light, 6: Lamp bracket, 7: Tightlock coupling, 8: Coupling electrical box, 9: Warning horns in coupling pocket. Details apply to all Class 357 sub-classes. The Class 357s are the only 'Electrostar' product to sport this front end design, all others incorporate gangway fitted ends.* **CJM**

Right Upper: *The 'East Thameside' or London Tilbury and Southend lines received Class 357 'Electrostar' stock from Adtranz/Bombardier in 1999. The first 46 are Class 357/0 sets, funded by Porterbrook Leasing. These four-car sets seat 278 in the 2+3 style, with one universal access toilet in the PTSOL coach. Sets carry c2c white livery, off-set by blue bi-parting sliding-plug doors and c2c branding on each vehicle. Set No. 357042 is seen at Limehouse.* **CJM**

Right Below: *The second batch of Class 357s, consisted of 28 four-car sets, funded by Angel Trains, these were classified as 357/2. Internally they were the same as the original 357/0 batch, and have operated as one fleet. The entire Class 357 fleet is allocated to East Ham depot. Set No. 357207, is seen at Shadwell.*
Antony Christie

Key Facts

- First production build of Adtranz / Bombardier 'Electrostar' product.
- 357/0 fleet funded by Porterbrook, 357/2 fleet funded by Angel Trains.
- 17 Class 357/2s modified in 2015 as 357/3 by reducing seating, increasing standing space and fitting 'straps'. Branded 'Metro' sets.

Class 357 'Electrostar'

2023 Fleet

357/0	
357001	C2C
357002	C2C
357003	C2C
357004	C2C
357005	C2C
357006	C2C
357007	C2C
357008	C2C
357009	C2C
357010	C2C
357011	C2C
357012	C2C
357013	C2C
357014	C2C
357015	C2C
357016	C2C
357017	C2C
357018	C2C
357019	C2C
357020	C2C
357021	C2C
357022	C2C
357023	C2C
357024	C2C
357025	C2C
357026	C2C
357027	C2C
357028	C2C
357029	C2C
357030	C2C
357031	C2C
357032	C2C
357033	C2C
357034	C2C
357035	C2C
357036	C2C
357037	C2C
357038	C2C
357039	C2C
357040	C2C
357041	C2C
357042	C2C
357043	C2C
357044	C2C
357045	C2C
357046	C2C

357/2	
357201	C2C
357202	C2C
357203	C2C
357204	C2C
357205	C2C
357206	C2C
357207	C2C
357208	C2C
357209	C2C
357210	C2C
357211	C2C

357/3	
357312	C2C
357313	C2C
357314	C2C
357315	C2C
357316	C2C
357317	C2C
357318	C2C
357319	C2C
357320	C2C
357321	C2C
357322	C2C
357323	C2C
357324	C2C
357325	C2C
357326	C2C
357327	C2C
357328	C2C

Top: In 2015-2016 to address passenger growth, 17 Class 357/2 sets were modified by East Ham as 'Metro' units. Interiors were re-styled, with low-density 2+2 seating, leaving just 220 seats, per four-car unit. Large stand-back areas were formed by door pockets, with roof hanging 'straps' provided to support standing passengers. The operator believed that to increase standing space was the best option to address overcrowding. It is attempted to deploy these sets on shorter distance services, but in reality they turn up on any train. Modified units were reclassified as 357/3 and renumbered in the 357312-357328 series, keeping the last two digits of their original number. Set No. 357326 is seen at Limehouse. Note the mauve Metro signs by the passenger doors. **CJM**

Above Inset: for the 2023 Pride week, a number TOCs branded trains in rainbow LGBT colours. On c2c set No. 357318 was given a rather neat train length railbow band, runnng full length at cantrail height and on the cab side pillars. the set is seen at Limehouse on 7 July 2023. **Antony Christie**

Left Middle Upper & Lower: The interior of the Class 357s use a grey seat moquette with bright pink trim, handles and grab poles. The Class 357/0 and 357/2 sets sport 2+3 seating (upper picture) and are cramped, with a difficulty to walk through the train between seats. A Passenger Information System (PIS) is installed and automated announcements provided. The Class 357/3s with Metro interiors are fitted with 2+2 seating (lower picture), with a much wider space to walk along the vehicle interior. Large stand-back areas are provided by door pockets and ceiling mounted hanging 'straps' are installed. Both: **CJM**

Left: The driving cab installed on the Class 357 'Electrostar' is based on the 'Turbostar' design using a wrap-round style. The driving position is located on the left side of the cab in the direction of travel. The design uses the combined power and brake controller, located on the left side. **CJM**

Left: Metro branding by door position on Class 375/3, also indication of 'Quiet Zone' which is not usually followed. **CJM**

118 Rolling Stock Review : 2023-2024

For history and pictures of this class, see Modern Locomotives Illustrated Issue 178

'Desiro' — Class 360

Class:	360
Number range:	360101-360121
Built by:	Siemens Transportation - Vienna, Austria, and Duewag, Germany
Type:	Desiro
Years introduced:	2002-2003
Max speed:	110mph (177km/h)
Formation:	DMSO(A)+PTSO+TSO+DMSO(B)
Vehicle numbers:	DMSO(A) - 65551-65571
	PTSO - 72551-72571
	TSO - 74551-74571
	DMSO(B) - 68551-68571
Vehicle length:	66ft 9in (20.34m)
Height:	12ft 1½in (3.95m)
Width:	9ft 2in (2.79m)
Weight:	Total - 168 tonnes
	DMSO(A) - 45 tonnes
	PTSO - 43 tonnes
	TSO - 35 tonnes
	DMSO(B) - 45 tonnes
Internal layout:	2+2 / 2+3
Seating:	Total 272S (+9TU)
	DMSO(A) - 67S
	PTSO - 60S (+9TU)
	TSO - 78S
	DMSO(A) - 67S
Gangway:	Within set only
Toilets:	PTSO - 1
Brake type:	Air (regenerative)
Bogie type:	SGP SF5000
Power collection:	25kV ac overhead
Traction motor type:	4 x 1TB2016 - 0GB02 three phase
Horsepower:	2,680hp (2,000kW)
Coupling type:	Outer - Dellner 12, Inner - Semi-auto
Multiple restriction:	Within sub-class
Door type:	Bi-parting sliding plug
Construction:	Aluminium
Operator:	East Midlands Railway
Owner:	Angel Trains

Above, Right & Below: Siemens original UK Desiro product was for 21 four-car sets for First Great Eastern in 2002-2003. They seated 16 first and 256 standard class passengers, for longer-distance outer-suburban services. They went off lease in late 2020 and transferred to East Midland Railway for their 'Connect' services from St Pancras international to Corby, they are now based at Bedford depot. The sets sport a mix of 2+2 and 2+3 seating and are formed with two driving motor cars (DMSO) and two un-powered intermediate TSOs, one of which carries a pantograph. The fleet have all been refurbished by Arlington Fleet Services, Eastleigh and now carry EMR aubergine, with grey branding and passenger doors. In the top view is the interior based on the 3+2 layout, this view shows the PTOS coach. Above is the Pantograph Trailer Standard Open (PTSO) showing the aubergine livery with the 2023 applied Luton Airport branding. Below, set No. 360104 is shown from its DMSO(B), with the full Luton Airport Express branding applied from summer 2023. Inset, detail of branding. All: **Antony Christie (2) / CJM (2)**

2023 Fleet

360101 EMR	360108 EMR	360115 EMR
360102 EMR	360109 EMR	360116 EMR
360103 EMR	360110 EMR	360117 EMR
360104 EMR	360111 EMR	360118 EMR
360105 EMR	360112 EMR	360119 EMR
360106 EMR	360113 EMR	360120 EMR
360107 EMR	360114 EMR	360121 EMR

Rolling Stock Review : 2023-2024

Class 373

For history and pictures of this class, see *Modern Locomotives Illustrated* Issue 227

Class:	373 (e300)
Number range:	373007-373016/373205-373230, 373999 (spare PC)
Built by:	GEC Alsthom in UK and France
Years introduced:	1992-1996
Max speed:	186mph (300km/h)
Formation:	DM+MS+TS+TS+TS+TBK+TF+TF+TBF
Set length:	1,291ft 8in (393.72m)
Vehicle length:	DM - 72ft 8in (22.15m) MS - 71ft 8in (21.84m) TS, TBK, TF, TBF - 61ft 4in (18.69m)
Height:	12ft 4½in (3.77m)
Width:	9ft 3in (2.82m)
Weight:	Total - 816.1 tonnes DM - 68.5 tonnes, MS - 44.6 tonnes TS - 28.1-29.7 tonnes, TBK - 31.1 tonnes TF - 29.6 tonnes, TBF - 39.4 tonnes
Internal layout:	2+1F, 2+2S
Seating [refurb]:	Total - 103F/276S (half train) DM - 0, MS - 52S, TS - 56S, TS - 56S, TS - 56S TS - 56S, TBK - 0, TF - 39F, TF - 39F, TBF - 25F
Gangway:	Within set and at half set inner end
Toilets:	MS - 2, TBF - 1, TF - 1, TS - 1 or 2
Brake type:	Air
Power collection:	25kV ac overhead, 1.5kV dc overhead *
Horsepower:	25kV ac operation - 16,400hp (12,240kW) 1.5kV dc operation - 7,368hp (5,000kW)
Coupling type:	Outer - Schaku 10, Inner - Bar
Multiple restriction:	Two half sets only
Door type:	Single-leaf sliding plug
Construction:	Steel
Operator:	Eurostar UK, SNCF, SNCB
Notes:	

* Sets 373209-10/29-30 are fitted with French 1,500V overhead equipment to allow South of France running

Above & Below: *Of the original Class 373 fleet, introduced to work passenger services through the Channel Tunnel in the early 1990s, a fleet of just 8 trains (16 half-sets) remain, following introduction of Class 374s. Operational sets carry the new corporate Eurostar livery, and are officially known as the e300 fleet. The Class 373s are based at Temple Mills, London and Le Landy, Paris. During the early 2020s pandemic the sets were stored, but returned to traffic as passenger numbers increased from 2022. An e300 seats 206 first and 552 standard class passengers. Although built for multi-voltage operation, sets are now only maintained for 25kV AC and 1,500V DC use. In the above image an end of half set TBF coach is seen. In the image below, travelling at its full 186mph (300km/h), French set No. 373230+373229 is seen on HS1 passing Nashenden, with the 13.13 Paris Nord to London St Pancras International.*
Antony Christie / CJM

Below: *Class 373 driving cab layout, one of the most complex of any train in the world.* **CJM**

Key Facts

- Original fleet built to operate passenger services through the Channel Tunnel from London to Paris and Brussels.
- Each train formed of two 10-car semi-articulated ½ sets, making these the longest trains to operate in the UK.
- Trains can be split for evacuation of the Channel Tunnel between the power car and train and in the middle of each formation.
- 373/0 owned by Eurostar UK, and Class 373/2 owned by French Railway (SNCF).
- Sets maintained at Temple Mills, London. One spare power car No. 3999 is available to cover power car shortages or maintenance.

2023 Fleet

373/0
373007	EUS
373008	EUS
373015	EUS
373016	EUS

373/2
373205	EUS
373206	EUS
373209	EUS
373210	EUS
373211	EUS
373212	EUS
373219	EUS
373220	EUS
373221	EUS
373222	EUS
373229	EUS
373230	EUS

Spare PC
373399	EUS

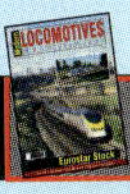

'Velaro' — Class 374

For history and pictures of this class, see Modern Locomotives Illustrated Issue 227

Class:	374 (e320)
Number range:	374001-374034 (half sets)
Introduced:	2015-2018*
Built by:	Siemens, Krefeld
Type:	Velaro
Formation:	Half train - DMF+TBF+MF+TS+TS+MS+TS+MSRB
Set length:	400m
Vehicle length:	DMF - 85ft 5in (26.03m), Intermediate - 81ft 3in (24.77m)
Height:	12ft 5in (3.80m)
Width:	9ft 4in (2.84m)
Seating:	Total - 110F, 336S DMF - 40F, TBF - 36F, MF - 34F, TS - 76S, TS - 76S, MS - 76S, TS - 76S, MSRB - 32
Internal layout:	2+1F, 2+2S
Gangway:	Within set and at half set inner end
Toilets:	TBF - 2, MFO - 1, TSO - 2, MSO - 2, MSORB - 2
Weight:	Total - 455 tonnes DMF - 58 tonnes, TBF - 59 tonnes, MFO - 59 tonnes, TSO - 53 tonnes, TSO - 53 tonnes, MSO - 58 tonnes, TSO - 57 tonnes, MSRB - 58 tonnes
Brake type:	Air, regenerative
Power collection:	25kV ac overhead, 1,500V dc overhead 3,000V dc overhead
Power rating:	ac operation - 8,000kW dc operation - 4,200kW
Max speed:	200mph (320km/h)
Coupling type:	Outer - Dellner 12, Inner - Bar
Multiple restriction:	Two half sets only
Door type:	Single-leaf sliding plug
Construction:	Aluminium
Operator/Operator:	Eurostar
Note:	* Sets 374001-002 built 2011 for design testing

Above & Below: Between 2015 and 2018 Eurostar took delivery of a fleet of 34 eight-vehicle half trains - forming 17 full length trains. Built by Siemens, they are part of their 'Velaro' platform and in the UK are classified 374 or referred to as e320 trains, reflecting their top speed in km/h. The tri-voltage sets are based at Temple Mills (Stratford) in the UK and operate the majority of Cross Channel services. The fleet is painted in the current Eurostar livery. Unlike the original Class 373 sets, these are not articulated with each vehicle having two bogies. Power pick-ups are mounted throughout the train. In the above view, car 99 70 3740118-2 a motor standard buffet. Below, sets Nos. 374026 and 374025 approach North Downs Tunnel in April 2022. **Antony Christie / CJM**

2023 Fleet

374001	EUS	374018	EUS
374002	EUS	374019	EUS
374003	EUS	374020	EUS
374004	EUS	374021	EUS
374005	EUS	374022	EUS
374006	EUS	374023	EUS
374007	EUS	374024	EUS
374008	EUS	374025	EUS
374009	EUS	374026	EUS
374010	EUS	374027	EUS
374011	EUS	374028	EUS
374012	EUS	374029	EUS
374013	EUS	374030	EUS
374014	EUS	374031	EUS
374015	EUS	374032	EUS
374016	EUS	374033	EUS
374017	EUS	374034	EUS

Key Facts

- Second generation Channel Tunnel stock, 34 half trains formed into 17 16-car sets.
- Allocated to Temple Mills depot, (Stratford) in the UK.
- Designed as part of the Siemens 'Velaro' family of high speed trains.
- Fleet able to operate on three power systems, 25kV AC, 1,500V and 3,000V DC.

Below: Class 374 (e320) driving cab layout. **Antony Christie**

Rolling Stock Review : 2023-2024

Class 375 'Electrostar'

For history and pictures of this class, see Modern Locomotives Illustrated Issue 194

Class:	375/3	375/6	375/7	375/8
Number range:	375301-375310	375601-375630	375701-375715	375801-375830
Built by:	Bombardier Derby	Adtranz/Bombardier Derby	Bombardier Derby	Bombardier Derby
Type:	Electrostar	Electrostar	Electrostar	Electrostar
Configuration:	Express	Express	Express	Express
Years introduced:	2001-2002	1999-2001	2001-2002	2003-2004
Max speed:	100mph (161km/h)	100mph (161km/h)	100mph (161km/h)	100mph (161km/h)
Formation:	DMSO+TSO+DMSO	DMSO(A)+MSO+PTSO+DMSO(B)	DMSO(A)+MSO+TSO+DMSO(B)	DMSO(A)+MSO+TSO+DMSO(B)
Vehicle numbers:	DMSO - 67921-67930 TSO - 74351-74360 DMSO - 67931-67940	DMSO(A) - 67801-67830 MSO - 74251-74280 PTSO - 74201-74230 DMSO(B) - 67851-67880	DMSO(A) - 67831-67845 MSO - 74281-74295 TSO - 74231-74245 DMSO(B) - 67881-67895	DMSO(A) - 73301-73330 MSO - 79001-79030 TSO - 78201-78230 DMSO(B) - 73701-73730
Vehicle length:	DMSO 66ft 11in (20.40m) TSO - 65ft 6in (19.99m)	DMSO - 66ft 11in (20.40m) MSO, PTSO - 65ft 6in (19.99m)	DMSO - 66ft 11in (20.40m) MSO, TSO - 65ft 6in (19.99m)	DMSO - 66ft 11in (20.40m) MSO, TSO - 65ft 6in (19.99m)
Height:	12ft 4in (3.78m)	12ft 4in (3.78m)	12ft 4in (3.78m)	12ft 4in (3.78m)
Width:	9ft 2in (2.80m)	9ft 2in (2.80m)	9ft 2in (2.80m)	9ft 2in (2.80m)
Weight:	Total - 124.1 tonnes DMSO - 43.8 tonnes TSO - 36.5 tonnes DMSO - 43.8 tonnes	Total - 173.6 tonnes DMSO(A) - 46.2 tonnes MSO - 40.5 tonnes PTSO - 40.7 tonnes DMSO(B) - 46.2 tonnes	Total - 158.1 tonnes DMSO(A) - 43.8 tonnes MSO - 36.4 tonnes TSO - 34.1 tonnes DMSO(B) - 43.8 tonnes	Total - 162.3 tonnes DMSO(A) - 43.3 tonnes MSO - 39.8 tonnes TSO - 35.9 tonnes DMSO(B) - 43.3 tonnes
Internal layout:	2+2	2+2	2+2	2+2
Seating:	Total - 176S DMSO - 60S TSO - 56S DMCO(B) - 60S	Total - 242S DMSO(A) - 60S MSO - 66S PTSO - 56S DMSO(B) - 60S	Total - 242S DMSO(A) - 60S MSO - 66S TSO - 56S DMSO(B) - 60S	Total - 242S DMSO(A) - 60S MSO - 66S TSO - 52S DMSO(B) - 64S
Gangway:	Throughout	Throughout	Throughout	Throughout
Toilets:	TSO - 1	PTSO, MSO - 1	TSO, MSO - 1	TSO, MSO - 1
Brake type:	Air (regenerative)	Air (regenerative)	Air (regenerative)	Air (regenerative)
Bogie type:	Power - Adtranz P3-25 Trailer - Adtranz T3-25	Power - Adtranz P3-25 Trailer - Adtranz T3-25	Power - Adtranz P3-25 Trailer - Adtranz T3-25	Power - Bombardier P3-25 Trailer - Bombardier T3-25
Power collection:	750V dc third rail	750V dc third rail, and 25kV ac overhead	750V dc third rail	750V dc third rail
Traction motor type:	4 x Adtranz	6 x Adtranz	6 x Adtranz	6 x Adtranz
Horsepower:	1,341hp (1,000kW)	2,012hp (1,500kW)	2,012hp (1,500kW)	2,012hp (1,500kW)
Coupling type:	Outer - Dellner 12, Inner - Bar	Outer - Dellner 12, Inner - Bar	Outer - Dellner 12, Inner - Bar	Outer - Dellner 12, Inner - Bar
Multiple restriction:	Class 375-378	Class 375-378	Class 375-378	Class 375-378
Door type:	Bi-parting sliding plug	Bi-parting sliding plug	Bi-parting sliding plug	Bi-parting sliding plug
Construction:	Aluminium, steel cabs	Aluminium, steel cabs	Aluminium, steel cabs	Aluminium, steel cabs
Operator:	Southeastern	Southeastern	Southeastern	Southeastern
Owner:	Eversholt Leasing	Eversholt Leasing	Eversholt Leasing	Eversholt Leasing
Note:	All sub classes to be fitted with Bombardier ETCS and EBICab 2000 ATP			

Above: Slam-door replacement stock for the South Eastern main line, saw a fleet of three and four-coach Adtranz/Bombardier 'Electrostar' units introduced, classified as 375. In 2023, five different sub-classes exist, all with slight variations. The 10 three-car sets, of Class 375/3, are used for branch line duties, or to strengthen longer-distance services. Set No. 375309 is recorded passing New Beckenham with empty stock for New Cross. These sets, in common with all Class 375s are now standard class only their first class section being removed in early 2023. Sets are painted in Southeastern dark blue with contrasting light blue passenger doors. **CJM**

'Electrostar' — Class 375

2023 Fleet

375/9
375901-375927
Bombardier Derby
Electrostar
Outer suburban
2003-2004
100mph (161km/h)
DMSO(A)+MSO+
 TSO+DMSO(B)
DMSO(A) - 73331-73357
MSO - 79031-79057
TSO - 79061-79087
DMSO(B) - 73731-73757
DMSO - 66ft 11in (20.40m)
MSO, TSO - 65ft 6in (19.99m)
12ft 4in (3.78m)
9ft 2in (2.80m)
Total - 161.9 tonnes
DMSO(A) - 43.4 tonnes
MSO - 39.3 tonnes
TSO - 35.8 tonnes
DMSO(B) - 43.4 tonnes
2+2, 2+3
Total - 274S
DMSO(A) - 71S
MSO - 73S
TSO - 59S
DMSO(B) - 71S
Throughout
TSO, MSO - 1
Air (regenerative)
Power - Bombardier P3-25
Trailer - Bombardier T3-25
750V dc third rail

6 x Adtranz
2,012hp (1,500kW)
Outer - Dellner 12,
Inner - Bar
Class 375-378
Bi-parting sliding plug
Aluminium, steel cabs
Southeastern
Eversholt Leasing

375/3		375/8	
375301	SET	375801	SET
375302	SET	375802	SET
375303	SET	375803	SET
375304	SET	375804	SET
375305	SET	375805	SET
375306	SET	375806	SET
375307	SET	375807	SET
375308	SET	375808	SET
375309	SET	375809	SET
375310	SET	375810	SET
		375811	SET
375/6		375812	SET
375601	SET	375813	SET
375602	SET	375814	SET
375603	SET	375815	SET
375604	SET	375816	SET
375605	SET	375817	SET
375606	SET	375818	SET
375607	SET	375819	SET
375608	SET	375820	SET
375609	SET	375821	SET
375610	SET	375822	SET
375611	SET	375823	SET
375612	SET	375824	SET
375613	SET	375825	SET
375614	SET	375826	SET
375615	SET	375827	SET
375616	SET	375828	SET
375617	SET	375829	SET
375618	SET	375830	SET
375619	SET		
375620	SET	**375/9**	
375621	SET	375901	SET
375622	SET	375902	SET
375623	SET	375903	SET
375624	SET	375904	SET
375625	SET	375905	SET
375626	SET	375906	SET
375627	SET	375907	SET
375628	SET	375908	SET
375629	SET	375909	SET
375630	SET	375910	SET
		375911	SET
375/7		375912	SET
375701	SET	375913	SET
375702	SET	375914	SET
375703	SET	375915	SET
375704	SET	375916	SET
375705	SET	375917	SET
375706	SET	375918	SET
375707	SET	375919	SET
375708	SET	375920	SET
375709	SET	375921	SET
375710	SET	375922	SET
375711	SET	375923	SET
375712	SET	375924	SET
375713	SET	375925	SET
375714	SET	375926	SET
375715	SET	375927	SET

Top: The Class 375s are fitted with a universal style toilet, suitable for disabled passengers. It is located in the Trailer Standard coach and the area is recognisable by a deep red cant rail stripe and window surround, plus a disabled sign on the bodyside, seen at the far end in this view. **CJM**

Above: Until early 2023, the intermediate motor coach of four-car Class 375/6, 375/7 and 375/8s was a Motor Composite with first class seating at one end for 16. However from early 2023 first class seating was withdrawn and sets have had the yellow first class markings removed, the seats remain the same. MS No. 74281 from set 375701 is seen from the former first class end. **Antony Christie**

Below: The Class 375/6 sub-class consists of 30 units, these emerged in 1999-2001 and are equipped for dual AC/DC operation, they carry standard Southeastern dark blue livery, off-set by powder blue doors. They have the three light front end light cluster arrangement. With its DMSO(B) nearest the camera, set No. 375627 is seen passing St Johns in south-east London on 25 April 2023. Above, is a Southeastern Class 707 'CityBeam' set. **Antony Christie**

Key Facts

- Replacement stock for Southeastern slam door fleet post privatisation.
- Built by Adtranz/Bombardier as part of the 'Electrostar' platform, built alongside the SouthCentral fleet, which was later reclassified as 377.
- Five sub-classes exist, the first to enter service were 375/6s fitted with dual AC/DC power equipment. All others units are DC power collection only.
- All sets refurbished and carry Southeastern blue livery. First class seating was removed from all sets in January 2023.

Rolling Stock Review : 2023-2024

Class 375   'Electrostar'

Top: *A fleet of 15 Class 375/7 sets were introduced in 2001-2002, at the time designated as 'Express' units. By 2020 all had been refurbished, with originally the first class being moved to an intermediate Motor Composite Open (MCO) vehicle, first class being removed from January 2023. No. 375707 is seen at Waterloo East, showing the three section lighting clusters fitted to this sub-class.* **Antony Christie**

Above: *In 2003-2004, 30 Class 375/8 sets were introduced. These sport the later design of two-section light cluster end, with a larger headlight and a joint marker/tail light. Set No. 375828 is seen at Waterloo East, with its disabled area at the lead end of the second coach.* **Antony Christie**

Left: *Class 375/8 interior, showing the layout in the DMSO vehicle viewed from the cab end. Seats are in the 2+2 style with the majority in the group layout around fixed tables. Luggage space is provided between seat backs and on above seat shelves.* **Antony Christie**

'Electrostar' — Class 375

Above: *The final batch of 27 sets were classified as 375/9, these were introduced in 2003-2004, and deemed 'outer suburban' sets, with increased seating. After refurbishment this sub-class retained first class in both driving vehicles until first was withdrawn in January 2023. The Trailer Standard Open (TSO), houses a universal access toilet and the disabled area. When originally constructed, the Class 375s were fitted with Tightlock couplings, these were exchanged for the Dellner type around 2003. By the time the '375/9s' were introduced, Dellner couplings were workshop fitted. Set No. 375906 is seen passing St Johns.* **Antony Christie**

Right Middle: *A few '375s' are named using 'sticker' plates. This plate is fitted to No. 375823.* **Antony Christie**

Right: *Electrostar driving cab layout. This basic design applies to all gangway fitted Electrostar sets, with some equipment in different positions to suit operational requirements. 1: Cab light switch, 2: Door controls left side, 3: Door close button, 4: Couple button, 5: Power/brake controller, 6: Master switch, 7: Drivers key socket, 8: AWS display 9: Line light, 10: Drivers reminder appliance, 11: Sand button, 12: Train fault warning, 13: Speedometer, 14: TPWS panel, 15: Main reservoir and brake cylinder gauge, 16: Speed set equipment, 17: Safety systems isolation warning, 18: AWS reset button, 19: Right side door controls, 20: Door interlock light, 21: Hazard warning button, 22: Emergency brake plunger, 23: Uncouple button (under flap), 24: Horn valve, 25: Windscreen wash/wipe controls.* **CJM**

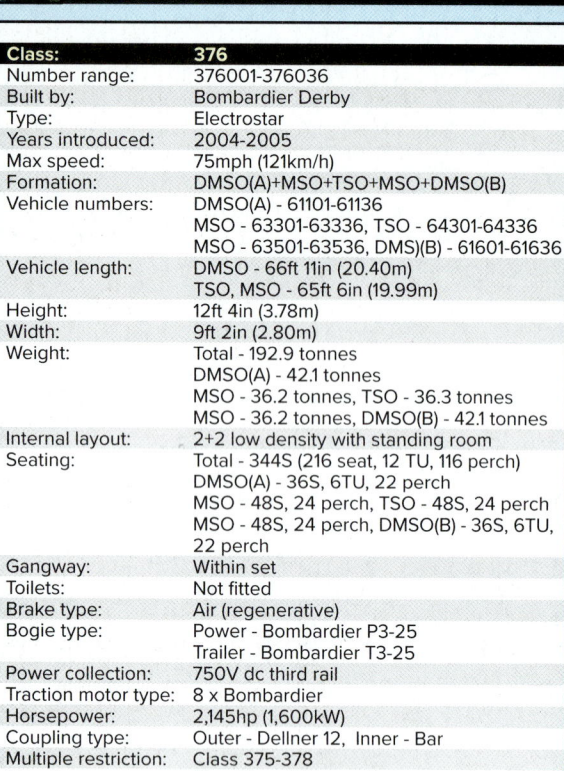

Class 376 'Electrostar'

Class:	376
Number range:	376001-376036
Built by:	Bombardier Derby
Type:	Electrostar
Years introduced:	2004-2005
Max speed:	75mph (121km/h)
Formation:	DMSO(A)+MSO+TSO+MSO+DMSO(B)
Vehicle numbers:	DMSO(A) - 61101-61136
	MSO - 63301-63336, TSO - 64301-64336
	MSO - 63501-63536, DMS)(B) - 61601-61636
Vehicle length:	DMSO - 66ft 11in (20.40m)
	TSO, MSO - 65ft 6in (19.99m)
Height:	12ft 4in (3.78m)
Width:	9ft 2in (2.80m)
Weight:	Total - 192.9 tonnes
	DMSO(A) - 42.1 tonnes
	MSO - 36.2 tonnes, TSO - 36.3 tonnes
	MSO - 36.2 tonnes, DMSO(B) - 42.1 tonnes
Internal layout:	2+2 low density with standing room
Seating:	Total - 344S (216 seat, 12 TU, 116 perch)
	DMSO(A) - 36S, 6TU, 22 perch
	MSO - 48S, 24 perch, TSO - 48S, 24 perch
	MSO - 48S, 24 perch, DMSO(B) - 36S, 6TU, 22 perch
Gangway:	Within set
Toilets:	Not fitted
Brake type:	Air (regenerative)
Bogie type:	Power - Bombardier P3-25
	Trailer - Bombardier T3-25
Power collection:	750V dc third rail
Traction motor type:	8 x Bombardier
Horsepower:	2,145hp (1,600kW)
Coupling type:	Outer - Dellner 12, Inner - Bar
Multiple restriction:	Class 375-378
Door type:	Bi-parting sliding pocket
Construction:	Aluminium with steel cabs
Operator:	Southeastern
Owner:	Eversholt Leasing
Note:	To be fitted with Bombardier ETCS and EBICab 2000 ATP

Right: Class 376 'Electrostar' front end. 1: Air vent, 2: High level marker light, 3: Destination indicator, 4: Headlight, 5: Combined white marker and red tail light, 6: Lamp bracket, 7: Dellner-12 coupling, 8: Warning horns (in coupling pocket), 9: Coupling electrical connections. **CJM**

Left: Class 376 driving cab. The design, not incorporating a gangway, permitted a much improved and more spacious driving cab, giving better visibility, with equipment spread out over two-thirds of the body width. Equipment is based on the standard 'Electrostar' design, with which the sets are capable of operating in multiple (dc sets only). **CJM**

2023 Fleet

376001	SET
376002	SET
376003	SET
376004	SET
376005	SET
376006	SET
376007	SET
376008	SET
376009	SET
376010	SET
376011	SET
376012	SET
376013	SET
376014	SET
376015	SET
376016	SET
376017	SET
376018	SET
376019	SET
376020	SET
376021	SET
376022	SET
376023	SET
376024	SET
376025	SET
376026	SET
376027	SET
376028	SET
376029	SET
376030	SET
376031	SET
376032	SET
376033	SET
376034	SET
376035	SET
376036	SET

Above & Inset: In 2004-2005, a fleet of 36 five-car low-density, high-capacity 'Metro' units were introduced on Southeastern, for inner-suburban routes. The non-gangwayed sliding door sets, have reduced seating, but increased standing and 'perch' areas, catering for the high level of passenger demand in south-east London. The sets are basic, they do not have air conditioning, toilets or plug doors, providing a 'tram' like travelling experience. Four of the five vehicles are powered, the middle coach, being a Trailer Standard Open (TSO). The fleet is based at Slade Green and deployed on Metro services. They carry Southeastern white livery with a grey base band, off-set by yellow passenger doors. Above, set No. 376035 is seen at Waterloo East. The inset right, shows the spacious interior, with lots of standing space. Both: **CJM**

Key Facts

- Part of the Bombardier 'Electrostar' family.
- Five-car high-capacity no-frills sets, built to cope with traffic on Kent suburban network.
- Fitted with bi-parting sliding pocket doors, rather than the plug type.
- Gangwayed only within set, full width driving cab, no air conditioning or toilets.

Left: The outer intermediate vehicles of each set are Motor Standard Open (MSO) with seating for 48, plus 24 'perch' seats. The pocket sliding doors on this design, were installed to reduce build costs, they are quite noticeable when viewed against the usual sliding plug doors of other 'Electrostar' products, MS No. 63323 from set 376023 is shown. **CJM**

'Electrostar' — Class 377

For history and pictures of this class, see Modern Locomotives Illustrated Issue 194

2023 Fleet

Above & Inset: The 'Electrostar' product for the Southern franchise, now part of GTR, emerged in 2001-2002 with a batch of 28 three-car Class 377/3s, this was followed by several follow-on orders. The Class 377/1 sub-class, delivered in 2002-2004 covers 64 four-car sets. These seat 24 first and between 210-242 standard class passengers (depending on layout). '377/1s' sport three section light clusters. The first class area is located in the DMCO vehicles. Set No. 377122 is illustrated at Forest Hill. The inset image shows Motor Standard Open Lavatory (MSOL) No. 77129 from set 377129. Both: **CJM**

Below: A sub-class of 15 dual-voltage AC/DC '377/2s' emerged in 2002-2003, able to operate Cross-London services from Clapham Junction to the West Coast Main Line via Kensington. These sets seat 24 first and 222 standard class passengers, with 2+2 seating in the driving cars and 2+3 in intermediate vehicles. First class seating is inward of the driving cab. Set No. 377205 is seen passing South Kenton on the West Coast main line with a Watford-Croydon service. '377/2s' sport a large headlight and combined marker/tail light. **CJM**

Key Facts

- The '377' is largest fleet of Adtranz/Bombardier 'Electrostar' stock, operated by GTR Southern and Southeastern.
- Seven sub-classes exist, all but 377/5s operated by GTR.
- Class 377/2, 377/5 and 377/7s are fitted for dual voltage AC/DC operation.
- Class 377/3 are three-car sets.
- Class 377/6 and 377/7 are high-capacity 'Metro' style units.
- Members of Class 377/6 and 377/7 have seperate window frames rather than ribbon glazing.
- As delivered, Class 377/1-377/4 were classified as Class 375, fitted with Tightlock couplings, reclassified when Dellner couplings installed.

Left: Class 377 front end. 1: High level marker light, 2: Gangway doors, 3: Headlight, 4: Combined white marker / red tail light, 5: Lamp bracket, 6: Track light, 7: Dellner coupling, 8: Electrical connection box, 9: Air horns in coupling pocket. **CJM**

377/1

377101	GTR	377306	GTR	377464	GTR
377102	GTR	377307	GTR	377465	GTR
377103	GTR	377308	GTR	377466	GTR
377104	GTR	377309	GTR	377467	GTR
377105	GTR	377310	GTR	377468	GTR
377106	GTR	377311	GTR	377469	GTR
377107	GTR	377312	GTR	377470	GTR
377108	GTR	377313	GTR	377471	GTR
377109	GTR	377314	GTR	377472	GTR
377110	GTR	377315	GTR	377473	GTR
377111	GTR	377316	GTR	377474	GTR
377112	GTR	377317	GTR	377475	GTR
377113	GTR	377318	GTR		
377114	GTR	377319	GTR	**377/5**	
377115	GTR	377320	GTR	377501	SET
377116	GTR	377321	GTR	377502	SET
377117	GTR	377322	GTR	377503	SET
377118	GTR	377323	GTR	377504	SET
377119	GTR	377324	GTR	377505	SET
377120	GTR	377325	GTR	377506	SET
377121	GTR	377326	GTR	377507	SET
377122	GTR	377327	GTR	377508	SET
377123	GTR	377328	GTR	377509	SET
377124	GTR			377510	SET
377125	GTR	**377/4**		377511	SET
377126	GTR	377401	GTR	377512	SET
377127	GTR	377402	GTR	377513	SET
377128	GTR	377403	GTR	377514	SET
377129	GTR	377404	GTR	377515	SET
377130	GTR	377405	GTR	377516	SET
377131	GTR	377406	GTR	377517	SET
377132	GTR	377407	GTR	377518	SET
377133	GTR	377408	GTR	377519	SET
377134	GTR	377409	GTR	377520	SET
377135	GTR	377410	GTR	377521	SET
377136	GTR	377411	GTR	377522	SET
377137	GTR	377412	GTR	377523	SET
377138	GTR	377413	GTR		
377139	GTR	377414	GTR	**377/6**	
377140	GTR	377415	GTR	377601	GTR
377141	GTR	377416	GTR	377602	GTR
377142	GTR	377417	GTR	377603	GTR
377143	GTR	377418	GTR	377604	GTR
377144	GTR	377419	GTR	377605	GTR
377145	GTR	377420	GTR	377606	GTR
377146	GTR	377421	GTR	377607	GTR
377147	GTR	377422	GTR	377608	GTR
377148	GTR	377423	GTR	377609	GTR
377149	GTR	377424	GTR	377610	GTR
377150	GTR	377425	GTR	377611	GTR
377151	GTR	377426	GTR	377612	GTR
377152	GTR	377427	GTR	377613	GTR
377153	GTR	377428	GTR	377614	GTR
377154	GTR	377429	GTR	377615	GTR
377155	GTR	377430	GTR	377616	GTR
377156	GTR	377431	GTR	377617	GTR
377157	GTR	377432	GTR	377618	GTR
377158	GTR	377433	GTR	377619	GTR
377159	GTR	377434	GTR	377620	GTR
377160	GTR	377435	GTR	377621	GTR
377161	GTR	377436	GTR	377622	GTR
377162	GTR	377437	GTR	377623	GTR
377163	GTR	377438	GTR	377624	GTR
377164	GTR	377439	GTR	377625	GTR
		377440	GTR	377626	GTR
377/2		377441	GTR		
377201	GTR	377442	GTR	**377/7**	
377202	GTR	377443	GTR	377701	GTR
377203	GTR	377444	GTR	377702	GTR
377204	GTR	377445	GTR	377703	GTR
377205	GTR	377446	GTR	377704	GTR
377206	GTR	377447	GTR	377705	GTR
377207	GTR	377448	GTR	377706	GTR
377208	GTR	377449	GTR	377707	GTR
377209	GTR	377450	GTR	377708	GTR
377210	GTR	377451	GTR		
377211	GTR	377452	GTR		
377212	GTR	377453	GTR		
377213	GTR	377454	GTR		
377214	GTR	377455	GTR		
377215	GTR	377456	GTR		
		377457	GTR		
377/3		377458	GTR		
377301	GTR	377459	GTR		
377302	GTR	377460	GTR		
377303	GTR	377461	GTR		
377304	GTR	377462	GTR		
377305	GTR	377463	GTR		

Class 377 — 'Electrostar'

Class:	377/1	377/2	377/3	377/4
Number range:	377101-377164	377201-377215	377301-377328	377401-377475
Original numbers:	-	-	375311-375338	-
Built by:	Bombardier Derby	Bombardier Derby	Bombardier Derby	Bombardier Derby
Type:	Electrostar	Electrostar	Electrostar	Electrostar
Years introduced:	2002-2004	2002-2003	2001-2002	2004-2005
Max speed:	100mph (161km/h)	100mph (161km/h)	100mph (161km/h)	100mph (161km/h)
Formation:	DMCO(A)+MSOL+TSOL+DMCO(B)	DMCO(A)+MSOL+TSOL+DMCO(B)	DMSO+TSOL+DMCO	DMCO(A)+MSOL+TSOL+DMCO(B)
Vehicle numbers:	DMCO(A) - 78501-78564 MSOL - 77101-77164 TSOL - 78901-78964 DMCO(B) - 78701-78764	DMCO(A) - 78571-78585 MSOL - 77171-77185 PTSOL - 78971-78985 DMCO(B) - 78771-78785	DMSO - 68201-68228 TSOL - 74801-74828 DMCO - 68401-68428	DMCO(A) - 73401-73475 MSOL - 78801-78875 TSOL - 78601-78675 DMCO(B) - 73801-73875
Vehicle length:	DMCO - 66ft 11in (20.40m) MSOL, TSOL - 65ft 6in (19.99m)	DMCO - 66ft 11in (20.40m) MSOL, TSOL - 65ft 6in (19.99m)	DMSO/DMCO - 66ft 11in (20.40m) TSOL - 65ft 6in (19.99m)	DMCO - 66ft 11in (20.40m) MSOL, TSOL - 65ft 6in (19.99m)
Height:	12ft 4in (3.78m)	12ft 4in (3.78m)	12ft 4in (3.78m)	12ft 4in (3.78m)
Width:	9ft 2in (2.80m)	9ft 2in (2.80m)	9ft 2in (2.80m)	9ft 2in (2.80m)
Weight:	Total - 162.6 tonnes DMCO(A) - 44.8 tonnes MSOL - 39 tonnes TSOL - 35.4 tonnes DMCO(B) - 43.4 tonnes	Total - 168.3 tonnes DMCO(A) - 44.2 tonnes MSOL - 39.8 tonnes PTSOL - 40.1 tonnes DMCO(B) - 44.2 tonnes	Total - 122.4 tonnes DMCO(A) - 43.5 tonnes TSOL - 35.4 tonnes DMCO(B) - 43.5 tonnes	Total - 160.9 tonnes DMCO(A) - 43.1 tonnes MSOL - 39.3 tonnes TSOL - 35.3 tonnes DMCO(B) - 43.2 tonnes
Internal layout:	2+2F, and 2+2 & 2+3S	2+2F, and 2+2 & 2+3S	2+2F, and 2+2 & 2+3S	2+2F, and 2+2 & 2+3S
Seating:	Total - 24F/210 or 222 or 242S DMCO(A) - 12F/48-56S MSOL - 62S - 70S TSOL - 52S - 62S DMCO(B) - 12F/48-56S	Total - 24F/222S DMCO(A) - 12F/48S MSOL - 69S PTSOL - 57S DMCO(B) - 12F/48S	Total - 24F/163S DMSO - 60S TSOL - 56S DMCO - 12F/48S	Total - 20F/221S DMCO(A) - 10F/48S MSOL - 69S TSOL - 56S DMCO(B) - 10F/48S
Gangway:	Throughout	Throughout	Throughout	Throughout
Toilets:	MSOL, TSOL - 1	MSOL, PTSOL - 1	TSOL - 1	MSOL, TSOL - 1
Brake type:	Air (regenerative)	Air (regenerative)	Air (regenerative)	Air (regenerative)
Bogie type:	Power - Bombardier P3-25 Trailer - Bombardier T3-25	Power - Bombardier P3-25 Trailer - Bombardier T3-25	Power - Bombardier P3-25 Trailer - Bombardier T3-25	Power - Bombardier P3-25 Trailer - Bombardier T3-25
Power collection:	750V dc third rail	750V dc third rail, and 25kV ac overhead	750V dc third rail	750V dc third rail
Traction motor type:	6 x Bombardier	6 x Bombardier	4 x Bombardier	6 x Bombardier
Horsepower:	2,012hp (1,500kW)	2,012hp (1,500kW)	1,341hp (1,000kW)	2,012hp (1,500kW)
Coupling type:	Outer - Dellner 12, Inner - Bar	Outer - Dellner 12, Inner - Bar	Outer - Dellner 12, Inner - Bar	Outer - Dellner 12, Inner - Bar
Multiple restriction:	Class 375-378	Class 375-378	Class 375-378	Class 375-378
Door type:	Bi-parting sliding plug	Bi-parting sliding plug	Bi-parting sliding plug	Bi-parting sliding plug
Body structure:	Aluminium, steel ends	Aluminium, steel ends	Aluminium, steel ends	Aluminium, steel ends
Operator:	GTR	GTR	GTR	GTR
Owner:	Porterbrook Leasing	Porterbrook Leasing	Porterbrook Leasing	Porterbrook Leasing
Notes:	When delivered 377/1-377/4 were classified as Class 375 and fitted with Tightlock couplers All sub-classes to be fitted with Bombardier ETCS and EBICab 2000 ATP			

Left: The first fleet of 'Electrostars' to emerge in 2001-2002 was 28 three-car Class 375/3 sets, fitted with 2+2 seating and used on lower patronage routes, branch lines or give the ability to form 3, 6, 7, 9 or 11 car trains. These are third rail only units and incorporate the three light cluster front end style. Later re-classified as Class 377 after the fitting of Dellner couplings, the GTR sets are based at Selhurst depot. No. 377302 passes Forest Hill. These sets are scheduled to take over the Coastway services from Brighton. **CJM**

Left Lower & Inset: The most numerous sub-class of '377' covers the 75 four-car Class 377/4s, built in 2004-2005. These are DC only sets and have the revised front light cluster, with a large headlight and small combined marker/tail light. They seat just 10 first class passengers in each DTCO coach, with 48 standard class seats. The DTCOs have standard class seating in the 2+2 style, intermediate vehicles use the 2+3 style. A standard toilet compartment is located in the MSOL coach, while a universal toilet compartment is installed in the Trailer Standard Open Lavatory (TSOL), this vehicle also has wheelchair spaces. Set No. 377428 is seen passing Cooden Beach. The 377/4s are currently being refreshed, including the fitting of a track light on the front fairing and electronic display boards in passenger saloons. Inset shows the TSOL with a pantograph well. Both: **CJM**

'Electrostar' — Class 377

377/5	377/6	377/7
377501-377523	377601-377626	377701-377708
-		
Bombardier Derby	Bombardier Derby	Bombardier Derby
Electrostar	Electrostar	Electrostar
2008-2009	2013-2014	2013-2014
100mph (161km/h)	100mph (161km/h)	100mph (161km/h)
DMCO+MSOL+ PTSOL+DMSO	DMCO+MSOL+PTSOL+ MSO+DMSO	DMCO+MSOL+PTSOL+ MSO+DMSO
DMCO - 73501-73523	DMCO - 70101-70126	DMCO - 65201-65208
MSOL - 75901-75923	MSOL - 70201-70226	MSOL - 70601-70608
PTSOL - 74901-74923	PTSOL - 70301-70326	PTSOL - 65601-65608
DMSO - 73601-73623	MSO - 70401-70426	MSO - 70701-70708
	DMSO - 70510-70526	DMSO - 65401-65408
DMCO/DMSO - 66ft 11in (20.40m)	DMCO, DMSO - 66ft 11in (20.40m)	DMSO - 66ft 11in (20.40m)
MSOL, PTSOL - 65ft 6in (19.99m)	MSOL, TSOL - 65ft 6in (19.99m)	MSO,L PTSOL - 65ft 6in (19.99m)
12ft 4in (3.78m)	12ft 4in (3.78m)	12ft 4in (3.78m)
9ft 2in (2.80m)	9ft 2in (2.80m)	9ft 2in (2.80m)
Total - 168.9 tonnes	Total - 204.3 tonnes	Total - 212.3 tonnes
DMCO - 43.1 tonnes	DMCO - 44.7 tonnes	DMCO - 45.6 tonnes
MSOL - 40.3 tonnes	MSOL - 38.8 tonnes	MSOL - 41 tonnes
PTSOL - 40.6 tonnes	PTSOL - 37.8 tonnes	PTSOL - 40.9 tonnes
DMSO - 44.9 tonnes	MSO - 38.3 tonnes	MSO - 39.6 tonnes
	DMSO - 44.7 tonnes	DMSO - 45.2 tonnes
2+2F, and 2+2 & 2+3S	2+2F, 2+2S	2+2F, and 2+2S
Total - 10F/228S ·	Total - 24F/274S	Total - 24F/274S
DMCO(A) - 10F/48S	DMCO - 24F/36S	DMSO - 24F/36S
MSOL - 69S	MSOL - 64S	MSOL - 64S
PTSOL - 53S	PTSO - 46S, MSO - 66S	PTSO - 46S, MSO - 66S
DMSO - 5S	DMSO - 62S	DMSO - 62S
Throughout	Throughout	Throughout
MSOL, PTSOL - 1	MSOL, PTSOL - 1	MSOL, PTSOL - 1
Air (regenerative)	Air (regenerative)	Air (regenerative)
Power - Bombardier P3-25	Power - Bombardier P3-25	Power - Bombardier P3-25
Trailer - Bombardier T3-25	Trailer - Bombardier T3-25	Trailer - Bombardier T3-25
750V dc third rail	750V dc third rail	750V dc third rail
25kV ac overhead		25kV ac overhead
6 x Bombardier	8 x Bombardier	8 x Bombardier
2,012hp (1,500kW)	2,680hp (2,000kW)	2,680hp (2,000kW)
Outer - Dellner 12,	Outer - Dellner 12,	Outer - Dellner 12,
Inner - Bar	Inner - Bar	Inner - Bar
Class 375-378	Class 375-378	Class 375-378
Bi-parting sliding plug	Bi-parting sliding plug	Bi-parting sliding plug
Aluminium, steel ends	Aluminium, steel ends	Aluminium, steel ends
South Eastern	GTR	GTR
Porterbrook Leasing	Porterbrook Leasing	Porterbrook Leasing

Top & Above: Class 377 interior, the upper view shows the 2+3 style used in the standard class area. Seating on the three-seat side is very cramped. The lower image shows the first class 2+2 layout located just inward of the driving cab. This area offers no better seating than standard class, the only additions are hinged armrests and paper antimacassars. Both: **Antony Christie**

Left: In 2008-2009, a further batch of '377s' were built, when 23 dual-voltage Class 377/5s were delivered from Derby Litchurch Lane, originally ordered for Southern, the fleet was sub-leased to Thameslink, later replaced by Class 387s. After a short spell with GTR Southern, the sets were transferred to the Southeastern franchise and now sport Southeastern blue livery. The first class accommodation in one driving car has now been declassified. No. 377508 is seen at Waterloo East. **Antony Christie**

Right: In 2013-2014 the sub-classes of 377/6 and 377/7 were introduced, by this time the 'Electrostar' body design had changed to reduce costs, with the original ribbon glazing replaced by individual windows with frames. The 377/6 and 377/7 sets are virtually identical, five-car units for suburban 'Metro' use. The '377/6s' are DC only, while the '377/7s' are AC/DC, allowing operating from south to north London by way of Kensington Olympia. First class is provided for 24 in the DMCO vehicle, located at the inner end. The Trailer Standard Open (TSO), located in the middle of the formation, houses a universal toilet compartment. One of the Motor Standard Open (MSO) vehicles also houses a standard toilet. Class 377/6 No. 377617 passes Honor Oak Park with its DMCO nearest the camera. **CJM**

Rolling Stock Review : 2023-2024

Class 378   'Capitalstar'

Class:	378/1	378/2
Number range:	378135-378154	378201-378234, 378255-378257
Original number range:	-	378001-378024
Built by:	Bombardier Derby	Bombardier Derby
Type:	Capitalstar	Capitalstar
Years introduced:	2009-2010	2009-2011
Max speed:	75mph (121km/h)	75mph (121km/h)
Formation:	DMSO(A)+MSO(A)+TSO+MSO(B)+DMSO(B)	DMSO(A)+MSO(A)+PTSO+MSO(B)+DMSO(B)
Vehicle numbers:	DMSO(A) - 38035-38054 MSO - 38235-38254 TSO - 38335-38354 MSO(B) - 38435-38545 DMSO(B) - 38135-38154	DMSO(A) - 38001-38034, 38055-38057 MSO - 38201-38234, 38255-38257 PTSO - 38301-38334, 38355-38357 MSO(B) - 38401-38434, 38455-38457 DMSO(B) - 38101-38134, 38155-38157
Vehicle length:	DMSO - 67ft 2in (20.46m) MSO/TSO - 66ft 1in (20.14m)	DMSO - 67ft 2in (20.46m) MSO/PTSO - 66ft 1in (20.14m)
Height:	12ft 4½in (3.77m)	12ft 4½in (3.77m)
Width:	9ft 2in (2.80m)	9ft 2in (2.80m)
Weight:	Total - 199.6 tonnes DMSO(A) - 43.1 tonnes, MSO(A) - 39.3 tonnes TSO - 34.3 tonnes, MSO(B) - 40.2 tonnes DMSO(B) - 42.7 tonnes	Total - 205.7 tonnes DMSO(A) - 43.4 tonnes, MSO(A) - 39.6 tonnes PTSO - 39.2 tonnes, MSO(B) - 40.4 tonnes DMSO(B) - 43.1 tonnes
Internal layout:	Longitudinal	Longitudinal
Seating:	Total - 186S DMSO(A) - 36S, MSO(A) - 40S TSO - 34S, MSO(B) - 40S, DMSO(B) - 36S	Total - 186S DMSO(A) - 36S, MSO(A) - 40S PTSO - 34S, MSO(B) - 40, DMSO(B) - 36S
Gangway:	Within set, emergency end door	Within set, emergency end door
Toilets:	None	None
Brake type:	Air (Regenerative)	Air (Regenerative)
Bogie type:	Powered - Bombardier P3-25 Trailer - Bombardier T3-25	Powered - Bombardier P3-25 Trailer - Bombardier T3-25
Power collection:	750V dc third rail	25kV ac overhead, 750V dc third rail
Traction motor type:	8 x 200kW Bombardier asynchronous	8 x 200kW Bombardier asynchronous
Horsepower:	2,145hp (1,600kW)	2,145hp (1,600kW)
Coupling type:	Outer - Dellner 12, Inner - Bar	Outer - Dellner 12, Inner - Bar
Multiple restriction:	Class 375-378	Class 375-378
Door type:	Bi-parting sliding	Bi-parting sliding
Construction:	Aluminium	Aluminium
Operator:	London Overground (TfL)	London Overground (TfL)
Owner:	QW Rail Leasing	QW Rail Leasing
Note:	All sub-classes to be fitted with Bombardier ETCS and EBICab 2000 ATP	

For history and pictures of this class, see Modern Locomotives Illustrated Issue 194

Below: Introduced specifically for London Overground 'Metro' operations, the Class 378 Bombardier 'Capitalstar' is part of the 'Electrostar' family. The product is similar to the Class 376, with wide sliding pocket doors. The main difference, is that the '378' has only longitudinal seating, similar to that used on London Underground stock. This configuration gives more standing room to cut down overcrowding and reduced station dwell times. The Class 378s entered service in July 2009, originally in three-car form, they were augmented to four and later five-car sets. There are two sub-classes, 378/1 are 750V DC only, and 378/2 which are dual voltage 25kV AC /750V DC powered sets. DC only sets were built with a pantograph well and can be retro-fitted. The fleet is based at New Cross Gate depot, but also receives maintenance at Willesden. DC only set No. 378135 is seen departing from Honor Oak Park, displaying the latest black, white and blue colours, off-set by orange passenger doors. Each set seats 186, but has room for at least five times that number to stand. **CJM**

Left: Class 378 front end equipment. 1: High level marker light, 2: Destination indicator, 3: Front end emergency door, 4: Tunnel light, 5: Headlight, 6: Combined red tail and white marker light, 7: Emergency door release handle, 8: Dellner 12 coupling, 9: Electrical coupling box. 10: Forward facing camera, 11: Lamp bracket. **CJM**

Right: Coupled to each Class 378 driving car is a Motor Standard Open, seating 40. Although these sets are fully air Conditioned, each coach has some hopper windows (usually locked shut). Car No. 38250 from set No. 378150 is seen at Highbury & Islington. **CJM**

'Capitalstar' — Class 378

2023 Fleet

378/1
378135	LOG
378136	LOG
378137	LOG
378138	LOG
378139	LOG
378140	LOG
378141	LOG
378142	LOG
378143	LOG
378144	LOG
378145	LOG
378146	LOG
378147	LOG
378148	LOG
378149	LOG
378150	LOG
378151	LOG
378152	LOG
378153	LOG
378154	LOG

378/2
378201	LOG
378202	LOG
378203	LOG
378204	LOG
378205	LOG
378206	LOG
378207	LOG
378208	LOG
378209	LOG
378210	LOG
378211	LOG
378212	LOG
378213	LOG
378214	LOG
378215	LOG
378216	LOG
378217	LOG
378218	LOG
378219	LOG
378220	LOG
378221	LOG
378222	LOG
378223	LOG
378224	LOG
378225	LOG
378226	LOG
378227	LOG
378228	LOG
378229	LOG
378230	LOG
378231	LOG
378232	LOG
378233	LOG
378234	LOG
378255	LOG
378256	LOG
378257	LOG

Above: *The 37 five-car dual-voltage Class 378/2s have identical interiors as the '378/1s'. Dual voltage sets operate the Euston-Watford and Stratford-Richmond/Clapham Junction routes. The sets can also operate on DC only lines. No. 378210, in original livery, is seen at Highbury & Islington on the AC system.* **CJM**

Left: *In July 2023, set No. 378205 emerged in a multi-coloured 'London for Everyone' livery, based on the rainbow stripes, with high-quality graphics. A very eye-catching livery. It is seen at Clapham Junction on 7 July 2023.* **Antony Christie**

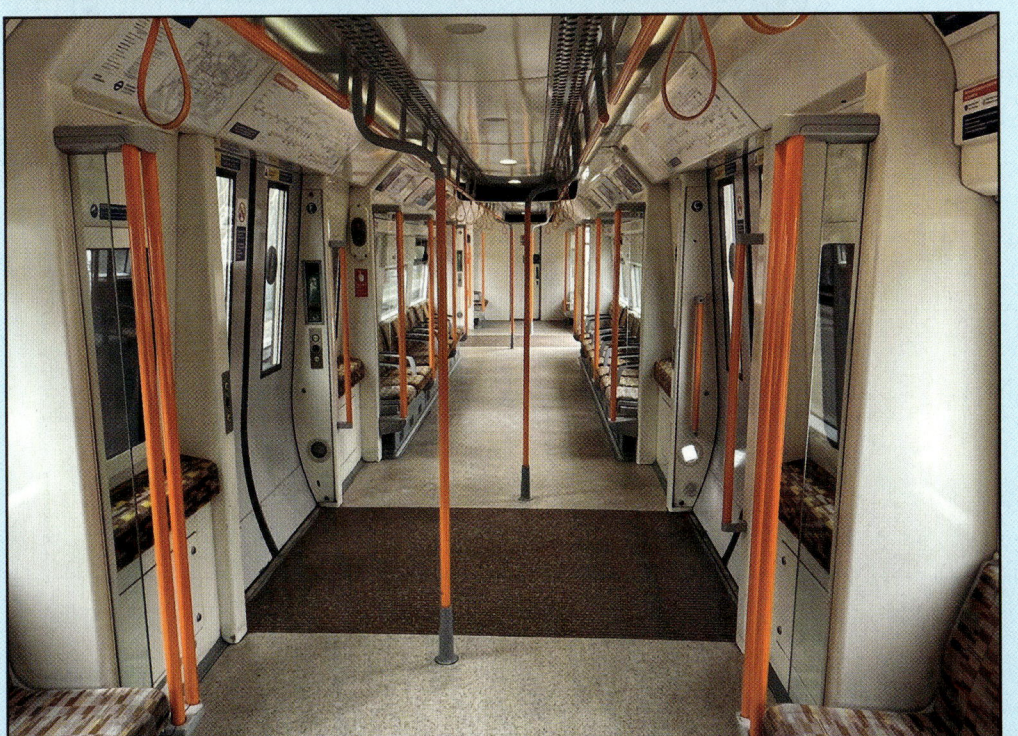

Below: *Class 387 interior has all longitudinal seating with an open and spacious layout. The sets have wide between-vehicle gangways, giving good visibility through the train, increasing security. Ceiling mounted straps are provided for standees, as are vertical grab poles, located throughout the train. Passenger doors are crew released and locally operated.* **CJM**

Key Facts

- Designed for TfL London Overground 'Metro' network.

- Sets 378201-234 built as three-car units, strengthened to four and then five vehicles to cope with passenger growth.

- Two sub-classes exist, 378/1 DC third rail and 378/2 dual voltage AC/DC sets. Interiors are identical.

- Fitted with London Underground style longitudinal seating.

- Emergency end doors fitted to allow emergency tunnel egress.

- Both sub-classes based at a purpose-built depot at New Cross.

Rolling Stock Review : 2023-2024

Class 380 'Desiro'

For history and pictures of this class, see Modern Locomotives Illustrated Issue 225

Class:	380/0	380/1
Number range:	380001-380022	380101-380116
Years introduced:	2010-2011	2010-2011
Built by:	Siemens Transportation Duewag, Germany	Siemens Transportation Duewag, Germany
Type:	Desiro	Desiro
Max speed:	100mph (161km/h)	100mph (161km/h)
Formation:	DMSO(A)+PTSOL+DMSO(B)	DMSO(A)+PTSOL+TSOL+DMSO(B)
Vehicle numbers:	DMSO(A) - 38501-38522 PTSOL - 38601-38622 DMSO(B) - 38701-38722	DMSO(A) - 38551-38566 PTSOL - 38651-38666 TSOL - 38551-38866 DMSO(B) - 38751-38766
Vehicle length:	75ft 5in (23.00m)	75ft 5in (23.00m)
Height:	12ft 4in (3.78m)	12ft 4in (3.78m)
Width:	9ft 2in (2.80m)	9ft 2in (2.80m)
Weight:	Total - 132.6 tonnes DMSO(A) - 45 tonnes PTSOL - 42.7 tonnes DMSO(B) - 44.9 tonnes	Total - 167.4 tonnes DMSO(A) - 45 tonnes PTSOL - 42.7 tonnes TSOL - 34.8 tonnes DMSO(B) - 44.9 tonnes
Internal layout:	2+2	2+2
Seating:	Total - 191S (+29 TU) DMSO(A) - 70S (+12TU) PTSOL - 57S (+12TU) DMSO(B) - 64S (+5TU)	Total - 265S (+29 TU) DMSO(A) - 70S (+12TU) PTSOL - 57S (+12TU) TSOL - 74S DMSO(B) - 64S (+5TU)
Gangway:	Throughout	Throughout
Toilets:	PTSOL - 1	PTSO - 1, TSOL - 1
Brake type:	Air (regenerative)	Air (regenerative)
Bogie type:	SGB SF5000	SGB SF5000
Power collection:	25kV ac overhead	25kV ac overhead
Traction motor type:	4 x Siemens 1TB2016-0GB02	4 x Siemens 1TB2016-0GB02
Horsepower:	1,341hp (1,000kW)	1,341hp (1,000kW)
Coupling type:	Outer - Voith, Inner - Bar	Outer - Voith, Inner - Bar
Multiple restriction:	Within class only	Within class only
Door type:	Bi-parting sliding plug	Bi-parting sliding plug
Construction:	Aluminium with steel ends	Aluminium with steel ends
Operator:	ScotRail	ScotRail
Owner:	Eversholt Leasing	Eversholt Leasing

2023 Fleet

380/0							
380001	ASR	380011	ASR	380022	ASR	380109	ASR
380002	ASR	380012	ASR			380110	ASR
380003	ASR	380013	ASR	**380/1**		380111	ASR
380004	ASR	380014	ASR	380101	ASR	380112	ASR
380005	ASR	380015	ASR	380102	ASR	380113	ASR
380006	ASR	380016	ASR	380103	ASR	380114	ASR
380007	ASR	380017	ASR	380104	ASR	380115	ASR
380008	ASR	380018	ASR	380105	ASR	380116	ASR
380009	ASR	380019	ASR	380106	ASR		
380010	ASR	380020	ASR	380107	ASR		
		380021	ASR	380108	ASR		

Top, Above, Left & Below: *A fleet of three- and four-car Siemens 'Desiro' units were delivered to ScotRail in 2010-2011 for the Ayrshire and Inverclyde routes. The three-car sets seat 191, and the four-car units 265 standard class passengers. The PTSOL vehicle (above) houses a universal toilet. Four-car sets, with an extra TSOL, have an additional standard toilet compartment. Left, is three-car set No. 380012 at Paisley Gilmore Street. The '2' on the front indicates the position of the pantograph and disabled area. On three car sets both ends carry a '2', on four car sets one end carries a '2' and the other a '3'. The view top shows the interior style, using the 2+2 layout with a good mix of group and airline styles. Airline seats have a good quality fold down table on the rear of the seat ahead. Power sockets are provided by each seat. A fleet of 16 four-car Class 380/1s are in traffic, along with the '380/0s', these are based at Glasgow Shields depot. The need to include a front gangway, gave for a very ugly end design. Sets carry Scottish Railways blue and white livery. Below, set No. 380108 is seen at Glasgow Central with its DMSO(A) seating 70 nearest the camera. All:* **CJM**

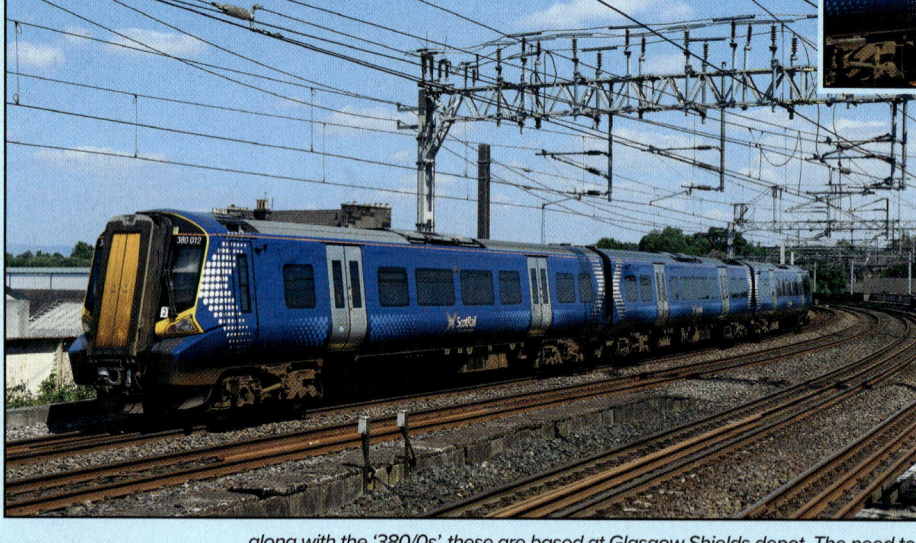

Key Facts

- A member of the Siemens 'Desiro' platform, designed for use by ScotRail.
- Formed as both three- and four-car formations.
- Fitted with ugly front end gangway connection, when compared with other 'Desiro' fleets.
- Installed with low-density 2+2 seating and usually used on outer suburban services.

132 — Rolling Stock Review : 2023-2024

'AT200' — **Class 385**

For history and pictures of this class, see *Modern Locomotives Illustrated* Issue 225

Class:	385/0	385/1
Number range:	385001-380046	385101-385124
Years introduced:	2017-2019	2017-2019
Built by:	Hitachi, Kasado, Japan and Newton Aycliffe UK	Hitachi, Kasado, Japan and Newton Aycliffe UK
Type:	AT200	AT200
Max speed:	100mph (161km/h)	100mph (161km/h)
Formation:	DMSO(A)+PTSO+DMSO(B)	DMCO+PTSO+TSO+DMSO
Vehicle numbers:	DMSO(A) - 441001-441046 PTSO - 442001-442046 DMSO(B) - 444001-444046	DMCO - 441101-441124 PTSO - 442101-442124 TSO - 443101-443124 DMSO - 444101-444124
Vehicle length:	Driving - 76ft 4in (23.18m) Intermediate - 72ft 4in (22.08m)	Driving - 76ft 4in (23.18m) Intermediate - 72ft 4in (22.08m)
Height:	12ft 2in (3.7m)	12ft 2in (3.7m)
Width:	8ft 10in (2.74m)	8ft 10in (2.74m)
Weight:	Total: 127.5 tonnes DMSO(A) - 44.6t, PTSO - 38.4t DMSO(B) - 44.5t	Total: 159.1t DMCO - 44.7t, PTSO - 38.4t TSO - 31.5t, DMSO - 44.5t
Internal layout:	2+2	2+1F / 2+2S
Seating:	Total - 190S (+14TU) DMSO(A) - 48S (+9TU) PTSO - 80S DMSO(B) - 62S (+5TU)	Total - 20F / 237S (+14TU) DMCO - 20F/15S (+9TU) PTSO - 80S TSO - 80S DMSO - 62S (+5TU)
Gangway:	Throughout	Throughout
Toilets:	DMSO - 1	DMCO - 1
Brake type:	Air (regenerative)	Air (regenerative)
Bogie type:	Hitachi	Hitachi
Power collection:	25kV ac overhead	25kV ac overhead
Traction motor type:	6 x Hitachi 335 hp (250kW)	8 x Hitachi 335 hp (250kW)
Horsepower:	2,010hp (1,499kW)	2,680hp (2,000kW)
Coupling type:	Outer - Dellner, Inner - Bar	Outer - Dellner, Inner - Bar
Multiple restriction:	Within class only	Within class only
Door type:	Bi-parting sliding plug	Bi-parting sliding plug
Construction:	Aluminium	Aluminium
Operator:	ScotRail	ScotRail
Owner:	Caledonian Rail Leasing	Caledonian Rail Leasing

Left: Class 385 front end equipment. 1: High level marker light, 2: Gangway door, 3: Headlight, 4: Forward facing camera, 5: Lamp bracket, 6: Combined marker/tail light, 7: Dellner coupling, 8: electrical connection box. **CJM**

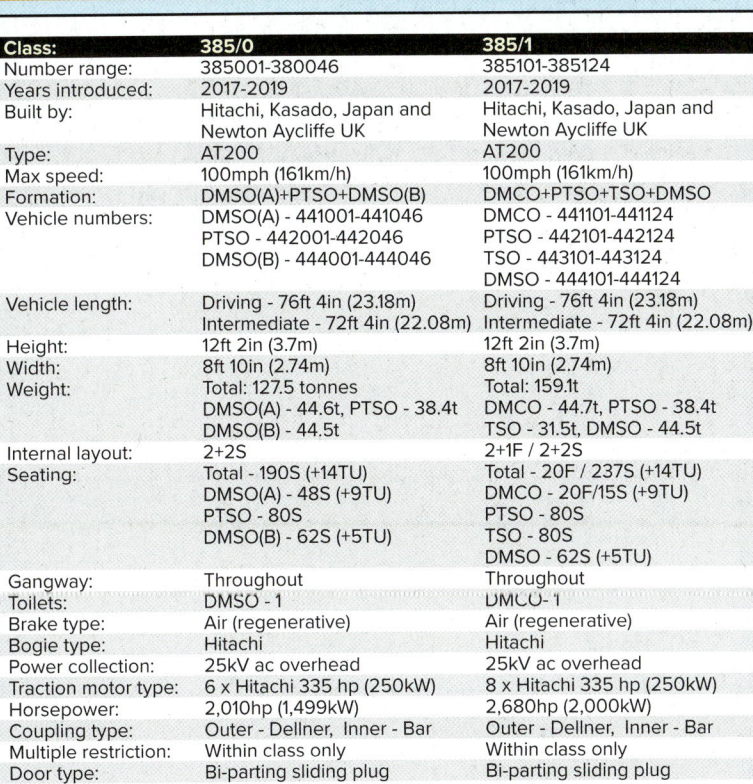

Right: The Class 385 has a very cramped driving cab, this was required due to the need of an end gangway. Forward vision to the right side is also restricted. **CJM**

Left Top: A total fleet of 70 Hitachi-built '385s' operate on Scottish Railways, based at the Hitachi-operated Craigentinny depot, Edinburgh. The 46 three-car Class 385/0s are standard class only, seating 190. Units are finished in Scottish Railways blue and white and carry the 'express' legend on the cab side. No. 385040 is seen arriving at Motherwell. **CJM**

Left Below: The 24 four-car Class 385/1 sets, have 20 first class seats in the DMCO vehicle, positioned inward of the driving cab. At the other end of the coach 15 standard class seats and a universal toilet are fitted. No. 385122 with its DMCO nearest the camera, is seen at Edinburgh Haymarket. Note the yellow first class banding around the windows and the cant rail stripe. **CJM**

Bottom Left & Bottom Right: First class seating in the DMCO (below left) with 20 seats in the 2+1 style around tables, sliding doors separate this area from the standard class seating. Window blinds and power sockets are provided by each seat. Below right is the standard class seating in the 2+2 style, with a mix of airline and group layouts. Power sockets are provided by each seat pair. This view shows an area of 2+2 group seating. Seats have hinged armrests, a luggage shelf is provided above and a good passenger information system is installed. Both: **CJM**

2023 Fleet

385/0
385001	ASR
385002	ASR
385003	ASR
385004	ASR
385005	ASR
385006	ASR
385007	ASR
385008	ASR
385009	ASR
385010	ASR
385011	ASR
385012	ASR
385013	ASR
385014	ASR
385015	ASR
385016	ASR
385017	ASR
385018	ASR
385019	ASR
385020	ASR
385021	ASR
385022	ASR
385023	ASR
385024	ASR
385025	ASR
385026	ASR
385027	ASR
385028	ASR
385029	ASR
385030	ASR
385031	ASR
385032	ASR
385033	ASR
385034	ASR
385035	ASR
385036	ASR
385037	ASR
385038	ASR
385039	ASR
385040	ASR
385041	ASR
385042	ASR
385043	ASR
385044	ASR
385045	ASR
385046	ASR

385/1
385101	ASR
385102	ASR
385103	ASR
385104	ASR
385105	ASR
385106	ASR
385107	ASR
385108	ASR
385109	ASR
385110	ASR
385111	ASR
385112	ASR
385113	ASR
385114	ASR
385115	ASR
385116	ASR
385117	ASR
385118	ASR
385119	ASR
385120	ASR
385121	ASR
385122	ASR
385123	ASR
385124	ASR

Key Facts

- Some sets built in Japan, most at the Hitachi Newton Aycliffe factory.
- Fleet maintained under a ten year contract by Hitachi at Craigentinny depot, Edinburgh.
- First example of Hitachi AT200 product platform to operate in the UK.
- Fully air conditioned.

Rolling Stock Review : 2023-2024

Class 387 'Electrostar'

Class:	387/1			387/2	387/3
Number range:	GTR - 387101-387129/172-174 GWR - 387130-387171 (HEX 387130-387141)			387201-387227	387301-387306
Built by:	Bombardier Derby			Bombardier Derby	Bombardier Derby
Type:	Electrostar			Electrostar	Electrostar
Introduced:	2014-2018			2016-2017	2017
Max speed:	110mph (177km/h)			110mph (177km/h)	110mph (177km/h)
Formation:	DMC(S)O+MSOL+PTSOL+DMSO			DMCO+MSOL+PTSOL+DMSO	DMSO(A)+MSOL+PTSOL+DMSO(B)
Vehicle numbers:	DMC(S)O - 421101-421174 MSOL - 422101-422174 PTSOL - 423101-423174 DMSO - 424101-424174			DMCO - 421201-421227 MSOL - 422201-422227 PTSOL - 423201-423227 DMSO - 424201-424227	DMSO(A) - 421301-421306 MSOL - 422301-422306 PTSOL - 423301-423306 DMSO(B) - 424301-424306
Vehicle length:	DMCO/DMSO - 66ft 11in (20.40m) MSOL, PTSOL - 65ft 6in (19.99m)			DMCO/DMSO - 66ft 11in (20.40m) MSOL, PTSOL - 65ft 6in (19.99m)	DMSO - 66ft 11in (20.40m) MSOL, PTSOL - 65ft 6in (19.99m)
Height:	12ft 4in (3.78m)			12ft 4in (3.78m)	12ft 4in (3.78m)
Width:	9ft 2in (2.80m)			9ft 2in (2.80m)	9ft 2in (2.80m)
Weight:	Total - 175 tonnes DMC(S)O - 46.0 tonnes MSOL - 41.3 tonnes PTSOL - 41.8 tonnes DMSO - 45.9 tonnes			Total - 175 tonnes DMCO - 46.0 tonnes MSOL - 41.3 tonnes PTSOL - 41.8 tonnes DMSO - 45.9 tonnes	Total - 175 tonnes DMSO(A) - 46.0 tonnes MSOL - 41.3 tonnes PTSOL - 41.8 tonnes DMSO(B) - 45.9 tonnes
Internal layout:	2+2	2+2	2+1 / 2+2	2+2	2+2
Seating:	GTR Total - 22F/201S DMCO - 22F/34S MSOL - 62S PTSOL - 45S DMSO - 60S	GWR Total - 223S DMSO - 56S MSOL - 62S PTSOL - 45S DMSO - 60S	HEX Total - 22F/187S DMCO - 22F/26S MSOL - 54S PTSOL - 39S DMSO - 52S	Total - 22F/199S DMCO - 22F/34S MSOL - 60S PTSOL - 45S DMSO - 60S	Total - 223S DMSO(A) - 56S MSOL - 62S PTSOL - 45S DMSO - 60S
Gangway:	Throughout			Throughout	Throughout
Toilets:	MSOL, PTSOL - 1			MSOL, PTSOL - 1	MSOL, PTSOL - 1
Brake type:	Air (regenerative)			Air (regenerative)	Air (regenerative)
Bogie type:	Power - Bombardier P3-25 Trailer - Bombardier T3-25			Power - Bombardier P3-25 Trailer - Bombardier T3-25	Power - Bombardier P3-25 Trailer - Bombardier T3-25
Power collection:	750V dc third rail and 25kV ac overhead			750V dc third rail and 25kV ac overhead	750V dc third rail and 25kV ac overhead
Traction motor type:	6 x Bombardier			6 x Bombardier	6 x Bombardier
Horsepower:	2,012hp (1,500kW)			2,012hp (1,500kW)	2,012hp (1,500kW)
Coupling type:	Outer - Dellner 12, Inner - Bar			Outer - Dellner 12, Inner - Bar	Outer - Dellner 12, Inner - Bar
Multiple restriction:	Within class			Within class	Within class
Door type:	Bi-parting sliding plug			Bi-parting sliding plug	Bi-parting sliding plug
Construction:	Aluminium, steel ends			Aluminium, steel ends	Aluminium, steel ends
Operator:	GTR / GWR			GTR (Gatwick Express)	GTR
Owner:	Porterbrook Leasing			Porterbrook Leasing	Porterbrook Leasing
Note:	All sub-classes to be fitted with Bombardier ETCS and EBICab 2000 ATP				

Right Upper & Right Lower: The Class 387 'Electrostar' fleet were delivered in three sub-classes. Class 387/1 is then divided into three groups, Nos. 387101-387129/172-174, are operated by Govia Thameslink (GTR) on Great Northern services, seating 22 first and 201 standard class. First class is located in the Driving Motor Composite Open (DMCO). The next batch, Nos. 387130-387171 are operated by Great Western. The GW batch, Nos. 387130-387141 are modified for Heathrow Express operation see page 136. The domestic fleet are all standard class. The '387's are similar to the later Class 377 and 379 builds, with framed bodyside windows rather than ribbon glazing. They incorporate a large headlight and a combined marker/tail light on either side of the front end, they also have a track light. Sets have a top speed of 110mph (177km/h). In the upper view, No. 387107 is seen from its DMCO end at King's Cross. Below, Great Western set No. 387152 is seen at arriving at West Ealing with a Didcot to Paddington service. **Antony Christie / CJM**

Key Facts

- Final refinement of 'Electrostar' platform.

- Class 387/1 fleet divided into three groups, sets 101-129/172-174 operated by GTR, sets 130-141 by Heathrow Express and sets 142-171 on Great Western on domestic duties. The two GW groups have different seating layouts.

- Class 387/2s are dedicated to Gatwick Express services, and carry red GatX livery and branding, with extra luggage space.

- Six Class 387/3s originally with c2c, now with GTR.

'Electrostar' — Class 387

Left: The 30 domestic Great Western Class 387s are standard class only with traditional style 'Electrostar' seats in the 2+2 layout, with a high number in the airline style. This view shows the disabled seating and wheelchair space in the PTSOL vehicle. Seats are finished in the Great Western green and grey moquette, vehicles are fully carpeted, with coach furniture finished in dark green. **CJM**

Above: The Class 387/1s operated by GTR have first class seating for 22 located at the inner end of the DMCO vehicle. The area is isolated from standard class by a pair of passenger activated sliding doors. The first class seats are in the 2+2 style with hinged armrests, some are in the facing style, others are of the airline layout. **Antony Christie**

Above: Standard class seating on GTR Class 387s is in the 2+2 style, finished in turquoise moquette with green trim for grab handles, poles and the edge of the above seat luggage rack. Seats are a mix of airline and facing layout. Airline seats have fold down tables attached to the seat ahead. **CJM**

Above: The Pantograph Trailer Standard Open Lavatory (PTSOL) vehicle has a universal toilet compartment at one end, with a semi rotary door. Disabled space is adjacent. The toilet compartment has open/close buttons as well as available, engaged and out of use indicators. Set 387111 is shown. **CJM**

Above: The Gatwick Express sets have a similar interior to the Class 387/1s, but seating is finished with a red moquette. Seating is for 22 first (in the DMCO) and 199 standard. Blue trim is used for grab handles, poles and luggage rack edges. The first area is seperated by bi-parting doors. **Antony Christie**

Below Left: A fleet of 27 Gatwick Express Class 387/2s are based at Stewarts Lane and deployed on the London Victoria to Gatwick Airport and Brighton route, working a 15min interval service between London and the Airport. Trains are formed of eight or 12 carriages. The sets are finished in red, with Gatwick Express branding in white. Illustrated is set 387205. In 2020-2021, due to changes in passenger demand during and after the Covid pandemic, several Gatwick Express sets were transferred to the GTR Great Northern and Great Western routes, most have now returned to the GatX line as services have been increased. **Antony Christie**

2023 Fleet

387/1		387156	GWR
387101	GTR	387157	GWR
387102	GTR	387158	GWR
387103	GTR	387159	GWR
387104	GTR	387160	GWR
387105	GTR	387161	GWR
387106	GTR	387162	GWR
387107	GTR	387163	GWR
387108	GTR	387164	GWR
387109	GTR	387165	GWR
387110	GTR	387166	GWR
387111	GTR	387167	GWR
387112	GTR	387168	GWR
387113	GTR	387169	GWR
387114	GTR	387170	GWR
387115	GTR	387171	GWR
387116	GTR	387172	GTR
387117	GTR	387173	GTR
387118	GTR	387174	GTR
387119	GTR		
387120	GTR	**387/2**	
387121	GTR	387201	GTR
387122	GTR	387202	GTR
387123	GTR	387203	GTR
387124	GTR	387204	GTR
007125	CTN	387205	GTR
387126	GTR	387206	GTR
387127	GTR	387207	GTR
387128	GTR	387208	GTR
387129	GTR	387209	GTR
387130	HEX	387210	GTR
387131	HEX	387211	GTR
387132	HEX	387212	GTR
387133	HEX	387213	GTR
387134	HEX	387214	GTR
387135	HEX	387215	GTR
387136	HEX	387216	GTR
387137	HEX	387217	GTR
387138	HEX	387218	GTR
387139	HEX	387219	GTR
387140	HEX	387220	GTR
387141	HEX	387221	GTR
387142	GWR	387222	GTR
387143	GWR	387223	GTR
387144	GWR	387224	GTR
387145	GWR	387225	GTR
387146	GWR	387226	GTR
387147	GWR	387227	GTR
387148	GWR		
387149	GWR	**387/3**	
387150	GWR	387301	GTR
387151	GWR	387302	GTR
387152	GWR	387303	GTR
387153	GWR	387304	GTR
387154	GWR	387305	GTR
387155	GWR	387306	GTR

Rolling Stock Review : 2023-2024

Class 387 'Electrostar'

Right & Inset: In 2022, the six Class 387/3 sets owned by Porterbrook Leasing and officially leased to c2c, had a period working on Great Western to cover for a fleet shortage caused by Class 80x unavailability. From mid-summer 2022 the six sets transferred to GTR Great Northern and absorbed into their core fleet. Currently, the sub-class are finished in white livery have mauve passenger doors, but this is likely to be changed to GTR white and blue. Set No. 387301 is seen operating on the Great Northern. The inset image shows the seating layout using the standard 2+2 arrangement.
Antony Christie / Jamie Squibbs

Right, Right Middle, Right Below & Below: In early 2021, Heathrow Express underwent a major upgrade, with the fleet of Class 332s withdrawn and replaced by a batch of 12 modified Class 387/1s, incorporating revised interiors to meet the demands of airport traffic, with additional luggage racks. A newly classified Driving Motor Composite Open (DMCO) was formed incorporating a Business First section with 22 first and 26 standard class seats. All sets carry a distinctive Heathrow Express livery of silver/grey and mauve with airline branding. Set No. 387138 is seen right passing West Ealing. Trains are usually formed of two four-car sets. Below right is MSOL coach No. 4222140, these vehicles seat 54 in the 2+2 style with a standard type toilet compartment at one end. Mid-2023 marked the 25th anniversary of the Heathrow Express service, this was marked by bodyside branding on the intermediate vehicles of set No. 387140. From late 2021, the Hex sets were named after major destinations served by Heathrow Airport. No. 387141 *Prague* is shown under the Brunel roof at Paddington. Below, is the revised Heathrow Express interior, the upper image shows the 2+1 Business First seating, while the lower shows the standard class 2+2 seating. Electronic display boards are provided in both classes, with information on airport terminals, connections and services. Onward connections are provided on the approach to Paddington.
CJM (2) / Antony Christie (3)

'Pendolino' — Class 390

Key Facts

- Introduced 2001-2005 by Virgin Trains as replacement stock for West Coast franchise.

- Incorporating tilting technology allowing higher speeds on curves.

- Extra vehicles and four additional trains delivered in 2010-2011.

- A three-year full refurbishment project commenced in July 2021 for the fleet, with work carried out at the Alstom Widnes facility. Work will include a full technical overhaul and upgrade, plus interior redesign. On 11-car sets, one first class vehicle is reconfigured for standard class, while all sets will have an upgraded buffet/bar area. New lighting, extra luggage space and power sockets fitted at each seat.

For history and pictures of this class, see Modern Locomotives Illustrated Issue 232

Class:	390
Number range:	9-car sets - 390001/002/005/006/008-011/013/016/020/023/035/038-040/042-047/049-050 11-car sets - 390103/104/107/112/114/115/117-119/121/122/124-134/136/137/141/148/151-157
Built by:	390001-390053 - Alstom, Washwood Heath, body shells from Italy 390154-390157 - Alstom, Savigliano, Italy§
Type:	Pendolino
Introduced:	390001-390053 - 2001-2005, 390154-157 - 2010-2011§
Max speed:	125mph (201km/h)
Formation:	9-car sets - DMRF+MF+PTF+MS+TS+MS+PTSRMB+MS+DMSO 11-car sets - DMRF+MF+PTF+MS+TS+MS+PTSRMB+MS+MS+MS+DMSO DMRF - 69101-69157 MF - 69401-69457 PTF - 69501-69557 MS - 69601-69657 TS - 65303-65357§ (11-car sets only) MS - 68903-69957§ (11-car sets only) TS - 68801-68857 MS - 69701-69757 PTSRMB - 69801-69857 MS - 69901-69957 DMSO - 69201-69257
Vehicle length:	DMRF, DMSO - 81ft 3in (24.80m), MF, PTF, TS, MS, PTSRMB - 78ft 4in (23.90m)
Height:	11ft 6in (3.56m)
Width:	8ft 9in (2.73m)
Weight:	Total - 9-car - 459.7 tonnes, 11-car - 556.9 tonnes DMRF - 55.6 tonnes MF - 52 tonnes PTF - 50.1 tonnes MF - 51.8 tonnes (MS on 9-car sets) TS - 45.5 tonnes (11-car sets only) MS - 50 tonnes (11-car sets only) TS - 45.5 tonnes MS - 51.7 tonnes PTSRMB - 52 tonnes MS - 51.7 tonnes DMSO - 51 tonnes
Internal layout:	2+1F/2+2S
Seating:	Total - 9-car - 99F/364S, 11-car - 145F/438S DMRF - 18F, MF - 37F, PTF - 44F, MS - 46U (MS on 9-car sets 74S), TS - 74S, MS - 74S, TS - 74S, MS - 62S, PTSRMB - 48S, MS - 60S, DMSO - 46S
Gangway:	Within set
Toilets:	MF, PTF, TS, MF, MSO, TS, MF(D) - 1
Brake type:	Air (regenerative)
Bogie type:	Fiat/SIG tilting
Power collection:	25kV ac overhead
Traction motor type:	9-car - 12 x Alstom Onix 800, 11-car 14 x Alstom Onix 800
Horsepower:	9-car - 6,839hp (5,100 kW), 11-car 9,120hp (6,803kW)
Coupling type:	Outer - Dellner 12, Inner - Bar
Multiple restriction:	No multiple facility, operable with selected Class 57/3s fitted with drop head Dellner
Door type:	Single-leaf sliding plug
Construction:	Aluminium
Operator:	Avanti West Coast
Owner:	Angel Trains

Above: *The core traction operating on Avanti West Coast is a fleet of either 9- or 11-car Class 390 'Pendolino' sets. These were introduced from 2001, and include tilt technology, enabling trains to take curves at higher speeds, reducing journey times, while maintaining a comfortable ride. Operated by Virgin Trains until December 2019, the fleet is now finished in Avanti West Coast livery. Nine car Class 390/0 No. 390042 is shown from its first class end at Carlisle.* **CJM**

Right: *In 2010-2011, four additional trains were built as 11-car formations and at the same time 35 9-car sets were lengthened to 11-car, easing overcrowding on the busy West Coast route. The lengthened sets were reclassified as 390/1, with the fourth digit of the number being changed from 0 to 1. With its standard class end nearest the camera, 11-car set No. 390104 is seen near Kilsby Tunnel on 14 February 2023 with the 15.13 Euston to Manchester Piccadilly.* **CJM**

Rolling Stock Review : 2023-2024

Class 390 'Pendolino'

390001	AWC
390002	AWC
390103	AWC
390104	AWC
390005	AWC
390006	AWC
390107	AWC
390008	AWC
390009	AWC
390010	AWC
390011	AWC
390112	AWC
390013	AWC
390114	AWC
390115	AWC
390016	AWC
390117	AWC
390118	AWC
390119	AWC
390020	AWC
390121	AWC
390122	AWC
390123	AWC
390124	AWC
390125	AWC
390126	AWC
390127	AWC
390128	AWC
390129	AWC
390130	AWC
390131	AWC
390132	AWC
390134	AWC
390135	AWC
390136	AWC
390137	AWC
390138	AWC
390039	AWC
390040	AWC
390141	AWC
390042	AWC
390043	AWC
390044	AWC
390045	AWC
390046	AWC
390047	AWC
390148	AWC
390049	AWC
390050	AWC
390151	AWC
390152	AWC
390153	AWC
390154	AWC
390155	AWC
390156	AWC
390157	AWC

Top & Above: Each Class 390 has two pantograph fitted intermediate vehicles, formed as the third vehicle from each end of the train. Only one pantograph is used at a time, usually the rear in direction of travel. Top, with a Pantograph Trailer First (PTF) on the far left a MF vehicle from set No. 390142 is shown at Carlisle. These vehicles seat 37 first class passengers. Above is Motor Standard 69642, a vehicle seating 76. Both: **CJM**

Above Right: In 2023 a major upgrade to Class 390 passenger accommodation was ongoing, with many enhancements. Pre-2022 each nine car set accommodated 99 first and 370 standard class, while an 11-car seated 145 first and 444 standard. Due to the design tapering at the top to allow for tilting, the interior, with high-back seats, looks cramped. Standard seating is mainly in the airline style, with limited group seats per coach. The original moquette was deep red or dark blue in standard class with a blue fleck carpet. **CJM**

Right Upper & Right Lower: The major facelift currently being undertaken to the Pendolino fleet by Alstom at Widnes sees several changes, the 11-car sets were upgraded first, followed by the nine-car sets. Work has seen a revised interior for both standard and first class and the introduction of a new Standard Premium coach. The overall number of first class coaches has been reduced by one, this was previously done on nine-car sets. A new shop, enlarged luggage racks, upgraded toilets and at seat power supply is included in the project. While the original seats are retained, new moquette is applied as are new carpets. The image right upper shows the upgraded standard class, with a mainly airline style layout with a small number of group seats. The image right below, shows the new Standard Premium coach, using first class seating, with new moquette, table tops and carpets. A number of new glazed divider screens have been installed. Both: **Jamie Dyke**

Left: Class 390 'Pendolino' driving cab layout. With such a major advance in train design, came a totally re-styled cab. Designed in partnership between Alstom, Virgin Trains, drivers and the driver's trade union ASLE&F. The result is this spacious futuristic design. Power and brake control is on the left side with all displays and controls in sensible positions and within easy reach of the driver. A full Train Management System (TMS) is located on the drivers left side. **CJM**

Below: 'Stick-on' style nameplate Flying Scouseman applied to No. 390148. **Antony Christie**

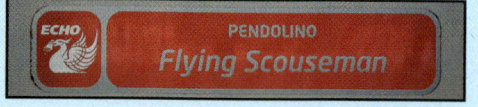

For history and pictures of this class, see Modern Locomotives Illustrated Issue 223

'Javelin' — Class 395

Class:	395
Number range:	395001-395029
Built by:	Hitachi, Kasado, Japan
Type:	Javelin (AT300)
Introduced:	2007-2009
Max speed:	HS1 - 140mph (225km/h)
	Classic - 100mph (161km/h)
Formation:	PDTSA+MS1+MS2+MS3+MS4+PDTSB
Vehicle numbers:	PDTSA - 39011-39291
	MS1 - 39012-39292
	MS2 - 39013-39293
	MS3 - 39014-39294
	MS4 - 39015-39295
	PDTSB - 39016-39296
Vehicle length:	Driving - 68ft 5in (20.88m)
	Intermediate - 65ft 6in (20m)
Height:	12ft 6in (3.81m)
Width:	9ft 2in (2.81m)
Weight:	Total - 276.2 tonnes
	PDTS - 46.7 tonnes
	MS - 45.7 tonnes
Internal layout:	2+2
Seating:	Total - 340S (+12TU)
	PDTSA - 28S (+12TU)
	MS1 - 66S
	MS2 - 66S
	MS3 - 66S
	MS4 - 66S
	PDTSB - 48S
Gangway:	Within unit only
Toilets:	PDTS - 1
Brake type:	Air (regenerative)
Bogie type:	Hitachi
Power collection:	25kV ac overhead & 750V dc third rail
Traction motor type:	16 x IGBT Converter, three-phase
Horsepower:	2,253hp (1,680 kW)
Coupling type:	Outer: Scharfenburg, Inner - Bar
Multiple restriction:	Within class only (2-unit max)
Door type:	Single-leaf sliding
Construction:	Aluminium
Operator:	South Eastern
Owner:	Eversholt Leasing

2023 Fleet

395001	SET
395002	SET
395003	SET
395004	SET
395005	SET
395006	SET
395007	SET
395008	SET
395009	SET
395010	SET
395011	SET
395012	SET
395013	SET
395014	SET
395015	SET
395016	SET
395017	SET
395018	SET
395019	SET
395020	SET
395021	SET
395022	SET
395023	SET
395024	SET
395025	SET
395026	SET
395027	SET
395028	SET
395029	SET

Top: *A fleet of 29 six-car Hitachi-built AT300 or UK Class 395 'Javelin' sets, were introduced to operate domestic services over High Speed 1 (HS1), as well as working over classic dc electrified lines. They are fitted with TVM430 cab signalling as well as AWS and TPWS, they are based at the Hitachi depot in Ashford (Kent). Trains are standard class only, branded in SouthEastern dark blue livery with sky blue passenger doors. Set Nos. 395025 is recorded arriving at Margate in summer 2022.* **CJM**

Above: *Between the two driving cars are four Motor Standard (MS) coaches, each seating 66, two single leaf sliding doors are on each side, feeding direct into the passenger saloon. Each '395' seats 340 in the low-density 2+2 style. MS No. 39254 from set 395025 is shown. Both bogies of each MS vehicles are powered, driving cars are unpowered.* **CJM**

Above: *Class 395 interior, all vehicles are set out in the 2+2 low-density style, with a mix of airline and group seating. Sets have recently completed a major fleet refresh, with new seat swabs, an interior repaint, new tables and USB sockets fitted.* **Antony Christie**

Left: *The driving cab of the Class 395 is one of the most complex found on any modern electric multiple unit, having facilities to operate using TVM430 cab signalling as well as UK conventional systems, with AWS and TPWS installed. A combined power/brake controller is operated by the drivers left hand. The TVM430 signalling display is located at the top of the central section of the desk.* **CJM**

Key Facts

- Dual voltage stock, designed to operate domestic trains over HS1 at a max speed of 140mph (225km/h).
- Able to work over high speed and classic tracks, fitted with TVM430 cab signalling, as well as conventional AWS/TPWS.
- Named 'Javelin' sets after being used to provide a special service between St Pancras International and Stratford for the 2012 London Olympic Games.
- Major fleet refurbishment started in 2023, with set No. 395012 returning to traffic in August with upgraded seats and new equipment.

Class 397 'Civity'

For history and pictures of this class, see *Modern Locomotives Illustrated* Issue 243

Class:	397
Number range:	397001-397012
Years introduced:	2018-2020
Built by:	CAF, Beasain, Spain
Type:	Civity
Max speed:	125mph (201km/h)
Formation:	DMF+PTS(A)+MS+PTS(B)+DMS
Vehicle numbers:	DMF - 471001-471012
	PTS(A) - 472001-472012
	MS - 473001-473012
	PTS(B) - 474001-474012
	DMS - 475001-475012
Vehicle length:	Driving: 78ft 10in (24.02m)
	Intermediate: 76ft 4in (23.35m)
Train length:	387ft 2in (118m)
Height:	12ft 6in (3.80m)
Width:	8ft 10in (2.71m)
Internal Layout:	2+1F / 2+2S
Seating:	Total - 24F/264S (+8TU)
	DMF - 24F, PTS(A) - 76S
	MS - 68S, PTS(B) - 76S
	DMS - 44S (+8TU)
Gangway:	Within set
Toilets:	DMF-1, MS-2, DMS-1
Brake type:	Air (regenerative)
Bogie type:	CAF
Power collection:	25kV ac overhead
Traction Motor Type:	TSA
Horsepower:	3,540hp (2,640kW)
Coupling type:	Outer - Dellner 12, Inner - Bar
Multiple restriction:	Within class
Door type:	Sliding plug
Construction:	Aluminium with steel cabs
Operator:	TransPennine Trains
Owner:	Eversholt Leasing

Right: CAF-built Class 397 TransPennine Express 'Civity' unit front end design. 1: High level marker light, 2: Destination indicator, 3: Forward facing camera, 4: Headlight, 5: Combined white marker and red tail light, 6: Screen wash water jet. 7: Hinged nose cone, opens to reveal Dellner 12 coupling, 8: Coupling guide. Front end of DMS 475006 from set No. 397006 shown. **CJM**

Below & Right Middle: In 2016 TransPennine Express ordered 12 five-car EMUs from CAF, classified 397, to replace Class 350/4 sets on Liverpool, Manchester to Scotland via WCML services. The '397s' are formed with three powered vehicles; both driving cars and the middle coach, with two un-powered trailers. Owned by Eversholt, they entered service in 2020. On the right is Pantograph Trailer Standard PTS 472008, with its pantograph at the far end. Below, set No. 397005 with its DMS vehicle leading approaches Carlisle. DMS vehicles seat 44 with one standard toilet at the inner end. Both: **Antony Christie**

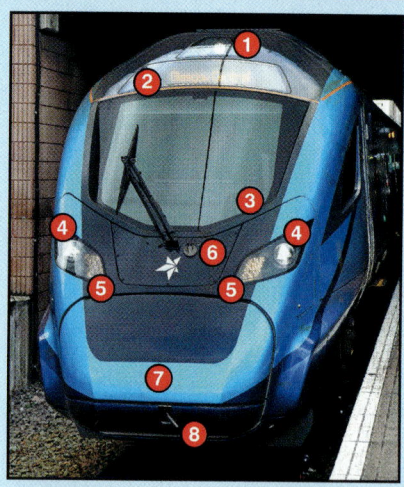

397001	TPT
397002	TPT
397003	TPT
397004	TPT
397005	TPT
397006	TPT
397007	TPT
397008	TPT
397009	TPT
397010	TPT
397011	TPT
397012	TPT

Below Left & Below Right: Class 397 interiors are spacious and bright. In first class, located in one driving car, seats in 2+1 style accommodate 24 (shown left). each seat has a fixed table. In standard class (shown right) seats are in the 2+2 layout with a mix of group and airline. All vehicles are carpeted throughout, are fitted with passenger information displays, automated announcements and coaches include an electronic seat reservation system. Both: **Nathan Williamson**

'Citylink' — Class 398

Class:	398
Number range:	398001-398036
Built by:	Vossloh, Valencia, Spain
Type:	CityLink (TramTrain)
Built:	2021-2024
Max speed:	62mph (97km/h)
Formation:	DMSO(A)+TSO+DMSO(B)
Vehicle numbers:	DMSO(A) - 999051-999086 TSO - 999151-999186 DMSO(B) - 999251-999286
Length:	40m
Height:	3.71m
Width:	2.65m
Weight:	tba
Internal layout:	2+2
Seating:	Total: 104S DMSO - 34S, TSO - 36S
Gangway:	Within unit only
Toilets:	Not fitted
Brake type:	Air (regenerative), Emergency track brake
Underframe type:	Bo-2-2-Bo
Power collection:	25kV ac overhead
Traction motor type:	4 x VEM 194hp (145kW)
Horsepower:	776hp (580 kW)
Coupling type:	Albert (emergency)
Multiple restriction:	Not fitted
Door type:	Bi-parting sliding plug
Construction:	Steel

Above & Below: As part of the ongoing modernisation of the Cardiff Valleys network, a fleet of 36 three-section 'Tram-Trains' are currently under construction/delivery by Stadler to Transport for Wales (TfW), to operate on Cardiff Valley routes and the Cardiff City line including the route to Cardiff Bay and possible street extensions in the future. The electric and battery sets are currently on delivery to a new purpose-built depot at Taffs Well. The sets carry the Transport for Wales grey livery with red passenger doors with the 'Metro' brand name on the side. The design was first shown off to the public at Innotrans, Berlin in September 2022, where set No. 398004 is shown. Since these illustration were taken, a lower yellow warning end has been applied.
Both: **CJM**

2023 Fleet

398001	TFW	398019	TFW
398002	TFW	398020	TFW
398003	TFW	398021	TFW
398004	TFW	398022	TFW
398005	TFW	398023	TFW
398006	TFW	398024	TFW
398007	TFW	398025	TFW
398008	TFW	398026	TFW
398009	TFW	398027	TFW
398010	TFW	398028	TFW
398011	TFW	398029	TFW
398012	TFW	398030	TFW
398013	TFW	398031	TFW
398014	TFW	398032	TFW
398015	TFW	398033	TFW
398016	TFW	398034	TFW
398017	TFW	398035	TFW
398018	TFW	398036	TFW

Right Upper: Class 398 'Tram-Train' front end. These sets are fitted with both rail and street running lights. The lower front end is liftable to reveal the coupling. **CJM**

Right Middle: The interior is based on the low-density 2+2 seating style with a mix of airline and group layouts. The interior layout is basic as would be expected on a tram-train design, but adequate for short to medium journeys. Phone/lap-top charging points are provided as is a wi-fi connection. A wide between vehicle gangway is provided. No luggage racks or above seat shelf is provided. **CJM**

Right Lower: The driving cab, with the driving position in the middle of the cab uses the now standard combined power brake controller on the left side. Screens provide the driver with good all round visibility of the vehicle to ensure passenger safety especially during street running. The sets with automatically change from 'train' to 'street' systems. **CJM**

Left: As the Stadler Class 398s are designed for street running, although this will not be used when introduced, the wheelsets and underframe is shrouded, to avoid anyone coming into contact with moving parts. This view also shows the Metro bodyside branding and vehicle set number high up behind the cab. **CJM**

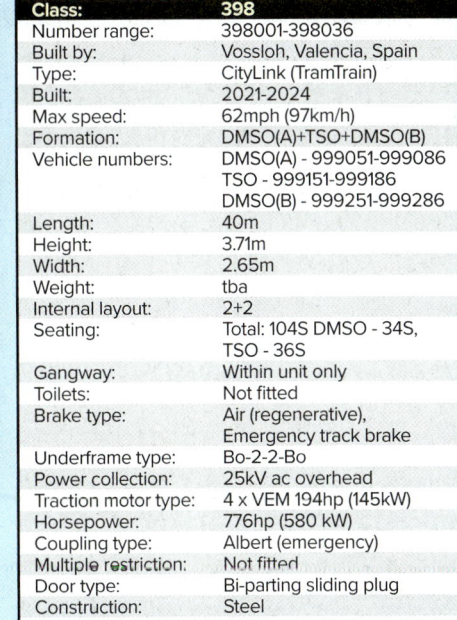

Rolling Stock Review : 2023-2024

Class 399 'Citylink'

For history and pictures of this class, see Modern Locomotives Illustrated Issue 243

Class:	399
Number range:	399201-399207
Built by:	Vossloh, Valencia, Spain
Type:	CityLink (TramTrain)
Built:	2015-2016 (Introduced 2017-2018)
Max speed:	60mph (97km/h)
Formation:	DMSO(A)+MSO+DMSO(B)
Vehicle numbers:	DMSO(A) - 999001-999007
	MSO - 999101-999107
	DMSO(B) - 999201-999207
Length:	122ft 1in (37.2m)
Height:	12ft 2in (3.71m)
Width:	8ft 7in (2.65m)
Weight:	64 tonnes
Internal layout:	2+2
Seating:	Total: 88S DMSO - 22S, MSO - 44S
Gangway:	Within unit only
Toilets:	Not fitted
Brake type:	Air (regenerative), Emergency track brake
Underframe type:	Bo-2-Bo-Bo
Power collection:	750V dc overhead
Traction motor type:	6 x VEM 194hp (145kW) units
Horsepower:	1,166hp (870 kW)
Coupling type:	Albert (emergency)
Multiple restriction:	Not fitted
Door type:	Bi-parting sliding plug
Construction:	Steel
Owner/operator:	Sheffield Super Tram

Above & Below Left: *A fleet of seven Vossloh-built 'CityLink' 'tram-trains' operate on the Sheffield to Rotherham Parkgate route, of South Yorkshire Super Tram system for which they are given the TOPS classification of '399', as they work over Network Rail trackage between Tinsley North Junction and Rotherham Parkgate. They also operate over the tram network. The three section vehicles and carry Super Tram colours. Above, set No. 399204 is seen at Cathedral in Sheffield city centre with a 'TT' service to Rotherham Parkgate. Below left is No. 399203 operating over the Super Tram network at Arena heading towards the city centre. Both:* **CJM**

Right: *Class 399 'tram-train' front end. The lighting on these vehicles conforms to both the Network Rail requirement and the tram requirements for street running, with suitable head and marker lights as well as street operating running lights and direction indicators. An emergency coupling is carried behind the front lower valance.* **CJM**

Left: *Class 399 driving cab. This design has a central position for the driver, with the combined power/brake controller is on the raised section on the left side of the desk, side mounted screens give the driver a clear view of the train/platform interface. The equipment on the right side is the change over from 'train' to 'tram' running.* **CJM**

Key Facts

- First tram-trains in the UK, designed to operate on both South Yorkshire Super Tram and Network Rail between Tinsley North Junction and Rotherham Parkgate.
- Owned and operated by South Yorkshire Super Tram, maintained at Nunnery depot.
- Services over Network Rail commenced in late 2018 as trial of TramTrain technology still running in 2023.
- Only sets modified P1 wheel profiles permitted to operate over NR metals.

Right Middle: *Each tram-train is mounted on four bogies, with seating for 88 standard class passengers, supported by a lot of room for standees. Driving cars have 22 seats and the middle section 44 seats. Seats are a mix of group, facing and longitudinal. Passenger doors are under the control of the driver, with each train having a conductor for revenue purposes. The vehicles provide level access for disabled travellers.* **CJM**

Right Lower: *Detail of the intermediate Motor Standard Open (MSO) vehicle, housing the 750V DC pantograph and traction equipment. Note the bogie/underframe shrouds to protect the public from moving parts.*
Antony Christie

2023 Fleet

399201	SST	399205	SST
399202	SST	399206	SST
399203	SST	399207	SST
399204	SST		

'Desiro' Class 444

For history and pictures of this class, see Modern Locomotives Illustrated Issue 178

Class:	444
Number range:	444001-444045
Built by:	Siemens Transportation SGP, Austria
Type:	Desiro
Introduced:	2003-2004
Max speed:	100mph (161km/h)
Formation:	DMSO+TSO(A)+TSO(B)+TSO(C)+DMCO
Vehicle numbers:	DMSO - 63801-63845
	TSO(A) - 67101-67145
	TSO(B) - 67151-67195
	TSO(C) - 67201-67245
	DMCO - 63851-63895
Vehicle length:	77ft 3in (23.57m)
Height:	12ft 1½in (3.7m)
Width:	8ft 9in (2.74m)
Weight:	Total - 221.8 tonnes
	DMSO - 51.3 tonnes, TSO(A) - 40.3 tonnes
	TSO(B) - 36.8 tonnes, TSO(C) - 42.1 tonnes
	DMCO - 51.3 tonnes
Internal layout:	As built - 2+2S, 2+2F
Seating:	Total - Original - 32F/327S
	DMSO - 76S, TSO(A) - 76S, TSO(B) - 76S
	TSO(C) - 59S, DMCO - 32F/40S
Gangway:	Throughout
Toilets:	TSO(A), TSO(B) - 1, TSO(C) - 2
Brake type:	Air (regenerative)
Bogie type:	Siemens SGB SF5000
Power collection:	750V dc third rail
Traction motor type:	4 x Siemens 1TB2016-0GB02 3-phase
Horsepower:	2,682hp (2,000kW)
Coupling type:	Outer - Dellner 12, Inner - Semi-auto
Multiple restriction:	444 and 450
Door type:	Single-leaf sliding plug
Construction:	Aluminium
Operator:	South Western Railway
Owner:	Angel Trains

Above: South West Trains ordered a fleet of 45 five-car Siemens 'Desiro' sets as part of the modernisation of the South Western main line routes, emerging in 2003-2004. The low-density fleet now seat 32 first and 327 standard class passengers. The sets have two powered driving cars and three intermediate Trailer Standards. Above is TS No. 67155, a vehicle seatsing 76 in the 2+2 style with one standard design toilet at one end. The vehicle displays the latest SWR grey and blue livery. **Antony Christie**

Above: Class 444 'Desiro' front end equipment, also applicable to Class 450 and 350 fleets. 1: Transmitter, 2: High level marker light, 3: Destination display, 4: Gangway door, 5: Forward facing camera, 6: White marker light, 7: Red tail light, 8: Headlight, 9: Anti-climber plate, 10: Warning Horns, 11: Dellner coupling with electrical connection. **CJM**

Above: In normal operation, it is usual to find the first class end Driving Motor Composite Open (DMCO) at the London end of formations, these currently offer 32 first class sets in the 2+2 style, originally, this coaches were set out in the 2+1 style but to increase capacity a revised layout is now fitted. With its DMCO coach nearest the camera, set No. 444027 passes Worting Junction, Basingstoke heading towards Waterloo in early 2023. Note the yellow above the three leading windows and to the side of the door. Class 444s have a single leaf plug door at either end of each coach, feeding a transverse walkway. **CJM**

Below: In mid-2023, the repainting of Class 444 sets into the latest South West Railway colours of grey and blue was ongoing, with several sets still in service carrying their original 'Stagecoach' blue, white, orange and red colours, with SW Railway branding, as displayed on set No. 444004 at St Johns, Woking on 25 April 2023. **CJM**

2023 Fleet

444001	SWR	444024	SWR
444002	SWR	444025	SWR
444003	SWR	444026	SWR
444004	SWR	444027	SWR
444005	SWR	444028	SWR
444006	SWR	444029	SWR
444007	SWR	444030	SWR
444008	SWR	444031	SWR
444009	SWR	444032	SWR
444010	SWR	444033	SWR
444011	SWR	444034	SWR
444012	SWR	444035	SWR
444013	SWR	444036	SWR
444014	SWR	444037	SWR
444015	SWR	444038	SWR
444016	SWR	444039	SWR
444017	SWR	444040	SWR
444018	SWR	444041	SWR
444019	SWR	444042	SWR
444020	SWR	444043	SWR
444021	SWR	444044	SWR
444022	SWR	444045	SWR
444023	SWR		

Key Facts

- Part of Siemens 'Desiro' family.
- 45 sets designed for use on Waterloo to Weymouth and Portsmouth line as replacement for Class 442 stock.
- Driving cabs are the same as on Class 450 and 350 stock.
- Main line sets with a single-leaf passenger door feeding a transverse walkway rather than a door opening direct into seating saloon.
- The original TSRMB coach with a pantograph well, have been rebuilt as a TS with universal toilet compartment.

Rolling Stock Review : 2023-2024

Class 450   'Desiro'

For history and pictures of this class, see Modern Locomotives Illustrated Issue 178.

Class:	450
Number range:	450001-450127
Built by:	Siemens Duewag, Germany, and Siemens SGP, Austria
Introduced:	2002-2007
Max speed:	100mph (161km/h)
Formation:	DMCO(A)+TSO(A)+TSO((B)+DMCO(B)
Vehicle numbers:	DMCO(A) - 63201-300/701-710/901-917
	TSO(A) - 64201-300/851-860/921-937
	TSO(B) - 68101-200/801-810/901-917
	DMCO(B) - 63601-700/751-760/921-937
Vehicle length:	66ft 9in (20.4m)
Height:	12ft 1½in (3.7m)
Width:	9ft 2in (2.7m)
Weight:	Total - 172.2 tonnes
	DMCO(A) - 48.0 tonnes
	TSO(A) - 35.8 tonnes
	TSO(B) - 39.8 tonnes
	DMCO(B) - 48.6 tonnes
Internal layout:	2+2F/2+2 & 2+3S
Seating:	Total - 16F/254S
	DMCO(A) - 8F/62S
	TSO(A) - 69S (+4TU)
	TSO(B) - 61S (+9TU)
	DMCO(B) - 8F/62S
Gangway:	Throughout
Toilets:	TSO - 1
Brake type:	Air (regenerative)
Bogie type:	Siemens SGP SF5000
Power collection:	750V dc third rail
Traction motor type:	4 x 1TB2016 0GB02 three-phase
Horsepower:	2,682hp (2,000kW)
Coupling type:	Outer - Dellner 12, Inner - Semi-auto
Multiple restriction:	444 and 450
Door type:	Bi-parting sliding plug
Construction:	Aluminium
Operator:	South Western Railway
Owner:	Angel Trains

Above & Below: Slam-door replacement stock for South West Trains, included a fleet of 127 four-car outer-suburban Siemens 'Desiro' sets, introduced from 2002. Originally, each set accommodated 24 first and 242 standard class passengers. The cab end style being the same as the Class 444s. The '450s' were originally painted in outer-suburban SWT blue with red contrasting bi-parting doors. From 2019 sets were facelifted and between 2020-2023 sets have been repainted in South Western Railway grey/blue livery. These two images show the current livery. Above is Trailer Standard Open (A) No. 64221 from set 450066. All '450' vehicles have two bi-parting plug doors at the 1/3 and 2/3 positions. Below are two four-car sets led by No. 450027, showing the current position for first class inwards of the cab at both ends, the first class area is identified by a yellow cant rail band and a yellow surround to the doors. **Antony Christie / CJM**

Right & Below: Although while not set up for overhead power collection, the '450s' have a pantograph well on the roof of their Trailer Standard B [TS(B)] coach, this is illustrated right. This vehicle also houses the universal toilet compartment. During the height of the Covid Pandemic, South Western Railway branded set 450067 with 'Thank You Key Workers', wording on both driving cars. The thoughtful embellishment remained in 2023. The set is seen near St Johns, Woking in April 2023. All Class 450s are based at the Siemens-operated Northam depot near Southampton, but can also be found stabled throughout the South Western Railway network. Both: **CJM**

'Desiro' — Class 450

2023 Fleet

450001 SWR	450065 SWR
450002 SWR	450066 SWR
450003 SWR	450067 SWR
450004 SWR	450068 SWR
450005 SWR	450069 SWR
450006 SWR	450070 SWR
450007 SWR	450071 SWR
450008 SWR	450072 SWR
450009 SWR	450073 SWR
450010 SWR	450074 SWR
450011 SWR	450075 SWR
450012 SWR	450076 SWR
450013 SWR	450077 SWR
450014 SWR	450078 SWR
450015 SWR	450079 SWR
450016 SWR	450080 SWR
450017 SWR	450081 SWR
450018 SWR	450082 SWR
450019 SWR	450083 SWR
450020 SWR	450084 SWR
450021 SWR	450085 SWR
450022 SWR	450086 SWR
450023 SWR	450087 SWR
450024 SWR	450088 SWR
450025 SWR	450089 SWR
450026 SWR	450090 SWR
450027 SWR	450091 SWR
450028 SWR	450092 SWR
450029 SWR	450093 SWR
450030 SWR	450094 SWR
450031 SWR	450095 SWR
450032 SWR	450096 SWR
450033 SWR	450097 SWR
450034 SWR	450098 SWR
450035 SWR	450099 SWR
450036 SWR	450100 SWR
450037 SWR	450101 SWR
450038 SWR	450102 SWR
450039 SWR	450103 SWR
450040 SWR	450104 SWR
450041 SWR	450105 SWR
450042 SWR	450106 SWR
450043 SWR	450107 SWR
450044 SWR	450108 SWR
450045 SWR	450109 SWR
450046 SWR	450110 SWR
450047 SWR	450111 SWR
450048 SWR	450112 SWR
450049 SWR	450113 SWR
450050 SWR	450114 SWR
450051 SWR	450115 SWR
450052 SWR	450116 SWR
450053 SWR	450117 SWR
450054 SWR	450118 SWR
450055 SWR	450119 SWR
450056 SWR	450120 SWR
450057 SWR	450121 SWR
450058 SWR	450122 SWR
450059 SWR	450123 SWR
450060 SWR	450124 SWR
450061 SWR	450125 SWR
450062 SWR	450126 SWR
450063 SWR	450127 SWR
450064 SWR	

Top & Above: *After refurbishment, the first class area was moved to both driving cars, taking the space behind the cab to the first door pocket. 2+2 seating is provided, illustrated in the top image. The view above shows the standard class seating using a mix of 3+2 and 2+2 layouts, using both airline and group styles. Seating now used the standard SWR blue moquette, with leatherette headrests in standard class, leather covered seats are used in first class. Grab handles are blue and grab poles orange. A Passenger Information System (PIS) is installed, with both automated and manual announcements. Doors are under the control of the guard and opened locally by passengers.* Both: **CJM**

Key Facts

- Introduced as South West Trains as slam-door replacement stock, allocated to Northam depot.
- Operate outer suburban and main line SWR services.
- A batch of 28 (450043-070) were modified with a high-capacity layout, classified as 450/5, now returned to normal configuration.
- First class seating now housed in driving cars.
- Can operate in multiple with Class 444 fleet.

Right: *Class 450 driving cab, also applicable to the Class 444 fleet. The cab position occupies just a one third width of the front end and is very cramped, visibility is also limited by the presence of the gangway connection. A combined power/brake controller is used, operated by the driver's left hand.* **CJM**

Rolling Stock Review : 2023-2024

Class 455

2023 Fleet

455/7			
455701	SWR	455732	SWR	455867	SWR
455702	SWR	455733	SWR	455868	SWR
455703	SWR	455734	SWR	455869	SWR
455705	SWR	455735	SWR	455870	SWR
455706	SWR	455737	SWR	455871	SWR
455707	SWR	455738	SWR	455872	SWR
455708	SWR	455739	SWR	455873	SWR
455709	SWR	455741	SWR	455874	SWR
455710	SWR	455742	SWR		
455711	SWR	455750	SWR	**455/9**	
455712	SWR			455901	SWR
455713	SWR	**455/8**		455902	SWR
455714	SWR	455848	SWR	455903	SWR
455715	SWR	455849	SWR	455904	SWR
455716	SWR	455850	SWR	455905	SWR
455717	SWR	455851	SWR	455906	SWR
455718	SWR	455852	SWR	455908	SWR
455719	SWR	455853	SWR	455909	SWR
455720	SWR	455854	SWR	455910	SWR
455721	SWR	455856	SWR	455911	SWR
455722	SWR	455857	SWR	455912	SWR
455723	SWR	455858	SWR	455913	SWR
455724	SWR	455859	SWR	455914	SWR
455725	SWR	455860	SWR	455915	SWR
455726	SWR	455861	SWR	455916	SWR
455727	SWR	455862	SWR	455917	SWR
455728	SWR	455863	SWR	455919	SWR
455729	SWR	455864	SWR	455920	SWR
455730	SWR	455865	SWR		
455731	SWR	455866	SWR		

Class:	455/7	455/8	455/9
Number range:	455701-455750	455848-455874	455901-455920
Built by:	BREL York	BREL York	BREL York
Refurbished by:	Bombardier, Ashford	Bombardier, Ashford	Bombardier, Ashford
Introduced:	1984-1985	1982-1984	1985
Refurbished:	2003-2007	2003-2008	2004-2007
Max speed:	75mph (121km/h)	75mph (121km/h)	75mph (121km/h)
Formation:	DTSO(A)+MSO+ TSO+DTSO(B)	DTSO(A)+MSO+ TSO+DTSO(B)	DTSO(A)+MSO+ TSO+DTSO(B)
Vehicle numbers:	DTSO(A) - 77727-77811 (odds) MSO - 62783-62825 TSO - 71526-71568 DTSO(B) - 77728-77812 (evens)	DTSO(A) - 77579-77725 (odds) MSO - 62709-62782 TSO - 71637-71710 DTSO(B) - 77580-77726 (evens)	DTSO(A) - 77813-77852 (odds) MSO - 62826-62845 TSO - 71714-71733 DTSO(B) - 77814-77852 (evens)
Vehicle length:	DTSO - 65ft 0½in (19.83m) MSO/TSO - 65ft 4½in (19.92m)	DTSO - 65ft 0½in (19.83m) MSO/TSO - 65ft 4½in (19.92m)	DTSO - 65ft 0½in (19.83m) MSO/TSO - 65ft 4½in (19.92m)
Height:	DTSO/MSO - 12ft 1½in (3.7m) TSO - 11ft 6½in (3.58m) (ex-Class 508)	12ft 1½in (3.7m)	12ft 1½in (3.7m)
Width:	9ft 3¼in (2.82m)	9ft 3¼in (2.82m)	9ft 3¼in (2.82m)
Weight:	Total: 133.4 tonnes DTSO(A) - 30.8 tonnes MSO - 45.7 tonnes TSO - 26.1 tonnes DTSO(B) - 30.8 tonnes	131.7 tonnes DTSO(A) - 29.5 tonnes MSO - 45.6 tonnes TSO - 27.1 tonnes DTSO(B) - 29.5 tonnes	Total - 131.8 tonnes DTSO(A) - 30.7 tonnes MSO - 46.3 tonnes TSO - 28.3 tonnes DTSO(B) - 26.5 tonnes
Internal layout:	2+3	2+3	2+3
Seating:	Total - 236S + 8 tip-up DTSO(A) - 50S (+4TU) MSO - 68S, TSO - 68S DTSO - 50S (+4TU)	Total - 236S (+8 TU) DTSO(A) - 50S (+4 TU) MSO - 68S, TSO - 68S DTSO(B) - 50S (+4 TU)	Total - 236S + 9 tip-up DTSO(A) - 50S (+4 TU) MSO - 68S, TSO - 68S DTSO(B) - 50S (+5 TU)
Gangway:	Throughout	Throughout	Throughout
Toilets:	Not fitted	Not fitted	Not fitted
Brake type:	Air (regen)	Air (regen)	Air (Westcode)
Bogie type:	DTSO - BREL BT13 MSO - BREL BP27 TSO - BREL BX1	DTSO - BREL BT13 MSO - BREL BP20 TSO - BREL BT13	DTSO - BREL BT13 MSO - BREL BP20 TSO - BREL BT13
Power collection:	750V dc third rail	750V dc third rail	750V dc third rail
Traction motor type:	4 X TSA TSA010163 of 322hp	4 X TSA TSA010163 of 322hp	4 X TSA TSA010163 of 322hp
Horsepower:	1,288hp (960kW)	1,288hp (960kW)	1,288hp (960kW)
Coupling type:	Outer - Tightlock, Inner - Bar	Outer - Tightlock, Inner - Bar	Outer - Tightlock, Inner - Bar
Multiple restriction:	Class 455	Class 455	Class 455
Door type:	Bi-parting sliding	Bi-parting sliding	Bi-parting sliding
Construction:	Steel (TSO - Aluminium)	Steel	Steel
Operator:	South Western Railway	South Western Railway	South Western Railway
Owner:	Porterbrook Leasing	Porterbrook Leasing	Porterbrook Leasing

Notes: Set 455912 has vehicle 67400 and set 455913 has vehicle 67301, both are former Class 210 vehicles

For history and pictures of this class, see *Modern Locomotives Illustrated* Issue 220

Above: Class 455 front end layout, applicable to all sub-classes. 1: Destination indicator, 2: Main reservoir pipe (yellow), 3: Multiple jumper cable, 4: Multiple jumper socket, 5: Lamp bracket, 6: Gangway door, 7: Headlight, 8: Combined marker/tail light, 9: Warning horns, 10: Tightlock coupling, 11: Electrical connection 'drum switch'. **CJM**

Key Facts

- The remaining Class 455s only have a short time left in service. They will be replaced when the Alstom Class 701 fleet is introduced.

- Two different front end designs are found, 455/8 have the original 'box' style, while 455/7 and 455/9 use a rounded roof profile.

- All remaining sets allocated to East Wimbledon and operate suburban services out of Waterloo. Can only operate in multiple with Class 455 stock.

Above & Below: The Class 455s were introduced for BR Southern Region in 1982-1983, when a batch of 74 Class 455/8s were built by BREL York, as fleet replacement SUB and EPB stock on the South Western division and also replace the 508s which had been loaned to the Southern from Merseyrail, 455/8 sets later migrated to the Central section. Two follow-on Class 455 orders were made, for 43 Class 455/7s and 20 Class 455/9s. In mid-2023, South Western Railway operated a fleet of 26 of the original design Class 455/8s, not all are operational with several stored. This batch has the earlier 'standard' front end, with the warning horns mounted at roof height. All sets are based at East Wimbledon depot and members of all three sub-classes operate as a common pool, all are refurbished and seat 236 in a mix of airline and group seating. In the 2010s, a major traction replacement project was undertaken to install a Kiepe AC traction package and Traction Systems Austria (TSA) traction motors. Set No. (45)5856 is seen at Earlsfield. All Class 455s carry just the last four digits of their number. To the right is an intermediate TSO coach, these seat 68 in the 2+2 layout. All vehicles have two pairs of bi-parting sliding doors on each side, crew controlled and passenger operated. Both: **CJM**

Rolling Stock Review : 2023-2024

Class 455

Above & Left: The second batch of Class 455s were the 43 Class 455/7s, built by BREL York in 1984-1985 as three car sets, on arrival on the Southern, they received a Trailer Standard Open (TSO) from a Class 508, thus making a four-car set. The Class 508 body profile was very different, and thus this batch have always had one odd profile vehicle, the coach being the third in this view of set No. 5717 passing Vauxhall, en route to Waterloo. The Class 455/7s were built with the later design rounded cab roof profile, with horns repositioned just below coupling level to the left side. The image right shows the ex-Class 508 TSO, No. 71557, formed in 455 set No. 5718. Note the near empty underframe. **CJM / Antony Christie**

Above: The final batch of Class 455s were 20 Class 455/9s, fitted with convection heating and thus have slightly different roof vents, this sub-class also has the revised rounded cab roof profile. Th '455/9s' are always recognisable in having vehicles all of the same body profile, plus the rounded cab roof design. All sub-classes of Class 455 are formed with a Motor Standard Open (MSO), housing traction equipment (third vehicle in train). Traction current (at 750V DC) is collected from shoes at the outer ends of DTSO vehicles, which is passed to the MSO via a train line. On either end of the MSO coach, a power jumper is provided for attachment to depot traction trolley systems. Class 455/9 No. 455910, is seen at Southampton Central. **Antony Christie**

Left & Right: The interior '455s' have been refurbished and feature a low-density 2+2 layout, with high levels of standing room, especially in the door pocket positions. A number of 'perch' seats are also provided. A Passenger Information System (PIS) is also fitted. **Antony Christie / CJM**

Rolling Stock Review : 2023-2024

Class 458 'Juniper'

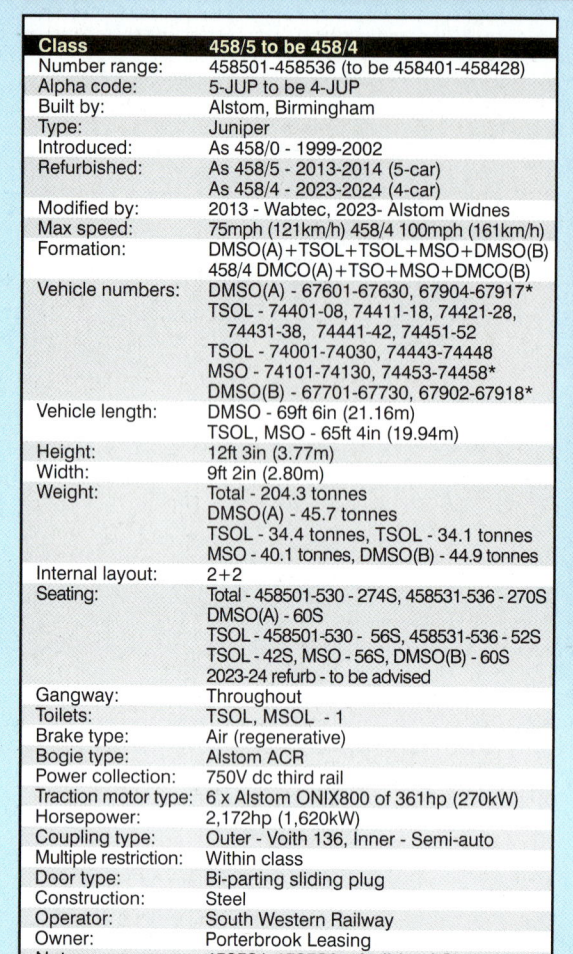

For history and pictures of this class, see Modern Locomotives Illustrated Issue 223

Class	458/5 to be 458/4
Number range:	458501-458536 (to be 458401-458428)
Alpha code:	5-JUP to be 4-JUP
Built by:	Alstom, Birmingham
Type:	Juniper
Introduced:	As 458/0 - 1999-2002
Refurbished:	As 458/5 - 2013-2014 (5-car) As 458/4 - 2023-2024 (4-car)
Modified by:	2013 - Wabtec, 2023- Alstom Widnes
Max speed:	75mph (121km/h) 458/4 100mph (161km/h)
Formation:	DMSO(A)+TSOL+TSOL+MSO+DMSO(B) 458/4 DMCO(A)+TSO+MSO+DMCO(B)
Vehicle numbers:	DMSO(A) - 67601-67630, 67904-67917* TSOL - 74401-08, 74411-18, 74421-28, 74431-38, 74441-42, 74451-52 TSOL - 74001-74030, 74443-74448 MSO - 74101-74130, 74453-74458* DMSO(B) - 67701-67730, 67902-67918*
Vehicle length:	DMSO - 69ft 6in (21.16m) TSOL, MSO - 65ft 4in (19.94m)
Height:	12ft 3in (3.77m)
Width:	9ft 2in (2.80m)
Weight:	Total - 204.3 tonnes DMSO(A) - 45.7 tonnes TSOL - 34.4 tonnes, TSOL - 34.1 tonnes MSO - 40.1 tonnes, DMSO(B) - 44.9 tonnes
Internal layout:	2+2
Seating:	Total - 458501-530 - 274S, 458531-536 - 270S DMSO(A) - 60S TSOL - 458501-530 - 56S, 458531-536 - 52S TSOL - 42S, MSO - 56S, DMSO(B) - 60S 2023-24 refurb - to be advised
Gangway:	Throughout
Toilets:	TSOL, MSOL - 1
Brake type:	Air (regenerative)
Bogie type:	Alstom ACR
Power collection:	750V dc third rail
Traction motor type:	6 x Alstom ONIX800 of 361hp (270kW)
Horsepower:	2,172hp (1,620kW)
Coupling type:	Outer - Voith 136, Inner - Semi-auto
Multiple restriction:	Within class
Door type:	Bi-parting sliding plug
Construction:	Steel
Operator:	South Western Railway
Owner:	Porterbrook Leasing
Notes:	458531-458536 rebuilds of Class 460s * vehicles not in order

2023 Fleet

458501	SWR	458513	SWR	458525	SWR
458502	SWR	458514	SWR	458526	SWR
458503	SWR	458515	SWR	458527	SWR
458504	(S)	458516	SWR	458528	SWR
458505	(S)	458417	SWR	458529	(S)
458506	(S)	458518	SWR	458530	SWR
458407	SWR	458519	SWR	458531	SWR
458508	SWR	458520	SWR	458532	SWR
458509	SWR	458521	SWR	458533	SWR
458510	SWR	458522	SWR	458534	SWR
458511	SWR	458523	(S)	458535	(S)
458512	SWR	458524	SWR	458536	SWR

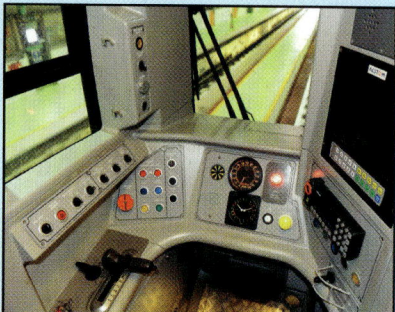

Above: *Class 458 front end layout. 1: High level marker light, 2: Warning horns (behind grille), 3: Gangway door, 4: Dual white marker and red tail light, 5: Headlight, 6: Anti-climber plates, 7: Voith coupling, 8: Electrical connection box.* **CJM**

Above Right: *The very cramped Class 458 driving cab, occupying one third of the vehicle end and offering restricted forward vision due to the gangway.* **CJM**

Key Facts

- Rebuilt Class 458/0 'Juniper' stock, extended from 4 to 5 cars by inserting vehicle from disbanded Class 460 stock. New front ends fitted.

- Six extra sets formed of rebuilt Class 460 (Gatwick Express) stock.

- Major changes announced in April 2021, will see 28 sets (458401-428) reformed with four-cars, upgraded for 100mph running and fitted with new seating, for Portsmouth line services. Project ongoing in 2023 at Alstom, Widnes. First refurbished set, classified as 458/4 returned to SWR in July 2023.

Below & Inset Right: *The original 30 Class 458 Alstom 'Juniper' sets emerged in 1999-2002. In 2013 the fleet was rebuilt at Wabtec Doncaster, augmented to five-car formation by adding one former Class 460 (Gatwick Express) vehicle to each set, plus, six additional five-car sets were formed using spare '460' vehicles. The rebuild also saw a new front end fitted. The original '460' vehicles can be identified by having ribbon glazing. The sets carry outer suburban blue livery with South Western Railway branding. A project commenced in mid 2022 by Alstom at Widnes to refurbish 28 sets as four-car units for use on the Waterloo - Portsmouth route. No set had emerged by late summer 2023. Set No. 458510 is seen at Clapham Junction. Inset right shows 2+2 standard class interior of an unrefurbished set.* Both: **CJM**

Left: *The first of the fully refurbished four-car Class 458/4 sets (No. 458417) emerged from the Alstom Widnes factory in July 2023 and moved to Bournemouth for commissioning. The set has lost its inherited Class 460 coach and now has first class seating behind both driving cabs. The disabled area and universal toilet is retained in the pantograph well fitted TS. The sets now with a new interior is painted in standard South Western Railway livery. The train is seen on its delivery run at Eastleigh.* **Neil Walkling**

Rolling Stock Review : 2023-2024

'Networker' — Class 465

For history and pictures of this class, see Modern Locomotives Illustrated Issue 223

Key Facts

- BR/NSE South Eastern division replacement stock for 'EPB' fleet.
- Built by BREL/ABB and Metro Cammell as part of a dual sourcing policy.
- Class 465/0 and 465/1 sets have been upgraded with Hitachi traction equipment.
- Some minor structural differences exist between the builds.
- Front ends redesigned to remove foot steps and handrails to reduce 'train surfing'.
- Storing of sets started in summer 2021 with a number stored at Worksop, later Ely.

Class:	465/0	465/1	465/9
Number range:	465001-465050	465151-465197	465901-456934
Former number range:	-	-	465201-465234
Built by:	BREL/ABB York	BREL/ABB York	Metro-Cammell, Birmingham
Type:	Networker	Networker	Networker
Years introduced:	1991-1993	1993	1991-1993
Modified by:	-	-	Wabtec, Doncaster
Year modified:	-	-	2005
Max speed:	75mph (121km/h)	75mph (121km/h)	75mph (121km/h)
Formation:	DMSO(A)+TSO+TSOL+DMSO(B)	DMSO(A)+TSO+TSOL+DMSO(B)	DMCO(A)+TSOL+TSO+DMCO(B)
Vehicle numbers:	DMSO(A) - 64759-64808 TSO - 72028-72126 (even Nos.) TSOL - 72029-72126 (odd Nos.) DMSO(B) - 64809-64858	DMSO(A) - 65800-65846 TSO - 72900-72992 (even Nos.) TSOL - 72901-72993 (odd Nos.) DMSO(B) - 65847-65893	DMCO(A) - 65700-65733 SOL - 72719-72785 (odd Nos.) TSO - 72720-72786 (even Nos.) DMCO(B) - 65750-65783
Vehicle length:	DMSO - 68ft 6½in (20.89m) TSO, TSOL - 65ft 9¾in (20.06m)	DMSO - 68ft 6½in (20.89m) TSO, TSOL - 65ft 9¾in (20.06m)	DMCO - 68ft 6½in (20.89m) TSO, TSOL - 65ft 9¾in (20.06m)
Height:	12ft 4½in (3.77m)	12ft 4½in (3.77m)	12ft 4½in (3.77m)
Width:	9ft 3in (2.81m)	9ft 3in (2.81m)	9ft 3in (2.81m)
Weight:	Total - 133.6 tonnes DMSO(A) - 39.2 tonnes TSO - 27.2 tonnes TSOL - 28 tonnes DMSO(B) - 39.2 tonnes	Total - 133.6 tonnes DMSO(A) - 39.2 tonnes TSO - 27.2 tonnes TSOL - 28 tonnes DMSO(B) - 39.2 tonnes	Total - 138.2 tonnes DMCO(A) - 39.2 tonnes TSOL - 30.3 tonnes TSO - 29.5 tonnes DMCO(B) - 39.2 tonnes
Internal layout:	2+3 high density	2+3 high density	2+2F/2+3S
Seating:	Total - 327S (+7 TU) DMSO(A) - 86S TSO - 90S TSOL - 65S (+7 TU) DMSO(B) - 86S	Total - 327S (+7 TU) DMSO(A) - 86S TSO - 90S TSOL - 65S (+7 TU) DMSO(B) - 86S	Total - 24F/291S (+7 TU) DMCO(A) - 12F/68S TSOL - 65S (+7 TU) TSO - 90S DMCO(B) - 12F/68S
Gangway:	Within set	Within set	Within set
Toilets:	TSOL - 1	TSOL - 1	TSOL - 1
Brake type:	Air (regenerative)	Air (regenerative)	Air (regenerative)
Bogie type:	Powered - Adtranz P3 Trailer - Adtranz T3	Powered - Adtranz P3 Trailer - Adtranz T3	Powered - SRP BP62 Trailer - SRP BT52
Power collection:	750V dc third rail	750V dc third rail	750V dc third rail
Traction motor type:	8 x Hitachi asynchronous	8 x Hitachi asynchronous	8 x Alsthom G352BY
Control system:	IGBT	IGBT	GTO
Horsepower:	3,004hp (2,240kW)	3,004hp (2,240kW)	3,004hp (2,240kW)
Coupling type:	Outer - Tightlock, Inner - Semi-auto	Outer - Tightlock, Inner - Semi-auto	Outer - Tightlock, Inner - Semi-auto
Multiple restriction:	Class 465 and 466	Class 465 and 466	Class 465 and 466
Door type:	Bi-parting sliding plug	Bi-parting sliding plug	Bi-parting sliding plug
Construction:	Aluminium	Aluminium	Aluminium
Operator:	SouthEastern Trains	Off lease (stored)	SouthEastern Trains
Owner:	Eversholt Leasing	Eversholt Leasing	Angel Trains
Sub-class differences:	BREL/ABB phase 1 train	BREL/ABB phase 2 train	465/2 Modified with first class

Right: BR/Network SouthEast ordered a fleet of 147 four-car 'Networker' sets to replace slam door suburban stock on the South Eastern division, built by BREL/ABB (97 sets) and Metro-Cammell (50 sets). They entered service in 1991-1993, based at Slade Green, apart from a few small detail differences, all are of the same design. Privatisation saw the sets pass to Connex SouthEastern and are now operated by Southeastern. Sets are refurbished, sporting white and dark blue livery, off-set by powder blue passenger doors. The 465/0 and 456/1 sets are now installed with Hitachi AC traction equipment, with the original underframe skirts on driving cars removed. '465s' seat 327 passengers and operate in a common pool. No. 465034 arrives at Lewisham. **CJM**

Right Lower: In 1993 a follow-on order for 47 sets classified as Class 465/1 emerged, similar to the original 50, but with minor detail differences, mainly to the interior fittings. Set No. 465161 is shown at Lewisham. **CJM**

Below: Class 465/0 interior, showing vehicle 64819, a Driving Motor Standard Open (DMSO) from set 465011. These vehicles seat 86 in the 2+3 layout. Each coach has two pairs of bi-parting sliding plug doors on each side. One toilet is provided in each four-car set in a TS vehicle. **Antony Christie**

Rolling Stock Review : 2023-2024

Class 465   'Networker'

Above: The interior of the Class 465s and 466s is based on the high-density 2+3 style, with a mix of airline and group seating. Interiors are of the Metro style, with lino floors, seats are without armrests, with the middle seats on the three seat side especially cramped. Through various refurbishments a number of different interiors are now found. Including the Class 465/9 batch which were refurbished with first class seating in driving cars, this was removed from the end of 2022. The 465/9s have a revised door vestibule area with curved grab rails as shown on set No. 465915. **Antony Christie**

Left: Networker front end layout and equipment, applicable to all Class 465 and 466 stock. 1: Destination indicator, 2: White marker light, 3: Headlight, 4: Red tail light, 5: Tightlock coupling, 6: Warning horns, 7: Coupling electric connection box. When originally built, a flat step was located above the coupling and handrails were attached to the front end, these had to be angled or removed to reduce 'train-surfing'. **CJM**

Below: The original 50 Class 465/2 fleet, built by Metro-Cammell, is now reduced to 16 sets (465235-465250 - all stored) after modification of 34 units as Class 465/9s, to include 12 first class seats in each driving car. The sets were used on Southeastern outer-suburban routes, but from December 2022 first class was withdrawn by Southeastern and sets are now all declassified. Set No. 465920 is seen at Tonbridge. **CJM**

2023 Fleet

465/0
465001 SET	465175 SET
465002 SET	465176 SET
465003 SET	465177 SET
465004 SET	465178 SET
465005 SET	465179 SET
465006 SET	465180 SET
465007 SET	465181 SET
465008 SET	465182 SET
465009 SET	465183 SET
465010 OLS	465184 SET
465011 SET	465185 SET
465012 SET	465186 SET
465013 SET	465187 SET
465014 SET	465188 SET
465015 SET	465189 SET
465016 SET	465190 SET
465017 SET	465191 SET
465018 SET	465192 SET
465019 SET	465193 SET
465020 SET	465194 SET
465021 SET	465195 SET
465022 SET	465196 SET
465023 SET	465197 SET
465024 SET	
465025 SET	**465/2**
465026 SET	465235 OLS
465027 SET	465236 OLS
465028 SET	465237 OLS
465029 SET	465238 OLS
465030 SET	465239 OLS
465031 SET	465240 OLS
465032 SET	465241 OLS
465033 SET	465242 OLS
465034 SET	465243 OLS
465035 SET	465244 OLS
465036 SET	465245 OLS
465037 SET	465246 OLS
465038 SET	465247 OLS
465039 SET	465248 OLS
465040 SET	465249 OLS
465041 SET	465250 OLS
465042 SET	
465043 SET	**465/9**
465044 SET	465901 SET
465045 SET	465902 SET
465046 SET	465903 SET
465047 SET	465904 SET
465048 SET	465905 SET
465049 SET	465906 SET
465050 SET	465907 SET
	465908 SET
465/1	465909 SET
465151 SET	465910 SET
465152 SET	465911 SET
465153 SET	465912 SET
465154 SET	465913 SET
465155 SET	465914 SET
465156 SET	465915 SET
465157 SET	465916 SET
465158 SET	465917 SET
465159 SET	465918 SET
465160 SET	465919 SET
465161 SET	465920 SET
465162 SET	465921 SET
465163 SET	465922 SET
465164 SET	465923 SET
465165 SET	465924 SET
465166 SET	465925 SET
465167 SET	465926 SET
465168 SET	465927 SET
465169 SET	465928 SET
465170 SET	465929 SET
465171 SET	465930 SET
465172 SET	465931 SET
465173 SET	465932 SET
465174 SET	465933 SET
	465934 SET

Rolling Stock Review : 2023-2024

For history and pictures of this class, see *Modern Locomotives Illustrated* Issue 223

'Networker' — Class 466

Key Facts

- Two-car version of 'Networker', built by Metro-Cammell to allow formation of 10-car trains in peak periods or operate two-car branch line services.
- DTSO are only unpowered driving cars of Networker design.
- Front ends modified to reduce train surfing, similar to Class 465.
- In summer 2021 five sets were taken off-lease and stored.

Class:	466
Number range:	466001-466043
Built by:	Metro-Cammell, Birmingham
Type:	Networker
Introduced:	1992-1994
Max speed:	75mph (121km/h)
Formation:	DMSO+DTSO
Vehicle numbers:	DMSO - 64860-64902
	DTSO - 78312-78354
Vehicle length:	68ft 6½in (20.89m)
Height:	12ft 4½in (3.77m)
Width:	9ft 3in (2.81m)
Weight:	Total - 72 tonnes
	DMSO - 40.6 tonnes
	DTSO - 31.4 tonnes
Internal layout:	2+3 high density
Seating:	Total - 168S
	DMSO - 86S
	DTSO - 82S
Gangway:	Within set
Toilets:	DTSO - 1
Brake type:	Air (regenerative)
Bogie type:	Powered - Adtranz P3
	Trailer - Adtranz T3
Power collection:	750V dc third rail
Traction motor type:	2 x Alstom G352AY of 375hp (280kW)
Horsepower:	750hp (560kW)
Coupling type:	Outer - Tightlock, Inner - Semi-auto
Multiple restriction:	Class 465 and 466
Door type:	Bi-parting sliding plug
Construction:	Aluminium
Operator:	SouthEastern Trains
Owner:	Angel Trains

2023 Fleet

466001	SET
466002	SET
466003	SET
466004	OLS
466005	SET
466006	SET
466007	SET
466008	SET
466009	SET
466010	OLS
466011	SET
466012	SET
466013	SET
466014	SET
466015	SET
466016	OLS
466017	SET
466018	SET
466019	SET
466020	SET
466021	SET
466022	SET
466023	SET
466024	OLS
466025	SET
466026	SET
466027	SET
466028	SET
466029	SET
466030	SET
466031	SET
466032	SET
466033	OLS
466034	SET
466035	SET
466036	SET
466037	SET
466038	SET
466039	SET
466040	SET
466041	SET
466042	SET
466043	OLS

Below & Inset: *To allow two, six or ten-car 'Networker' trains to be formed, Network SouthEast ordered a fleet of 43 two-car units, classified 466, from Metro Cammell. These could either be used for low patronage routes on their own, or to supplement Class 465 sets. The Class 466s include the only non-powered driving cars of the 'Networker' family. The fleet does not comply with the latest regulations for passengers with reduced mobility, not having a universal toilet compartment. Set No. 466001 is seen at London Bridge from its Driving Motor Standard Open (DMSO) end. The Class 466 interior (shown inset), is the same as on the Metro-Cammell-built Class 465s, it sports 2+3 seating. A modern Passenger Information System (PIS) and full CCTV system is fitted.*
CJM / Antony Christie

Right: *As with all Metro-Cammell-built 'Networker' driving cars, the '466s' have been retro-fitted with a roof mounted cab air conditioning system, this can be seen by the extra box on the cab roof and a grille adjacent to the fan. No. 466019 clearly shows this fitting. The train is led by its Driving Trailer Standard Open (DTSO) coach at London Bridge.* **Antony Christie**

Rolling Stock Review : 2023-2024

Class 484

Class:	484
Number range:	484001-484005
Origan:	Ex-LUL D stock
Built by:	Metro-Cammell
Rebuilt by:	Vivarail
Introduced originally:	1979-1983
Introduced to Isle of Wight:	2021-2022
Max speed:	60mph (97km/h)
Formation:	DMSO(A)+DMSO(B)
Vehicle numbers:	DMSO(A) -131-135
	DMSO(B) - 231-235
Vehicle length:	DMSO - 60ft 3¼in (18.37m)
Height:	11ft 10in (3.60m)
Width:	9ft 4½in (2.85m)
Weight:	Total - 53.56 tonnes
	DMSO(A) - 27.46 tonnes
	DMSO(B) - 26.1 tonnes
Internal layout:	2+2, Longitudinal
Seating:	Total - 80S + 4 Tip-up
	DMSO(A) 40S + 2 Tip-up
	DMSO(B) 40S + 2 Tip-up
Gangway:	Within set
Toilets:	Not fitted
Brake type:	Air
Bogie type:	Bombardier Flexx 1000
Power collection:	660V dc third rail
Traction motor type:	TSA AC motors
Horsepower:	(awaited)
Coupling type:	Wedgelock
Multiple restriction:	Within type only
Door type:	Single leaf sliding
Construction:	Aluminium
Owner:	Lombard Finance
Operator:	South Western Railway (Island Line)

Right & Below: *The two-car Class 484s, converted from former London Underground 'D-Stock' by Vivarail, as part of the Class 230 project, entered service on Island Rail in late 2021. The five sets are based at Ryde St Johns Road depot and are formed of two DMSO vehicles, each seating 40 in the 2+2 style. Four single leaf doors are on each side of each vehicle. Sets carry an Island Line version of South Western Railway livery. Above, set No. 484004 is seen departing from Ryde St Johns towards Ryde Esplanade. Below, No. 484001 is seen on the approach to Shanklin. Both:* **CJM**

Right: *Class 484 front end equipment layout. 1: Whistle, 2: Air vents, 3: High level marker light, 4: Destination display, 5: Headlight, 6: Combined marker/tail light, 7: Over-ride protection plate, 8: Warning horns located in coupling pocket, 9: Wedgelock coupling, with air connections below, 10: Electrical connection boxes with roll-back covers either side of coupling.* **CJM**

2023 Fleet

484001	SWR
484002	SWR
484003	SWR
484004	SWR
484005	SWR

Key Facts

- Replacement stock for the Isle of Wight network, consisting of five 2-car Vivarail 'D-Stock' conversions.
- Sets entered passenger service in November 2021 and are based at Ryde St Johns Road depot.
- Sets differ from previous 'D-Stock' conversions with a recessed whistle on the end above the non driving window.
- Trains have not been affected by the closure of Vivarail.

Below Left & Below Right: *The interiors of the Class 484s are impressive, seating is mainly in the longitudinal (London Underground) style, with a pair of facing seat groups towards the middle of each coach. Provision is made for a wheelchair parking space in each vehicle, no toilets are installed. Doors are under the control of the guard with local passenger operation. A good quality passenger information system has been installed. Both:* **CJM**

Class 507 and 508

2023 Fleet

507001	MER
507002	MER
507003	MER
507004	MER
507005	MER
507007	MER
507009	MER
507010	MER
507011	MER
507012	MER
507013	MER
507014	MER
507015	MER
507016	MER
507017	MER
507018	MER
507020	MER
507021	MER
507023	MER
507026	MER
507028	MER
507029	MER
507030	MER
507031	MER
507032	MER
507033	MER
508103	MER
508104	MER
508108	MER
508111	MER
508112	MER
508114	MER
508115	MER
508117	MER
508120	MER
508124	MER
508125	MER
508126	MER
508127	MER
508128	MER
508130	MER
508131	MER
508136	MER
508137	MER
508138	MER
508139	MER
508141	MER

Class:	507
Number range:	507001-507033
Built by:	BREL York
Years introduced:	1978-1980
Max speed:	75mph (121km/h)
Formation:	BDMSO+TSO+DMSO
Vehicle numbers:	BDMSO - 64367-64399
	TSO - 71342-71374
	DMSO - 64405-64437
Vehicle length:	BDMSO/DMSO - 64ft 11½in (19.80m)
	TSO - 65ft 4¼in (19.92m)
Height:	11ft 6½in (3.58m)
Width:	9ft 3in (2.82m)
Weight:	Total - 98 tonnes
	BDMSO - 37 tonnes
	TSO - 25.5 tonnes
	DMSO - 35.5 tonnes
Internal layout:	2+2
Seating:	Total - 186S (+6TU)
	BDMSO - 56S (+3TU)
	TSO - 74S
	DMSO - 56S (+3TU)
Gangway:	Within set (emergency end doors)
Toilets:	Not fitted
Brake type:	Air (Westcode/rheostatic)
Bogie type:	RX1
Power collection:	750V dc third rail
Traction motor type:	8 x GEC G310AZ of 110hp (82.12kW)
Horsepower:	880hp (657kW)
Coupling type:	Outer - Tightlock, Inner - Bar
Multiple restriction:	Class 507 and 508/1 only
Door type:	Bi-parting sliding
Construction:	Body - aluminium, Frame - steel
Operator:	MerseyRail
Owner:	Angel Trains

Below: *The Class 507 and 508 fleets should have been withdrawn by the start of 2023, but protracted delivery and commissioning of new Class 777s has seen that timescale slip, which is now likely to be very early 2024. The '507s' and '508s' were introduced from 1978 as replacement stock for the Merseyrail electric system, with the Class 508s arriving via a period on BR Southern Region. After privatisation they were subject to a major refurbishment contract, which saw a significant upgrade to interiors, as well as modernisation to the front ends, with new angled light clusters, and the addition of a high-level marker light. The sets were later repainted with one side carrying silver/grey and the other yellow, with Mersey Travel branding and pictograms. Class 507 No. 507017 is seen at Hooton from its silver/grey side. All sets are fitted with trip-cock equipment for Liverpool tunnel working.* **CJM**

Above Left: *Coupled between the two Driving Motor cars, is a Trailer Standard Open (TSO), this has seating for 74 passengers in the 2+2 style. Passenger doors on these fleets are under the control of the guard, with no local passenger control. Vehicle No. 71356 from set No. 507015 is seen from its yellow side.* **CJM**

Above Right: *Class 508 No. 501127 is seen from its yellow painted side. Note the 'trip-cock' shoe beam on the leading bogie, the power collection shoe in on the inner bogie.* **CJM**

Class:	508
Number range:	508103-508143
Former number range:	508003-508043
Built by:	BREL York
Years introduced:	1979-1980
Introduced as 508/1:	1984-1985
Max speed:	75mph (121km/h)
Formation:	DMSO+TSO+BDMSO
Vehicle numbers:	DMSO - 64651-64691
	TSO - 71485-71525
	BDMSO - 64694-64734
Vehicle length:	DMSO/BDMSO - 64ft 11½in (19.80m)
	TSO - 65ft 4½in (19.92m)
Height:	11ft 6½in (3.58m)
Width:	9ft 3in (2.82m)
Weight:	Total - 99.3 tonnes
	DMSO - 36.0 tonnes
	TSO - 26.7 tonnes
	BDMSO - 36.6 tonnes
Internal layout:	2+2
Seating:	Total - 186S (+6TU)
	DMSO - 56S (+3TU), TSO - 74S
	BDMSO - 56S (+3TU)
Gangway:	Within set, emergency end doors
Toilets:	Not fitted
Brake type:	Air (Westcode/rheostatic)
Bogie type:	BX1
Power collection:	750V dc third rail
Traction motor type:	8 x GEC G310AZ of 110hp (82.12kW)
Horsepower:	880hp (657kW)
Coupling type:	Outer - Tightlock, Inner - Bar
Multiple restriction:	Class 507 and 508/1 only
Door type:	Bi-parting sliding
Construction:	Body - aluminium, Frame - steel
Operator:	MerseyRail
Owner:	Angel Trains

Below: *Class 507/508 front end equipment. 1: High level marker light, 2: Air Horns, 3: Front emergency door (for tunnel section working), 4: Destination indicator, 5: Headlight, 6: Combined white marker and red tail light, 7: Tightlock coupling, 8: Electrical connection box with roller cover, 9: Emergency un-coupling lever, 10: Emergency air connection, 11: Drum switch for electrical connection.* **CJM**

Above: *A number of '507s' and '508s' carry nameplates, Wilfred Owen MC is attached to set 508136.* **CJM**

Below: *Refurbished interior of Class 507/508 stock, showing the 2+2 seating with plenty of circulating space. As these trains are not fitted with air conditioning, opening hopper windows are provided throughout.* **CJM**

Class 555   'Tyne & Wear Metro'

Class:	555
Number range:	555001-555046
Built by:	Stadler, St Margrethen, Switzerland
Product type:	Metro
Years introduced:	2023-2024
Max speed:	50mph (80km/h)
Formation:	DMS(A)+MS(A)+MS(B)+MS(C)+DMS(B)
Vehicle numbers:	DMS(A) - 990101-990146
	MS(A) - 990201-990246
	MS(B) - 990301-990346
	MS(C) - 990401-990436
	DMS(B) - 990501-990536
Train length:	196ft 7in (59.9m)
Height:	11ft 4in (3.44m)
Width:	8ft 8in (2.65m)
Weight:	Tare - 98 tonnes
	Full capacity - 141 tonnes
Floor height:	3ft 1in (940mm)
Internal layout:	Longitudinal
Accommodation:	Total - 600, Seats 104
Gangway:	Within set
Toilets:	Not fitted
Brake type:	Air (EP), regen and magnetic
Bogie type:	Stadler
Power collection:	1,500V dc overhead
Traction system:	IGBT - VVVF
Traction motor type:	8 x TSA TMF 41-17-4 of 160hp (120kW)
Power output:	Max: 1,770hp (1,320kW)
	Cont: 1,263hp (942kW)
Coupling type:	Outer - Dellner, Inner - Bar
Multiple restriction:	Class 555 only
Door type:	Bi-parting sliding plug
Construction:	Aluminum
Operator:	Nexus (Tyne & Wear)

Right Upper: *Class 555 is the Network Rail identity given to the new METRO EMUs under construction by Stadler for the Tyne & Wear Metro. The fleet of 46 five-car sets will enter service from late 2023 replacing the original Metrocar fleet. Set No. 555003 was the first to arrive in the UK in spring 2023 and is seen inside the new depot at Gosforth.* **Nexus**

Right Middle: *Dynamic overnight testing of the new fleet commenced in summer 2023, set No. 555003 is seen on one of the first test runs at Four Lane Ends station. These sets show some resemblance to the Merseyrail Class 777s.* **Nexus**

Above: *Prior to delivery to the UK, a limited amount of test running was undertaken in Europe On 18 November 2022, set No. 555002 is seen stabled at Arth-Goldau station in Switzerland. These sets are allocated the Network Rail classification of 555 as they operate in part over Network Rail infrastructure.* **Phil Wormald**

Left: *The interior of the Class 555 is a total change from the previous Tyne & Wear fleet, with longitudinal seating, wide between vehicle gangways, CCTV, public address, a passenger information system and air conditioning. Seats are covered in a 'Metro' grey and yellow moquette and most on board furniture is finished in Tyne & Wear yellow. The driving cars have cycle and pram parking places, while the two outer intermediate coaches have wheelchair spaces.* **Nexus**

'Desiro City' — Class 700

Class:	700/0 (8-car) RLU	700/1 (12-car) FLU
Number range:	700001-700060	700101-700155
Built by:	Siemens Duewag, Germany	Siemens Duewag, Germany
Type:	Desiro City	Desiro City
Years introduced:	2015-2018	2015-2018
Max speed:	100mph (161km/h)	100mph (161km/h)
Formation:	DMCO(A)+PTSO+MSO+TSO+TSOW+MSO+ PTSO+DMCO(B)	DMCO(A)+PTSO+MSO+MSO+TSO+TSO+ TSO+TSO+MSO+MSO+PTSO+DMCO(B)
Vehicle numbers:	DMCO(A) - 401001-401060 PTSO - 402001-402060 MSO - 403001-403060 TSO - 406001-406060 TSOW - 407001-407060 MSO - 410001-410060 PTSO - 411001-411060 DMCO(B) - 412001-412060	DMCO(A) - 401101-401155 PTSO - 402101-402155 MSO - 403101-403155 MSO - 404101-404155 TSO - 405101-405155 TSO - 406101-406155 TSO - 407101-407155 TSO - 408101-408155 MSO - 409101-409155 MSO - 410101-410155 PTSO - 411101-411155 DMCO(B) - 412101-412155
Train length:	528ft 2in (162m)	794ft 0in (242m)
Vehicle length:	Driving - 67ft 4in (20.52m) Intermediate - 66ft 2in (20.16m)	Driving - 67ft 4in (20.52m) Intermediate - 66ft 2in (20.16m)
Height:	12ft 1in (3.7m)	12ft 1in (3.7m)
Width:	9ft 2in (2.8m)	9ft 2in (2.8m)
Floor height:	43.3in (1.1m)	43.3in (1.1m)
Internal layout:	2+2 metro and commuter	2+2 metro and commuter
Seating:	Total - 52F/364S (+17TU) DMCO(A) - 26F/16S, PTSO - 54S MSO - 64S, TSO - 56S, TSOW - 40S, MSO - 64S, PTSO - 54S, DMCO(B) - 26F/16S	Total - 52F/602S (+18TU) DMCO(A) - 26F/20S, PTSO - 54S MSO - 60S, MSO - 56S, TSO - 64S, TSO - 56S, TSO - 38S, TSO - 64S, MSO - 56S, MSO - 60S, PTSO - 54S DMC(B) - 26F/20S
Gangway:	Within set	Within set
Toilets:	3 (PTS, TSOW)	5 (PTS, MSO, TSOW)
Weight:	Total - 273.9 tonnes DMCO(A) - 38.5t, PTSO - 33.1t, MSO - 36.2t, TSO - 29.1t, TSOW - 29.1t, MSO - 36.2t, PTSO - 33.2t, DMCO(B) - 38.5t	Total - 400.6 tonnes DMCO(A) - 38.2t, PTSO - 34.4t, MSO - 36.0t, MSO - 35.8t, TSO - 26.8t TSO - 28.3t, TSO(W) - 28.7t, TSO - 27.9t MSO - 36.6t, MSO - 35.3t, PTSO - 34.4t DMCO(B) - 38.2t
Brake type:	Air (regenerative)	Air (regenerative)
Bogie type:	Siemens SGP SF7000 IF	Siemens SGP SF7000 IF
Power collection:	25kV ac overhead & 750V dc third rail	25kV ac overhead & 750V dc third rail
Traction motor type:	16 x Siemens of 200kW	24 x Siemens of 200kW
Horsepower:	4,291hp (3,200kW)	6,439hp (4,800kW)
Wheel diameter (New):	32.3in (820mm)	32.3in (820mm)
Coupling type:	Outer - Dellner 12, Inner - Semi-auto	Outer - Dellner 12, Inner - Semi-auto
Multiple restriction:	Within class and Class 717	Within class and Class 717
Door type:	Bi-parting sliding pocket	Bi-parting sliding pocket
Construction:	Aluminium	Aluminium
Operator:	GTR	GTR
Owner:	CLT/GTR	CLT/GTR
Sub class variations:	8-car set (Reduced Length Unit)	12-car set (Full Length Unit)

Key Facts

- First build of Siemens 'Desiro City' platform.
- Two sub-classes in use FLU - Full Length Unit (12-cars) and RLU - Reduced Length Unit (8-cars).
- Designed for 'Thameslink' services, using Automatic Train Operation (ATO) in central London 'core' to increase train paths.
- First class seating provided in each driving car.
- Fleet of 115 sets built, allocated to Siemens operated maintenance facilities at Hornsey (London) and Three Bridges (Sussex).

Right Upper: Class 700 'Desiro City' front end equipment positions. 1: High level marker light, 2: Destination indicator, 3: Combined white marker and red tail light, 4: Headlight, 5: Lamp bracket, 6: Coupling electrical connection box, 7: Dellner 12 coupling, 8: Obstacle deflector plate. **CJM**

Below: The Class 700 sets can operate from either 25kV AC overhead or from a 750V DC third rail power supply. Viewed against older stock DC stock, the third rail power collector shoes look very flimsy, especially mounted on the inside frame bogies used on this stock. **CJM**

Above: Govia Thameslink Railway (GTR) has a fleet of 115 Class 700 Siemens-built 'Desiro City' sets for its Thameslink route, 60 eight-car Reduced Length Units (RLU) and 55 12-car Full Length Units (FLU). These operate on services via central London on a north-south axis, with up to 24 trains per hour in each direction passing through the 'core' section between King's Cross and London Bridge, working under Automatic Train Operation (ATO). A RLU eight-coach set No. 700060 is seen calling at Harpenden operating from the 25kV AC overhead system. All Class 700s are based at the Siemens-operated maintenance depots at Three Bridges (Sussex) and Hornsey, (North London). **CJM**

Rolling Stock Review : 2023-2024

Class 700 'Desiro City'

2023 Fleet

700/0					
700001	GTR	700040	GTR	700118	GTR
700002	GTR	700041	GTR	700119	GTR
700003	GTR	700042	GTR	700120	GTR
700004	GTR	700043	GTR	700121	GTR
700005	GTR	700044	GTR	700122	GTR
700006	GTR	700045	GTR	700123	GTR
700007	GTR	700046	GTR	700124	GTR
700008	GTR	700047	GTR	700125	GTR
700009	GTR	700048	GTR	700126	GTR
700010	GTR	700049	GTR	700127	GTR
700011	GTR	700050	GTR	700128	GTR
700012	GTR	700051	GTR	700129	GTR
700013	GTR	700052	GTR	700130	GTR
700014	GTR	700053	GTR	700131	GTR
700015	GTR	700054	GTR	700132	GTR
700016	GTR	700055	GTR	700133	GTR
700017	GTR	700056	GTR	700134	GTR
700018	GTR	700057	GTR	700135	GTR
700019	GTR	700058	GTR	700136	GTR
700020	GTR	700059	GTR	700137	GTR
700021	GTR	700060	GTR	700138	GTR
700022	GTR			700139	GTR
700023	GTR	**700/1**		700140	GTR
700024	GTR	700101	GTR	700141	GTR
700025	GTR	700102	GTR	700142	GTR
700026	GTR	700103	GTR	700143	GTR
700027	GTR	700104	GTR	700144	GTR
700028	GTR	700105	GTR	700145	GTR
700029	GTR	700106	GTR	700146	GTR
700030	GTR	700107	GTR	700147	GTR
700031	GTR	700108	GTR	700148	GTR
700032	GTR	700109	GTR	700149	GTR
700033	GTR	700110	GTR	700150	GTR
700034	GTR	700111	GTR	700151	GTR
700035	GTR	700112	GTR	700152	GTR
700036	GTR	700113	GTR	700153	GTR
700037	GTR	700114	GTR	700154	GTR
700038	GTR	700115	GTR	700155	GTR
700039	GTR	700116	GTR		
		700117	GTR		

Above: The Class 700s are high-capacity sets, with a 12 car FLU train seating 52 first and 602 standard class passengers. First class seating for 26 is provided in each driving car, positioned in the area directly behind the cab. On a Class 700/1 12-car FLU set, five toilets are provided, the one located in the 407xxx coach is of the universal type, with the disabled seating area and wheelchair space adjacent. FLU set No. 700116 is seen passing Honor Oak Park operating on the third rail DC system. **CJM**

Left Upper & Left Lower: The reduced length eight-car sets have two intermediate Trailer Standard Open (TSO) vehicles, formed in the middle of the train. One (4070xx vehicle) is a TS with wheelchair spaces, seating 48 in the 2+2 style, with a universal access toilet and two wheelchair spaces, No. 407006 is shown (upper). Each set is formed with two Pantograph Trailer Seconds (PTS), formed as the second coach of the consist. These seat 54 with one standard toilet compartment (left end in picture). No. 411126 from full length set 700126 is seen at London Bridge. Both: **Antony Christie**

Above: The first class seating area in the DMCO, uses the 2+2 style, with each seat having a fixed table, armrests and antimacassars on seat backs. **Antony Christie**

Below: As with a number of passenger and freight operators, GTR Thameslink have shown their support for the LGBT community by branding the two driving cars of full length set No. 700155 in a 'rainbow' livery. The set is seen passing a Class 378 at Honor Oak Park. **CJM**

Below: Class 700 driving cab. Due to level of items required this is a very busy environment, with train/platform interface screens on the left, with train management screens in front of the driver. On the right is a communications package for both on train and train to signalbox connection. The combined power/brake controller is located on the main desk in front of the seat armrest, while the Automatic Train Operation (ATO) controls are on the desk in front of the right armrest. **CJM**

'Aventra' / 'Arterio' — Class 701

For history and pictures of this class, see *Modern Locomotives Illustrated* Issue 243

Class:	701/0	701/5
Number range:	701/0 - 701001 - 701060	701/5 - 701501 - 701530
Built by:	Bombardier, Derby	Bombardier, Derby
Years introduced:	2023-2024	2023-2024
Type:	Aventra 'Arterio'	Aventra 'Arterio'
Max speed:	100mph (161km/h)	100mph (161km/h)
Formation:	DMS+PMS+TS+MS+MS+ MS+MS+TS+MS+DMS	DMS+MS+TS+MS+DMS
Vehicle numbers:	DMS - 480001-480060, PMS - 481001-481060 TS - 482001-482060, MS - 483001-483060 MS - 484001-484060, MS - 485001-485060 MS - 486001-486060, TS - 487001-487060 MS - 488001-488060, DMS - 489001-489060	DMS - 480101-480130 MS - 481101-481130 TS - 482101-482130 MS - 483101-483130 DMS - 484101-484130
Vehicle length:	Driving cars 20.88m, Intermediate cars 19.90m	Driving cars 20.88m, Intermediate cars 19.90m
Height:	3.58m	3.58m
Width:	2.78m	2.78m
Internal layout:	2+2	2+2
Seating:	Total - 701/0 - 540S +10 TU DMS - 56S, PMS - 60S, TS - 34S + 10TU, MS - 60S	701/5 - 266S +10 TU DMS - 56S, PMS - 60S, TS - 34S + 10TU, MS - 60S
Gangway:	Within set	Within set
Toilets:	1 in each TS = 2	1 in TS = 1
Brake type:	Air (regenerative)	Air (regenerative)
Bogie type:	Flex B5000	Flex B5000
Power collection:	750V dc third rail	750V dc third rail
Traction Motors:	16 x Bombardier 250kW	8 x Bombardier 250kW
Output:	4,000kW	2,000kW
Coupling type:	Outer - Dellner 12, Inner - Semi-auto	Outer - Dellner 12, Inner - Semi-auto
Multiple restriction:	Within class	Within class
Door type:	Bi-parting sliding plug	Bi-parting sliding plug
Construction:	Aluminium	Aluminium
Operator:	South Western Railway	South Western Railway
Owner:	Rock Rail	Rock Rail

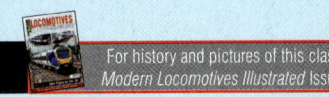

Above: Class 701 SWR 'Aventra' front end layout. 1: High level marker light, 2: Destination indicator, 3: Head, marker and tail light group, 4: Lamp bracket, 5: Dellner coupling, 6: Warning horns, 7: Electrical connection box. **Philip Sherratt**

2023 Fleet

701/0
701001 SWR, 701002 SWR, 701003 SWR, 701004 SWR, 701005 SWR, 701006 SWR, 701007 SWR, 701008 SWR, 701009 SWR, 701010 SWR, 701011 SWR, 701012 SWR, 701013 SWR, 701014 SWR, 701015 SWR, 701016 SWR, 701017 SWR, 701018 SWR, 701019 SWR, 701020 SWR, 701021 SWR, 701022 SWR, 701023 SWR, 701024 SWR, 701025 SWR, 701026 SWR, 701027 SWR, 701028 SWR, 701029 SWR, 701030 SWR, 701031 SWR, 701032 SWR, 701033 SWR, 701034 SWR, 701035 SWR, 701036 SWR, 701037 SWR, 701038 SWR, 701039 SWR, 701040 SWR, 701041 SWR, 701042 SWR, 701043 SWR, 701044 SWR, 701045 SWR, 701046 SWR, 701047 SWR, 701048 SWR, 701049 SWR, 701050 SWR, 701051 SWR, 701052 SWR, 701053 SWR, 701054 SWR, 701055 SWR, 701056 SWR, 701057 SWR, 701058 SWR, 701059 SWR, 701060 SWR

701/5
701501 SWR, 701502 SWR, 701503 SWR, 701504 SWR, 701505 SWR, 701506 SWR, 701507 SWR, 701508 SWR, 701509 SWR, 701510 SWR, 701511 SWR, 701512 SWR, 701513 SWR, 701514 SWR, 701515 SWR, 701516 SWR, 701517 SWR, 701518 SWR, 701519 SWR, 701520 SWR, 701521 SWR, 701522 SWR, 701523 SWR, 701524 SWR, 701525 SWR, 701526 SWR, 701527 SWR, 701528 SWR, 701529 SWR, 701530 SWR

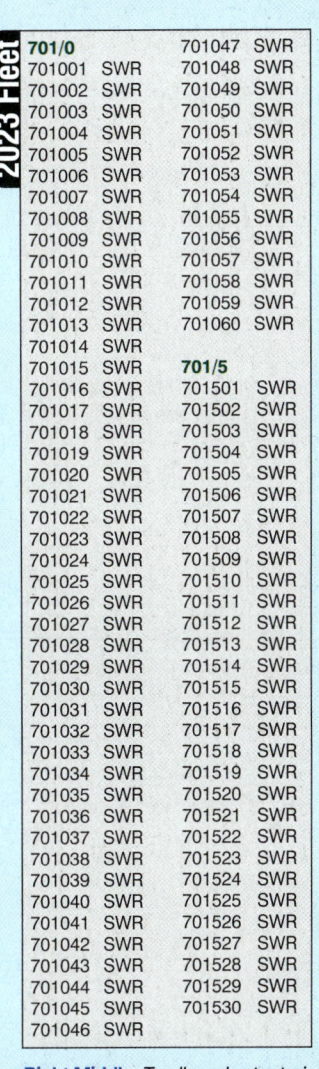

Above: Following the new South Western Railway franchise awarded to First Group (70%) and MTR Corporation (30%) in August 2017, 750 Bombardier, later Alstom, 'Aventra' vehicles, formed as 60 ten-car and 30 five-car sets were ordered. Trains have been on delivery since 2019, but by summer 2023 none were accepted for passenger use, when they are eventually accepted they will replace Classes 455 and 458s. Considerable test running is being undertaken. In full SWR livery, set No. 701024 is seen passing Vauxhall on a test run from Eastleigh to Staines via Waterloo in January 2023. **CJM**

Right Middle: To allow shorter train operation, a fleet of 30 Class 701/5s have been built, formed of five vehicles, all are powered except the middle vehicle which is a Trailer Standard Open, this vehicle also houses universal type toilet. Set No. 701512 is seen near Hook with a test train bound for Eastleigh on 7 March 2023. **CJM**

Right Below: Each 10-car Class 701 have five intermediate Motor Standard coaches, each seating 60 in the 2+2 style. MS No. 483004 from set 701004 is illustrated. **CJM**

Rolling Stock Review : 2023-2024

Class 707 'Desiro City'

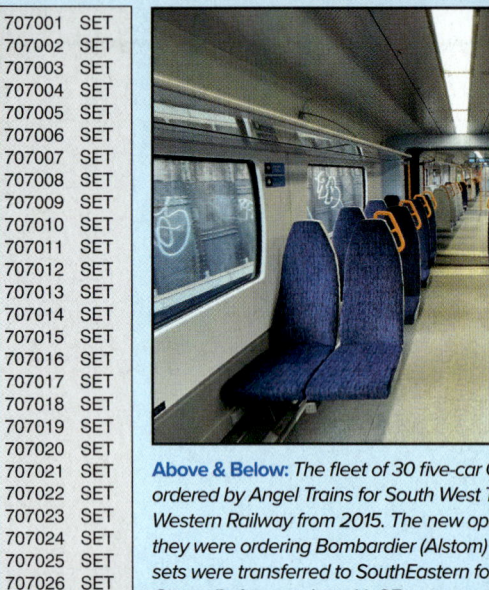

For history and pictures of this class, see *Modern Locomotives Illustrated* Issue 243

2023 Fleet

707001	SET
707002	SET
707003	SET
707004	SET
707005	SET
707006	SET
707007	SET
707008	SET
707009	SET
707010	SET
707011	SET
707012	SET
707013	SET
707014	SET
707015	SET
707016	SET
707017	SET
707018	SET
707019	SET
707020	SET
707021	SET
707022	SET
707023	SET
707024	SET
707025	SET
707026	SET
707027	SET
707028	SET
707029	SET
707030	SET

Above & Below: The fleet of 30 five-car Class 707 'Desiro City' sets, ordered by Angel Trains for South West Trains were delivered to South Western Railway from 2015. The new operator did not want the fleet, as they were ordering Bombardier (Alstom) Class 701 stock. In 2021-2022 sets were transferred to SouthEastern for Metro work, based at Slade Green. Before service with SE, sets were repainted in blue livery and given a SouthEastern blue interior. In the lower view, set No. 707009 is seen at New Cross. '707' interiors are of the low-density style using 2+2 seating in a mix of airline and group, wide passenger stand-back areas are provided by door pockets. Passenger information displays are provided throughout, giving train information and travel advice. The seat moquette uses SouthEastern blue with orange fixtures and fittings. The wide between vehicle gangways give an impression of space and helps with 'on-board' security. No toilets are provided. Both: **CJM**

Class:	707
Number range:	707001-707030
Built by:	Siemens Duewag, Germany
Type:	Desiro City
Years introduced:	2015-2018
Max speed:	100mph (161km/h)
Formation:	DMSO(A)+TSO+TSO+PTSO+DMSO(B)
Vehicle numbers:	DMSO(A) - 421001-421030
	TSO - 422001-422030
	TSO - 423001-423030
	PTSO - 424001-424030
	DMSO(B) - 425001-425030
Vehicle length:	Driving - 67ft 4in (20.52m)
	Intermediate - 66ft 2in (20.16m)
Height:	12ft 1in (3.7m)
Width:	9ft 2in (2.8m)
Floor height:	43.3in (1.1m)
Internal layout:	2+1, 2+2
Seating:	Total - 271S
	DMSO(A) - 46S, TSO - 64S, TSO - 53S + 4TU, PTSO - 62S, DMSO(B) - 46S
Gangway:	Within set
Toilets:	Not fitted
Weight:	Total - 160 tonnes
	DMSO(A) - 37.9t, TSO - 28.4t, TSO - 28.5t, PTSO - 27.6t, DMSO(B) - 37.6t
Brake type:	Air (regenerative)
Bogie type:	Siemens SGP SF7000 IF
Power collection:	750V dc third rail, wired for 25kV ac overhead
Traction motor type:	8 x Siemens of 150kW
Horsepower:	1,600hp (1,200kW)
Wheel diameter (New):	32.3in (820mm)
Coupling type:	Outer - Dellner 12, Inner - Semi-auto
Multiple restriction:	Within class only
Door type:	Bi-parting sliding pocket
Construction:	Aluminium
Operator:	South Eastern Railway
Owner:	Angel Trains

Below: In the middle of each train is a Trailer Standard Open (TSO), which has seating for 54 and two wheelchair parking spaces, this is marked on the outside of the vehicle by an orange surround to the window. For their SouthEastern use, the sets have been branded 'CityBeam'. In mid-2023 a handful of sets remained based at Wimbledon for use on South Western, to cover for the traction shortage caused by late delivery of Class 701 stock. **CJM**

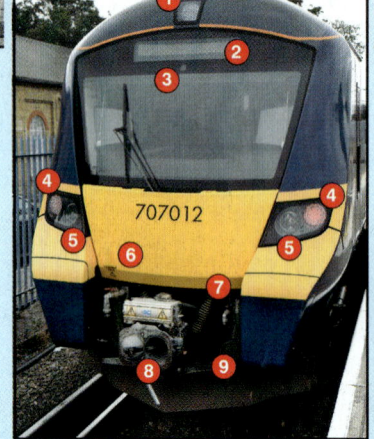

Above: Class 701 front equipment.
1: High level marker light, 2: Destination indicator, 3: Forward facing camera, 4: combined white marker / red tail light, 5: Headlight, 6: Lamp bracket, 7: Emergency air connection, 8: Dellner coupling, 9: Air horns in coupling pocket. **CJM**

Key Facts

- Ordered by Stagecoach, now operating for SouthEastern.
- Part of Siemens 'Desiro City' platform.
- Fitted with low-density 2+2 and 2+1 standard class 'Metro' style interior.
- No toilet compartment.
- Now operated as Citybeam' fleet.

Rolling Stock Review : 2023-2024

'Aventra' — Class 710

710/1
710101 LOL
710102 LOL
710103 LOL
710104 LOL
710105 LOL
710106 LOL
710107 LOL
710108 LOL
710109 LOL
710110 LOL
710111 LOL
710112 LOL
710113 LOL
710114 LOL
710115 LOL
710116 LOL
710117 LOL
710118 LOL
710119 LOL
710120 LOL
710121 LOL
710122 LOL
710123 LOL
710124 LOL
710125 LOL
710126 LOL
710127 LOL
710128 LOL
710129 LOL
710130 LOL

710/2
710256 LOL
710257 LOL
710258 LOL
710259 LOL
710260 LOL
710261 LOL
710262 LOL
710263 LOL
710264 LOL
710265 LOL
710266 LOL
710267 LOL
710268 LOL
710269 LOL
710270 LOL
710271 LOL
710272 LOL
710273 LOL

710/3
710374 LOL
710375 LOL
710376 LOL
710377 LOL
710378 LOL
710379 LOL

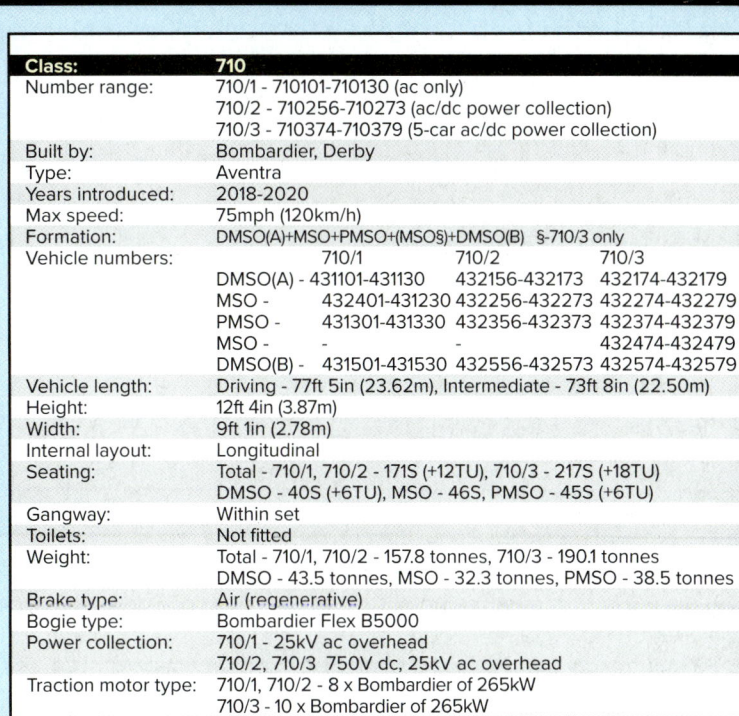

Class:	710
Number range:	710/1 - 710101-710130 (ac only)
	710/2 - 710256-710273 (ac/dc power collection)
	710/3 - 710374-710379 (5-car ac/dc power collection)
Built by:	Bombardier, Derby
Type:	Aventra
Years introduced:	2018-2020
Max speed:	75mph (120km/h)
Formation:	DMSO(A)+MSO+PMSO+(MSO§)+DMSO(B) §-710/3 only
Vehicle numbers:	710/1 / 710/2 / 710/3
DMSO(A) -	431101-431130 / 432156-432173 / 432174-432179
MSO -	432401-431230 / 432256-432273 / 432274-432279
PMSO -	431301-431330 / 432356-432373 / 432374-432379
MSO -	- / - / 432474-432479
DMSO(B) -	431501-431530 / 432556-432573 / 432574-432579
Vehicle length:	Driving - 77ft 5in (23.62m), Intermediate - 73ft 8in (22.50m)
Height:	12ft 4in (3.87m)
Width:	9ft 1in (2.78m)
Internal layout:	Longitudinal
Seating:	Total - 710/1, 710/2 - 171S (+12TU), 710/3 - 217S (+18TU)
	DMSO - 40S (+6TU), MSO - 46S, PMSO - 45S (+6TU)
Gangway:	Within set
Toilets:	Not fitted
Weight:	Total - 710/1, 710/2 - 157.8 tonnes, 710/3 - 190.1 tonnes
	DMSO - 43.5 tonnes, MSO - 32.3 tonnes, PMSO - 38.5 tonnes
Brake type:	Air (regenerative)
Bogie type:	Bombardier Flex B5000
Power collection:	710/1 - 25kV ac overhead
	710/2, 710/3 750V dc, 25kV ac overhead
Traction motor type:	710/1, 710/2 - 8 x Bombardier of 265kW
	710/3 - 10 x Bombardier of 265kW
Output:	710/1, 710/2 - 2,120kW, 710/3 - 1,650kW
Coupling type:	Outer - Dellner 12, Inner - Semi-auto
Multiple restriction:	Within class
Door type:	Bi-parting sliding plug
Construction:	Aluminium
Operator:	London Overground
Owner:	SMBC/Lombard

For history and pictures of this class, see *Modern Locomotives Illustrated* Issue 243

Key Facts

- Three fleets of Class 710 'Aventra' units operate with London Overground. 30 AC only units and 24 AC/DC sets.
- 710/1s operate on West Anglia lines from Liverpool Street to Chingford, Enfield and Cheshunt. 710/2 and 710/3s operate on Gospel Oak-Barking and Euston-Watford routes.
- All sets are based at Willesden depot.
- Seating is of the longitudinal 'London Underground' style.
- Sets do not carry yellow warning ends.

Top: *Class 710 front end equipment, applicable to all sub-classes. 1: High level marker light, 2: Destination indicator, 3: Combined white marker/red tail light, 4: Headlight, 5: Horns (behind grille), 6: Lamp bracket, 7: Dellner coupling, 8: Electrical connection box.* **CJM**

Above Inset: *The '710' fleet is fitted with a standard Bombardier/Alstom 'Aventra' cab. The screens on the left, display either the inside of the train or the train-platform interface for driver only operation. The combined power/brake controller is the 'swan neck' handle on the left side. A full train management system (TMS) is fitted. No. 710260 is shown.* **CJM**

Above: *The 30 four-car Class 710/1, AC powered sets operate on services from Liverpool Street to Enfield, Chingford and Cheshunt. Each seat 171, with standing room for over 500. The fleet carries London Overground black, white and orange livery with no yellow warning ends. Set No. 710127 is seen at Liverpool Street.* **Antony Christie**

Right: *The four-car Class 710s, have two Motor Standard intermediate coaches, one with a pantograph at the outer end. MSOs seat 46, with the pantograph vehicle seating 45. Vehicle No. 431327 seen from the non-pantograph end.* **Antony Christie**

Rolling Stock Review : 2023-2024

Class 710   'Aventra'

Left: Motor Standard Open (MSO) No. 732264 clearly shows the empty nature of these intermediate vehicle underframes, with just brake, heat and light equipment fitted. The far bogie has one axle powered, and that bogie has a power collection shoe. Seating in MSO coaches is for 52 in the longitudnal style, with two pairs of sliding plug doors on each side. **CJM**

Below: The interior of the Class 710 fleet is set out with 'London Underground' style longitudinal seating. This has not been welcome on the busy West Anglia routes. Wide between vehicle gangways are provided and a high quality passenger information system is installed giving train information and London travel advice. A very basic, tram like interior. **CJM**

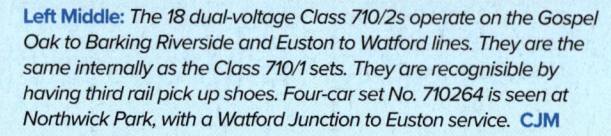

Left Middle: The 18 dual-voltage Class 710/2s operate on the Gospel Oak to Barking Riverside and Euston to Watford lines. They are the same internally as the Class 710/1 sets. They are recognisible by having third rail pick up shoes. Four-car set No. 710264 is seen at Northwick Park, with a Watford Junction to Euston service. **CJM**

Left Below: In mid-2021 six, five-car Class 710/3s, Nos, 710374-710379 started to enter service, but the full fleet of six were not all in traffic by mid-2023. These units have an additional 46 seat MSO and are deployed on the busy Euston to Watford all-stations services, set No. 710377 is seen approaching Willesden Junction low-level in March 2023. **Antony Christie**

Rolling Stock Review : 2023-2024

'Desiro City' — Class 717

For history and pictures of this class, see Modern Locomotives Illustrated Issue 243

2023 Fleet

717001	GTR
717002	GTR
717003	GTR
717004	GTR
717005	GTR
717006	GTR
717007	GTR
717008	GTR
717009	GTR
717010	GTR
717011	GTR
717012	GTR
717013	GTR
717014	GTR
717015	GTR
717016	GTR
717017	GTR
717018	GTR
717019	GTR
717020	GTR
717021	GTR
717022	GTR
717023	GTR
717024	GTR
717025	GTR

Right: Class 717 front end layout. 1: High level marker light, 2: Destination indicator, 3: Forward facing camera, 4: Fold out front emergency door/steps, 5: Combined white marker and red tail light, 6: Headlight, 7: Lamp bracket, 8: Coupling electric connection box, 9: Dellner 12 coupling, 10: Warning horns in coupling pocket, 11: Obstacle deflector plate. **CJM**

Class:	717
Number range:	717001-717025
Built by:	Siemens Duewag, Germany
Type:	Desiro City
Years introduced:	2018-2019
Max speed:	85mph (137km/h)
Formation:	DMSO(A)+TSO(A)+TSO(B)+MSO+PTSO+DMSO(B)
Vehicle numbers:	DMSO(A) - 451001-451025
	TSO(A) - 452001-452025
	TSO(B) - 453001-453025
	MSO - 454001-545025
	PTSO - 455001-455025
	DMSO(B) - 456001-456025
Vehicle length:	Driving - 67ft 4in (20.52m)
	Intermediate - 66ft 2in (20.16m)
Height:	12ft 1in (3.7m)
Width:	9ft 2in (2.8m)
Floor height:	43.3in (1.1m)
Internal layout:	2+2
Seating:	Total - 362S (+15TU)
	DMSO(A)-52 (+4TU), TSO(A)-68, TSO(B)-61 (+4TU), MSO-68, PTSO-61 (+3TU), DMSO(B)-52 (+4TU)
Gangway:	Within set, emergency end doors
Toilets:	Not fitted
Weight:	Total - 204.5 tonnes
	DMSO - 38.8t, TSO(A) - 28.8t, TSO(B) - 28.7t
	MSO - 35.5t, PTSO - 33.9t
Brake type:	Air (regenerative)
Bogie type:	Siemens SGP SF7000 IF
Power collection:	750V dc third rail and 25kV ac overhead
Traction motor type:	6 x Siemens of 200kW
Horsepower:	1,609hp (1,200kW)
Wheel diameter:	32.3in (820mm)
Coupling type:	Outer - Dellner 12, Inner - Semi-auto
Multiple restriction:	Within class and Class 700
Door type:	Bi-parting sliding pocket
Construction:	Aluminium
Operator:	GTR
Owner:	Rock Rail, Moorgate

Left: The Class 717 driving cab is very cramped, fitting adjacent to the front escape door. To include full cab equipment, the right hand panel is hinged. Recently all the Class 717s have been fitted with full ETCS cab signalling for operation on the Great Northern route from Moorgate. Space on the driving desk was provided for this during the original build. **CJM**

Right: A fleet of 25 six-car 'Desiro-City' Class 717s, owned by Rock Rail, are operated by Govia Thameslink on Great Northern services from Moorgate. Each set seats 362 standard class passengers in the 2+2 style. The sets are dual voltage, 25kV AC overhead and 750V DC third rail supply. The '717s' were the first of the 'Desiro City' design to include an emergency end door with hinged steps; these are automatically deployed, in seconds, if an emergency evacuation is needed in the tunnel section. Apart from the cab end, sets are very much like a Class 700. Set No. 717009 is seen at Palmers Green. **CJM**

Key Facts

- Introduced to replace GTR 'Great Northern' Class 313s.
- First 'Desiro City' design fitted with emergency end door and steps for tunnel evacuation.
- Sets do not have yellow warning ends.
- The '717s' can operate with Class 700s.
- Dual voltage sets, able to operate from 750V DC third rail or 25kV ac overhead.

Below: The interior style, based on the 2+2 layout, with large open gangways between vehicles and large standing areas helps to reduce station dwell times and speed up passenger boarding and departure times. The interiors are basic, the seating area of a DMSO is shown. **CJM**

Above: The Class 717s are formed of two driving - Driving Motor Standard Open (DMSO), this together with a mid-train Motor Standard Open (MSO), provides traction at 1,609hp (1,200kW). In this view view an intermeiate MSO is shown, these seat 68 and weigh 35.5 tonnes. Both bogies are powered. **CJM**

Rolling Stock Review : 2023-2024

Class 720 'Aventra'

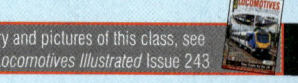

For history and pictures of this class, see Modern Locomotives Illustrated Issue 243

Class:	720/1 and 720/5		720/6
Number range:	720/1 - 720101 - 720144		720601-720612
	720/5 - 720501 - 720589		
Built by:	Bombardier, Derby		Bombardier, Derby
Type:	Aventra		Aventra
Years introduced:	2019-2023		2022-2023
Max speed:	100mph (161km/h)		100mph (161km/h)
Formation:	DMS+PMS+MS+MS+DTS	DMS+PMS+MS+MS+DTS	
Vehicle numbers:	720/1	720/5	
	DMS - 450101-450144	DMS - 450501-450589	DMS - 450601-450612
	PMS - 451101-451144	PMS - 451501-451589	PMS - 451601-451612
	MS - 452101-452144	MS - 452501-452589	MS - 452602-452612
	MS - 453101-453144	MS - 453501-453589	MS - 453601-453612
	DTS - 459101-459144	DTS - 459501-459589	DTS - 459601-459606
Train length:	400ft 3in (122m)		400ft 3in (122m)
Vehicle length:	Driving - 80ft 1in (24.47m)		Driving - 80ft 1in (24.47m)
	Intermediate - 79ft 5in (24.21m)		Intermediate - 79ft 5in (24.21m)
Height:	12ft 4in (3.87m)		12ft 4in (3.87m)
Width:	9ft 1in (2.78m)		9ft 1in (2.78m)
Internal layout:	3+2		3+2
Seating:	461S plus 18 Tip-up		tba
	DMS - 90S, PMS - 94S, MS - 110S, MS - 110S DTS - 57S		
Weight:	Total - 193.7 tonnes		tba
	DMS - 42.4t, PMS - 39.8t, MS(A) - 39.6t		
	MS(B) - 34.1t, DTS - 37.7t		
Gangway:	Within set		Within set
Brake type:	Air (regenerative)		Air (regenerative)
Bogie type:	Bombardier Flex B5000		Bombardier Flex B5000
Power collection:	25kV ac overhead		25kV ac overhead
Traction motor type:	8 x Bombardier 265kW		8 x Bombardier 265kW
Horsepower:	2,120kW		2,120kW
Toilets:	PMS, DTS 1		PMS, DTS 1
Coupling type:	Outer - Dellner 12, Inner - Semi-auto		Outer - Dellner 12, Inner - Semi-auto
Multiple restriction:	Within class		Within class
Door type:	Bi-parting sliding plug		Bi-parting sliding plug
Construction:	Aluminium		Aluminium
Operator:	Greater Anglia		c2c
Owner:	Angel Trains/CBA		Porterbrook

2023 Fleet

720/1
720101 GAR, 720102 GAR, 720103 GAR, 720104 GAR, 720105 GAR, 720106 GAR, 720107 GAR, 720108 GAR, 720109 GAR, 720110 GAR, 720111 GAR, 720112 GAR, 720113 GAR, 720114 GAR, 720115 GAR, 720116 GAR, 720117 GAR, 720118 GAR, 720119 GAR, 720120 GAR, 720121 GAR, 720122 GAR, 720123 GAR, 720124 GAR, 720125 GAR, 720126 GAR, 720127 GAR, 720128 GAR, 720129 GAR, 720130 GAR, 720131 GAR, 720132 GAR, 720133 GAR, 720134 GAR, 720135 GAR, 720136 GAR, 720137 GAR, 720138 GAR, 720139 GAR, 720140 GAR, 720141 GAR, 720142 GAR, 720143 GAR, 720144 GAR

720/5
720501 GAR, 720502 GAR, 720503 GAR, 720504 GAR, 720505 GAR, 720506 GAR, 720507 GAR, 720508 GAR, 720509 GAR, 720510 GAR, 720511 GAR, 720512 GAR, 720513 GAR, 720514 GAR, 720515 GAR, 720516 GAR, 720517 GAR, 720518 GAR, 720519 GAR, 720520 GAR, 720521 GAR, 720522 GAR, 720523 GAR, 720524 GAR, 720525 GAR, 720526 GAR, 720527 GAR, 720528 GAR, 720529 GAR, 720530 GAR, 720531 GAR, 720532 GAR, 720533 GAR, 720534 GAR, 720535 GAR, 720536 GAR, 720537 GAR, 720538 GAR, 720539 GAR, 720540 GAR, 720541 GAR, 720542 GAR, 720543 GAR, 720544 GAR, 720545 GAR, 720546 GAR, 720547 GAR, 720548 GAR, 720549 GAR, 720550 GAR, 720551 GAR, 720552 GAR, 720553 GAR, 720554 GAR, 720555 GAR, 720556 GAR, 720557 GAR, 720558 GAR, 720559 GAR, 720560 GAR, 720561 GAR, 720562 GAR, 720563 GAR, 720564 GAR, 720565 GAR, 720566 GAR, 720567 GAR, 720568 GAR, 720569 GAR, 720570 GAR, 720571 GAR, 720572 GAR, 720573 GAR, 720574 GAR, 720575 GAR, 720576 GAR, 720577 GAR, 720578 GAR, 720579 GAR, 720580 GAR, 720581 GAR, 720582 GAR, 720583 GAR, 720584 GAR, 720585 GAR, 720586 GAR, 720587 GAR, 720588 GAR, 720589 GAR

720/6
720601 c2c, 720602 c2c, 720603 c2c, 720604 c2c, 720605 c2c, 720606 c2c, 720607 c2c, 720608 c2c, 720609 c2c, 720610 c2c, 720611 c2c, 720612 c2c

Left: Class 720 'Aventra' front end equipment. 1: High level marker light, 2: Destination indicator, 3: Forward facing camera, 4: Front headlight/marker/tail light group, 5: Sign to indicate the position of wheelchair facilities, 6: Emergency lamp bracket, 7: Dellner coupling, 8: Electrical connection box. CJM

Below: The Bombardier (now Alstom)-built Class 720 'Aventra' units for Greater Anglia have been heavily delayed due to ongoing technical issues. The 665 vehicle order funded by Angel Trains, were to be formed as 22 ten-car and 89 five-car sets, but have been delivered in five vehicle formation. They are based at Ilford depot and replaced Class 317, 321, 360 and 379 units by spring 2023. The sets are stylish, well laid out, even though the seats are very cramped and hard, decent quality toilets are installed and a provision is made for disabled travellers. The disabled accommodation is marked by a blue sign on the front (see image left, with a blue banner above the side window). Set No. 720588 is seen at Rayleigh. Sets carry the last three digits of their running number on the front end, with full set numbers applied low down on the cab-side. The Class 720s are unusual as they have one powered and one trailer driving car, the powered one being at the pantograph end of the train. CJM

162 — Rolling Stock Review : 2023-2024

'Aventra' — Class 720

Right Above & Right Middle: *The three intermediate vehicles of Class 720s are powered, two are Motor Standard (MS) and one is a Pantograph Motor Standard (PMS). All passenger vehicles have two pairs of bi-parting sliding-plug doors on each side, released by train crew, and locally operated by passengers. In the upper view a MS coach is shown, the lower picture illustrates a Pantograph Motor Standard coach from its power collection end. Both:* **CJM**

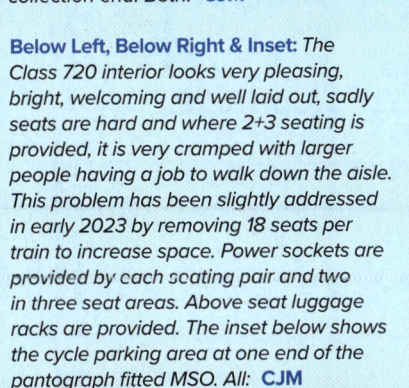

Below Left, Below Right & Inset: *The Class 720 interior looks very pleasing, bright, welcoming and well laid out, sadly seats are hard and where 2+3 seating is provided, it is very cramped with larger people having a job to walk down the aisle. This problem has been slightly addressed in early 2023 by removing 18 seats per train to increase space. Power sockets are provided by each seating pair and two in three seat areas. Above seat luggage racks are provided. The inset below shows the cycle parking area at one end of the pantograph fitted MSO. All:* **CJM**

Left: *From summer 2022, 12 five-car Class 720/6 sets were delivered by Alstom to Wembley for dynamic testing, these were destined for North Thameside operator c2c. Originally this order was for 10-car sets, but was amended during production for 12 five-cars. The sets are based at East Ham and entered service late summer 2023. Prior to hand over, a significant amount of main line testing was conducted on the West Coast main line, where set No. 720608 was recorded passing Daventry in February 2023.* **CJM**

Key Facts

- Originally ordered for Greater Anglia as a fleet of five and 10-car sets, contract altered for all five-vehicle trains.
- Delivery and commissioning heavily delayed due to 'Aventra' technical issues.
- Unusual configuration with one driving car unpowered.
- Sets delivered out of numeric sequence, with 80 per cent of fleet in service by mid-2023.
- Class 720/6s delivered mid-2022.

Class 730 'Aventra'

Class:	730/0	730/1
Number range:	730001-730048	730101-730136
Built by:	Alstom, Derby	Alstom, Derby
Type:	Aventra	Aventra
Years introduced:	2021-2022 - On delivery	2021-2022 - On delivery
Max speed:	90mph (145km/h)	110mph (177km/h)
Formation:	DMS(A)+PMS+DMS(B)	DMC+MS+PMS+MS+DMS
Vehicle numbers:	DMS(A) - 490001-490048	DMC - 490101-490136
	PMS - 492001-492048	MS - 491101-491136
	DMS(B) - 494001-494048	PMS - 492101-492136
		MS - 493101-493136
		DMS - 494101-494136
Train length:	tba	tba
Vehicle length:	Driving - 80ft 1in (24.47m)	Driving - 80ft 1in (24.47m)
	Intermediate - 79ft 5in (24.21m)	Intermediate - 79ft 5in (24.21m)
Height:	12ft 4in (3.87m)	12ft 4in (3.87m)
Width:	9ft 1in (2.78m)	9ft 1in (2.78m)
Internal layout:	2+2	2+2
Seating:	tba	tba
Weight:	tba	tba
Gangway:	Throughout	Throughout
Brake type:	Air (regenerative)	Air (regenerative)
Bogie type:	Bombardier Flex B5000	Bombardier Flex B5000
Power collection:	25kV ac overhead	25kV ac overhead
Traction motor type:	Bombardier 250kW	Bombardier 250kW
Horsepower:	tba	tba
Coupling type:	Outer - Dellner 12, Inner - Semi-auto	Outer - Dellner 12, Inner - Semi-auto
Multiple restriction:	Within class	Within class
Door type:	Bi-parting sliding plug	Bi-parting sliding plug
Construction:	Aluminium	Aluminium
Operator:	West Midlands Railway	West Midlands Railway
Owner:	Corelink Rail	Corelink Rail

Above: *Class 730 front end layout. 1: High level marker light, 2: Destination indicator, 3: Front gangway door, 4: Head, marker, tail light unit, 5: Lamp bracket, 6: Dellner coupling, 7: Air warning horns, 8: Coupling electrical connection box.* **Cliff Beeton**

Above & Below: *Another of the 'Aventra' platform of trains, currently on delivery and commissioning are the West Midlands Class 730 fleet. These come in two different sub-classes 730/0 and 730/1 and carry two different liveries. The 730/0s display West Midlands mauve and gold and will replace the Class 323s on Birmingham CrossCity services, while the 730/1s are finished in London North Western green for main line work from Euston to Liverpool, Crewe and Birmingham, replacing '350s'. Above, two Class 730/0 sets Nos. 730001 and 730004 are seen passing Tamworth. The inset left shows the intermediate PMS vehicle. Both:* **Antony Christie**

Right Upper & Right Lower: *In summer 2021, Alstom delivered the first Class 730/1s displaying London North Western grey and green livery for type testing. In mid-2023 the sets had not entered service. Set No. 730103 is seen at Crewe in November 2022. These five-car sets have two intermediate Motor Standard and one Pantograph Motor Standard vehicles. A Motor Standard is shown from set No. 730103. The intermediate vehicles are un-branded. Both:* **Cliff Beeton**

Key Facts

- A fleet of 48 Class 730/0 three-car sets for use on the CrossCity line in Birmingham to replace '323s' are on delivery, painted in mauve and orange livery.
- A fleet of 36 five-car sets to operate longer-distance London Northwestern services, carry LNW livery.
- First 'Aventra' platform trains to be fitted with end gangways.

2023 Fleet

730/0
730001 WMR
730002 WMR
730003 WMR
730004 WMR
730005 WMR
730006 WMR
730007 WMR
730008 WMR
730009 WMR
730010 WMR
730011 WMR
730012 WMR
730013 WMR
730014 WMR
730015 WMR
730016 WMR
730017 WMR
730018 WMR
730019 WMR
730020 WMR
730021 WMR
730022 WMR
730023 WMR
730024 WMR
730025 WMR
730026 WMR
730027 WMR
730028 WMR
730029 WMR
730030 WMR
730031 WMR
730032 WMR
730033 WMR
730034 WMR
730035 WMR
730036 WMR
730037 WMR
730038 WMR
730039 WMR
730040 WMR
730041 WMR
730042 WMR
730043 WMR
730044 WMR
730045 WMR
730046 WMR
730047 WMR
730048 WMR

730/1
730101 WMR
730102 WMR
730103 WMR
730104 WMR
730105 WMR
730106 WMR
730107 WMR
730108 WMR
730109 WMR
730110 WMR
730111 WMR
730112 WMR
730113 WMR
730114 WMR
730115 WMR
730116 WMR
730117 WMR
730118 WMR
730119 WMR
730120 WMR
730121 WMR
730122 WMR
730123 WMR
730124 WMR
730125 WMR
730126 WMR
730127 WMR
730128 WMR
730129 WMR
730130 WMR
730131 WMR
730132 WMR
730133 WMR
730134 WMR
730135 WMR
730136 WMR

'Flirt' — Class 745

For history and pictures of this class, see Modern Locomotives Illustrated Issue 243

Class:	745/0 InterCity	745/1 Stansted
Number range:	745/0 - 745001-745010	745/1 - 745101-745110
Built by:	Stadler, Bussnang, Switzerland	Stadler, Bussnang, Switzerland
Body shells:	Stadler, Bussnang, Switzerland & Szolnok, Hungary	Stadler, Bussnang, Switzerland & Szolnok, Hungary
Type:	Flirt	Flirt
Years introduced:	2020	2020
Max speed:	100mph (161km/h)	100mph (161km/h)
Formation:	DMF+PTF+TS(A)+TS(B)+TS(C)+MS(A)+MS(B)+ TS(D)+TS(E)+TS(F)+PTS+DMS	DMS+PTS+TS(A)+TS(B)+TS(C)+MS(A)+MS(B)+ TS(D)+TS(E)+TS(F)+PTS+DMS
Vehicle numbers:	745/0 - DMF - 413001-413010, PTF - 426001-426010 TS(A) - 332001-332010, TS(B) - 343001-343010 TS(C) - 341001-341010, MS(A) - 301001-301010 MS(B) - 302001-302010, TS(D) - 342001-342010 TS(E) - 344001-344010, TS(F) - 346001-346010 PTS - 322001-322010, DMS - 312001-312010	745/1 - DMS - 313001-313010, PTS - 326101-326110 TS(A) - 332101-332110, TS(B) - 343101-343110 TS(C) - 341101-341110, MS(A) - 301101-301110 MS(B) - 302101-302110, TS(D) - 342101-342110 TS(E) - 342101-342110, TS(F) - 346101-346110 PTS - 322101-322110, DMS - 312101-312110
Train length:	776ft 4in (236.6m)	776ft 4in (236.6m)
Vehicle length:	Driving - tba, Intermediate - tba	Driving - tba, Intermediate - tba
Height:	tba	tba
Width:	8ft 9in (2.72m)	8ft 9in (2.72m)
Floor height:	37.7in (960mm)	37.7in (960mm)
Internal layout:	2+1F, 2+2S	2+2S
Seating:	Total: 757 - 80F/624S/53TU Supplied by Fisa, Italy (Lean)	Total: 767S - 722S/45TU Supplied by Fisa, Italy (Lean)
Gangway:	Within set	Within set
Toilets:	5 - 1 universal, 4 standard	5 - 1 universal, 4 standard
Weight:	Total - tba	Total - tba
Brake type:	Air (regenerative)	Air (regenerative)
Bogie type:	Stadler, Jacob	Stadler, Jacob
Power collection:	25kV ac overhead	25kV ac overhead
Traction motor type:	8 x TSA (Traction Swiss Austria)	8 x TSA (Traction Swiss Austria)
Power at rail:	2600kW (3,487hp)	2600kW 2600kW (3,487hp)
Wheel diameter (New):	Motor bogie - 34.25in (870mm), Trailer bogie - 29.92in (760mm)	Motor bogie - 34.25in (870mm), Trailer bogie - 29.92in (760mm)
Coupling type:	Outer - Dellner 10, Inner - Special joint/articulated	Outer - Dellner 10, Inner - Special joint/articulated
Multiple restriction:	Not fitted for multiple operation	Not fitted for multiple operation
Door type:	Bi-parting sliding plug (one per car), 51.2in (1300mm opening)	Bi-parting sliding plug (one per car), 51.2in (1300mm opening)
Fittings:	Sanding & scrubber block	Sanding & scrubber block
Construction:	Aluminium	Aluminium
Operator:	Greater Anglia	Greater Anglia
Owner:	Rock Rail EA	Rock Rail EA

2023 Fleet

745/0		745/1	
745001	GAR	745101	GAR
745002	GAR	745102	GAR
745003	GAR	745103	GAR
745004	GAR	745104	GAR
745005	GAR	745105	GAR
745006	GAR	745106	GAR
745007	GAR	745107	GAR
745008	GAR	745108	GAR
745009	GAR	745109	GAR
745010	GAR	745110	GAR

Key Facts

- Greater Anglia replacement stock for Norwich Intercity services and Stansted Express route.
- Twelve-car trains formed as six 'pairs', can be split into two six-car sets for maintenance.
- Each 'pair' of vehicles are articulated.
- Class 745/0 sets are Intercity with first class seating at one end.
- Class 745/1 fleet are high capacity and dedicated to Stansted Airport services from Liverpool Street.

Above, Below & Inset: In 2019-2020, a fleet of 20 Stadler-built 12-car Class 745 'Flirt' units for Greater Anglia were delivered, formed as two sub-classes, marshalled as six semi-articulated 'pairs', formed into two six-car half trains. The 10 Class 745/0s, Nos. 745001-745010 are 12-car Intercity units, for the London Liverpool Street to Norwich route. Each train has 80 first and 624 standard class seats. As part of the Stadler 'Flirt' family, the sets look very similar to the Class 755s from the front end, but are recognisable by the omission of a multiple control connection box above the Dellner coupling. Sets carry the last three digits of their set number on the front end, a black triangle denotes the end with disabled facilities. Set No. 745004 passes Shenfield with its first class end, which is usually at the Liverpool Street end of the train, leading. All sets of both sub-classes are based at Norwich Crown Point. The inset image above, shows the retractable step plate, bridging the gap between train and platform, it extends to the required length, when the door open button is pressed, before the doors open. In the image right, the connection between the two two-vehicle sub-sets is seen. Units can be split in the middle for maintenance. One set of bi-parting plug doors are carried on each side of each coach, these are crew released, but operated locally by passengers. Coach letters are permanently applied either side of passenger doors. All: **CJM**

Rolling Stock Review : 2023-2024

Class 745   'Flirt'

Above Left & Above Right: *Interior of Intercity Class 745/0, left shows the standard class seating, using the 2+2 layout in a mix of airline and group style. On the right, is the first class, using 2+1 seating and located in one driving car and the attached trailer. Ambience in both saloons is good. A quality PIS and automated announcement system in fitted, as is an electronic seat reservation system. Both:* **CJM**

Upper Middle, Above & Left: *A fleet of 10 12-car Class 745/1s are dedicated to the Greater Anglia-operated Liverpool Street to Stansted Airport 'Stansted Express' service. These are standard class only sets, with accommodation for 722, with 45 tip-up seats. Based at Norwich Crown Point, sets also operate some London-Norwich services. Both '745' sub-classes are finished in the Greater Anglia two-tone grey, off-set by red panels by wheelsets and red passenger doors. In the middle view, set No. 745110 is seen at Liverpool Street. The view above shows an articulated bogie linking the two vehicles from a 'pair'. The green above window band indicated cycle space. On the left, the all standard class '745/1s' have seating in the 2+2 layout throughout the train. Each seat pair has a charging plug and a USB connection. The train has wide gangways between vehicles and seats have hinged armrests. The majority of seats are set out in the high-density airline style, with a limited number of group seats available. The interior of a TSO vehicle is shown. Note the gangway shape, giving the maximum space in the upper section, enabling a visual 'view' throughout the train. All:* **CJM**

'Flirt' — Class 755

For history and pictures of this class, see *Modern Locomotives Illustrated* Issue 243

Class:	755/3	755/4
Number range:	755/3 - 755325-755338	755/4 - 755401-755424
Built by:	Stadler, Bussnang, Switzerland & Siedlce, Poland. Some 'PPs' Valencia, Spain	
Body shells:	Stadler, Bussnang, Switzerland & Szolnok, Hungary	
Type:	Flirt	
Years introduced:	2019-2020	
Max speed:	100mph (161km/h)	
Formation:	755/3 - DMS(A)+PP+PTSW+DMS(B)	755/4 -DMS(A)+PTS+PP+PTSW+DMS(B)
Vehicle numbers:	DMS(A) - 911325-911338 PP - 971325-971338 PTSW - 981325-981338 DMS(B) - 912325-912338	DMS(A) - 911401-911424 PTS - 961401-961424 PP - 971401-971424 PTSW - 981401-981424 DMS(B) - 912401-912424
Train length:	755/3 - 213ft 2in (65m)	755/4 - 264ft 8in (80.7m)
Vehicle length:	DMS - 74ft 3in (20.81m), PTS 50ft (15.22m), PP - 21ft 9in (6.69m)	
Height:	13ft 0in (3.95m)	
Width:	DMS, PTS - 8ft 9in (2.72m), PP - 9ft 1in (2.82m)	
Floor height:	37.7in (960mm)	
Internal layout:	2+2S, Tip-up	
Seating:	Total - 755/3 - 144S + 23 Tip-up DMS(A) - 60S (+4TU), PTS - 60 (+4TU), PP - 0, PTSW - 32S (+7TU), DMS(B) - 52 (+12TU)	Total - 755/4 - 182 + 27 tip-up
Gangway:	Within set	
Toilets:	2, 1 universal, 1 standard	
Weight:	Total - 135.1 tonnes DMS(A) - 43.4 tonnes, PP - 25.4 tonnes PTSW - 24.2 tonnes, DMS(B) - 42.1 tonnes	Total - 163.5 tonnes DMS(A) - 41.4 tonnes, PTS - 25 tonnes PP - 28.5 tonnes, PTSW - 26.4 tonnes DMS(B) - 42.2 tonnes
Brake type:	Air (regenerative)	
Bogie type:	Stadler/Jacobs	
Power collection:	25kV ac overhead & Diesel	
Diesel Engine:	755/3 - 2 x Deutz V8/16 of 480kW (645hp)	755/4 - 4 x Deutz V8/16 of 480kW (645hp)
Engine output:	755/3 - 960kW (1,287hp)	755/4 - 1,920kW (2,574hp)
Fuel tank capacity:	506gal (2300lit)	
Traction motor type:	4 x 325kW (436hp) TSA (Traction Swiss Austria)	
Power at rail:	Electric - 2600kW (3,486hp)	
Wheel diameter (New):	Motor bogie - 34.25in (870mm), Trailer bogie - 29.92in (760mm)	
Coupling type:	Outer - Dellner 10, Inner - Jacobs	
Multiple restriction:	Within class only (up to three sets)	
Door type:	Bi-parting sliding plug (one per car), 51.2in (1300mm) opening)	
Fittings:	Sanding plus scrubber block. 755/3s fitted with Train Approach Warning System (TAWS)	
Construction:	Aluminium	
Operator:	Greater Anglia	
Owner:	Rock Rail	

Above: The front ends of the Stadler UK 'Flirt' design for both Class 745 and 755 are almost the same, except that on the Class 755 multiple operating equipment is located above the Dellner coupling. Equipment positions are. 1: High level marker light, 2: Forward facing camera, 3: Destination indicator, 4: Marker and headlights, 5: Red tail light, 6: Air horns in coupling pocket, 7: Electrical connection box, 8: Dellner coupling. **CJM**

Above & Inset: Introduced as part of the total fleet modernisation of Greater Anglia, The Class 755 three and four-car bi-mode Stadler 'Flirt' units replaced second generation DMUs, they are allocated to Norwich Crown Point. These were the first trains in the UK to include a 'power-pack' (PP) vehicle and these bi-mode sets operate from either the 25kV AC overhead, or from on-board diesel engines in the PP vehicle. The articulated sets operate all Greater Anglia local services in the Norwich area. Three-car set No. 755338 is shown from its DMS(B) vehicle at Bury St Edmunds. The inset shows the articulated bogie. **CJM**

Right: A fleet of 24 four-car Class 755/4 sets are in service, these sport an additional PTS formed between the Power Pack and DTS(A) vehicle. Set No. 755418 is seen at Norwich. **CJM**

Rolling Stock Review : 2023-2024

Class 755   'Flirt'

Left: *Pantograph Trailer Standard - Wheelchair (PTSW) No. 981418 shows the side detail of these vehicles, one PTSW is formed in each Class 745 three or four car unit. The pantograph is located at the left end (lowered), with the wheelchair area adjacent to the bi-parting doors, indicated by the blue band above the window, some sets have a wheelchair outline on the bodyside. The universal toilet area is also located in this coach, located by the section missing the window.* **CJM**

Below: *The Power Pack (PP) vehicle houses the diesel engine/alternator sets and are 4 inches (102mm) wider than the passenger stock. The power packs (two on a 755/3 and four on a Class 755/4, are housed either side of a narrow gangway, allowing passengers movement through the train. The PP coach does not have traction, this is provided by the leading bogie of each driving car. PP No. 971418 is shown.* **CJM**

Above Left & Left: *The Class 755 interior is based on the 2+2 configuration, in a mix of group and airline layouts. Above left is the interior of a PTS vehicle, note the passenger information system. The view left shows the seating at the end of the PTS leading through the Power-Pack coach, showing the narrow bi-parting door protected walkway. Note the seats are raised over the Jacobs bogie. Both:* **CJM**

Key Facts

- First Stadler 'Flirt' trains in the UK.
- Each train includes a wider profile 'Power-Pack' or PP coach, housing diesel engine/alternator sets.
- All vehicles articulated, only unit ends have traditional bogies, which are used for traction.
- Sets based at Norwich Crown Point.
- Three-car sets *could* be extended to four, fleet numbering allows for this change.

Left: *Class 755 driving cab, a well designed layout, based on the European 'Flirt' style, but with UK features.* **CJM**

755/3	755338 GAR	755412 GAR
755325 GAR		755413 GAR
755326 GAR	**755/4**	755414 GAR
755327 GAR	755401 GAR	755415 GAR
755327 GAR	755402 GAR	755416 GAR
755329 GAR	755403 GAR	755417 GAR
755330 GAR	755404 GAR	755418 GAR
755331 GAR	755405 GAR	755419 GAR
755332 GAR	755406 GAR	755420 GAR
755333 GAR	755407 GAR	755421 GAR
755334 GAR	755408 GAR	755422 GAR
755335 GAR	755409 GAR	755423 GAR
755336 GAR	755410 GAR	755424 GAR
755337 GAR	755411 GAR	

2023 Fleet

'Flirt' — Class 756

Class:	756/0	756/1
Number range:	756001-756007	756101-756117
Built by:	Stadler, Bussnang, Switzerland & Siedlce, Poland	
Body shells:	Stadler, Bussnang, Switzerland & Szolnok, Hungary	
Type:	Flirt	
Years introduced:	2022-2023	
Max speed:	75mph (121km/h)	
Formation:	755/3 - DMS(A)+PTSW+PP+DMS(B)	755/4 -DMS(A)+PTSW+PP+PTS+DMS(B)
Vehicle numbers:	DMS(A) - 911001-911007 PP - 971001-971007 PTSW - 981001-981007 DMS(B) - 912001-912007	DMS(A) - 911101-911117 PTS - 961101-961117 PP - 971101-971117 PTSW - 981101-981117 DMS(B) - 912101-912117
Train length:	756/0 - 213ft 2in (65m)	756/1 - 264ft 8in (80.7m)
Vehicle length:	DMS - 74ft 3in (20.81m), PTS 50ft (15.22m), PP - 21ft 9in (6.69m)	
Height:	13ft 0in (3.95m)	
Width:	DMS, PTS - 8ft 9in (2.72m), PP - 9ft 1in (2.82m)	
Floor height:	37.7in (960mm)	
Internal layout:	2+2S, Tip-up	
Seating:	Total - 756/0 - 118S + 18 tip-up DMS(A) - 40+12TU , PTS - 40+8TU, PP - 0, PTSW - 38+5TU , DMS(B) - 40+12TU	Total - 756/1 - 158 + 26 tip-up
Gangway:	Within set	
Toilets:	1 universal	
Weight:	Total - 140.3 - DMS(A) 42.3, PP 28.7, PTSW 26.6, DMS(B) 42.7	Total - 166.2 - DMS(A) 42.3, PTS 25.9, PP 28.7, PTSW 26.6, DMS(B) 42.7
Brake type:	Air (regenerative)	
Bogie type:	Stadler/Jacobs	
Power collection:	25kV ac overhead, Diesel and Battery [Diesel for battery chargiong]	
Diesel Engine:	756/0 - 1 x Deutz V8/16 of 480kW (645hp)	756/0 - 1 x Deutz V8/16 of 480kW (645hp)
Engine output:	756/0 - 480kW (645hp)	756/0 - 480kW (645hp)
Fuel tank capacity:	506gal (2300lit)	
Traction motor type:	4 x 325kW (436hp) TSA (Traction Swiss Austria)	
Power at rail:	Electric - 2600kW (3,486hp)	
Wheel diameter (New):	Motor bogie - 34.25in (870mm), Trailer bogie - 29.92in (760mm)	
Coupling type:	Outer - Dellner 10, Inner - Jacobs	
Multiple restriction:	Within class only (up to three sets)	
Door type:	Bi-parting sliding plug (one per car), 51.2in (1300mm) opening	
Fittings:	Sanding plus scrubber block.	
Construction:	Aluminium	
Operator:	Transport for Wales	
Owner:	SMB Equitix Leasing	

Above: *Class 756 front end equipment. The systems are the same as on the Class 231 (all diesel) sets, which the Class 765s can operate in multiple. The coupling is the Dellner 10 type, with electrical connection box above.* **CJM**

Above & Right: *In summer 2023, seven three-passenger vehicle and 17 four-passenger vehicle 'Flirt' tri-mode Class 756 sets were on delivery from plants in Switzerland, Poland and Hungary for Transport for Wales in Cardiff. Each train is powered either from the 25kV ac overhead, or battery, a diesel charges the battery as needed. A 'Power-Pack' or PP vehicle is located towards the middle of the formation, housing a Deutz diesel engine and battery packs. The PP is unpowered, with traction power provided to the driving cars. The trains are very similar in appearance to the Class 231 'Flirt' sets recently introduced. The '756s' carry Transport for Wales grey livery, off-set by red passenger doors. Compared to the similar Class 755s operated by Anglia, the '756s' have two pairs of bi-parting passenger doors on each side of each vehicle, except the PTSW which due it its internal layout with a universal access toilet only has one pair. Above three-car Class 756/0 No. 756001 is seen from its DMS(B) end. On the right four-car Class 756/1 No. 756103 is viewed from its DMS(A) end.* **Mark V. Pike / Antony Christie**

Rolling Stock Review : 2023-2024

Class 756   'Flirt'

Left: *Towards the centre of each Class 756 is a Power Pack or PP vehicle, the carries the auxiliary power systems in the way of Deutz diesel engine/generator units or traction batteries. The structure has two large opening doors on each side to access equipment. A narrow passenger corridor passes through the middle of the vehicle.* **CJM**

Above Right: *Bi-parting passenger doors, with retractable foot step below. When the door open button is activated, the step extends to bridge the train/platform interface, seconds before the doors open.* **CJM**

Left Middle: *The two PTS vehicles have different body profiles. The standard PTS has two pairs of doors on either side, with seating for 40 plus space for cycles and luggage. The PTSW with wheelchair space seats 38 and houses a universal toilet compartment and wheelchair area. Only one pair of bi-parting doors are fitted each side.* **Antony Christie**

Left Below: *Seating is provided in the 2+2 low-density style, with a good mix of airline and facing styles. Seats mounted at vehicle ends above the articulated bogies are on a slightly raised plinth to allow a level walkway through the train. All vehicles are carpeted throughout, seats have hinged armrests and each seat has access to a phone/laptop charging outlet, using both the UK three pin style and a USB connection.* **CJM**

Left Bottom: *The driving cab uses the now standard UK Flirt design, a very well styles and set out desk, based on the European Flirt product platform. The combined power/brake controller is on the left side. A full Train Management System (TMS) is provided and high-quality CCTV screens are able to show the interior of the train or the train/platform interface.* **CJM**

2023 Fleet

756/0	
756001	TFW
756002	TFW
756003	TFW
756004	TFW
756005	TFW
756006	TFW
756007	TFW

756/1	
756101	TFW
756102	TFW
756103	TFW
756104	TFW
756105	TFW
756106	TFW
756107	TFW
756108	TFW
756109	TFW
756110	TFW
756111	TFW
756112	TFW
756113	TFW
756114	TFW
756115	TFW
756116	TFW
756117	TFW

Key Facts

- Stadler 'Flirt' Class 756s are based on the Anglia Class 755s, but have 2 pairs of passenger doors either side on DMS and PTS coaches, one pair on PTSW.
- Sets commenced delivery to Cardiff in late 2022, should enter service in early 2024.
- All vehicles articulated, only unit ends have traditional bogies, which are used for traction.
- Sets fitted with automatic protruding steps to bridge train-platform interface.

Rolling Stock Review : 2023-2024

'Flex' Class 768 / 769

For history and pictures of this class, see Modern Locomotives Illustrated Issue 243

Class:	768 'Flex' Orion		
Number range:	768001-768010 (Potential conversion)	Internal layout:	Freight space
Former number range:	319010/009	Seating:	None
Built by:	As 319 - BREL York	Gangway:	Within set, emergency end doors
	As 769 - Wabtec, Loughborough	Toilets:	None
Years introduced:	As 319 - 1987, As 768 - 2021	Brake type:	Air (Westcode)
Max speed:	100 mph (161km/h)	Bogie type:	DTPMV, TPMV - BREL T3-7, PMPMV - BREL P7-4
Formation:	DTPMV+PMPMV+TPMV+DTPMV	Power collection:	25kV ac overhead or diesel
Vehicle numbers:	DTPMV - 77309, 77307 series	Traction motor type:	4 x GEC G315BZ
	PMPMV - 62900, 62899 series	Horsepower:	1,438hp (1,072kW)
	TPMV - 71781, 71780 series	Diesel engine:	2 x MAN D2876 LUE631 of 523hp (390kW)
	DTPMV - 77308, 77306 series	Diesel alternator:	ABB
Vehicle length:	DTPMV - 65ft ¾in (19.83m)	Coupling type:	Outer - Tightlock, Inner - Bar
	MPMV, PTPMV - 65ft 4¼in (19.93m)	Multiple restriction:	Within class and Class 319
Height:	11ft 9in (3.58m)	Door type:	Bi-parting sliding
Width:	9ft 3in (2.82m)	Construction:	Steel
Weight:	Total - 151.4 tonnes	Operator:	ROG
	DTPMV - 35.5 tonnes, MPMV - 49.6 tonnes,	Owner:	Porterbrook Leasing
	TPMV - 30 tonnes, DTPMV - 36.2 tonnes		

Above & Right: Rail Operations UK launched a light freight concept in 2021, using modified Class 319 stock, reclassified as Class 768, under the 'Project Orion' brand. The operator hopes to deploy a fleet of nine Class 319s and 10 Class 768s, with the ability for the Class 768s to haul all-electric Class 319s away from the electric network. Major engineering to upgrade the interiors to carry freight has been undertaken at Arlington Fleet Services, Eastleigh, with the diesel/alternator and traction modifications to driving cars undertaken at Wabtec, Loughborough. Painted in full ROUK Orion livery, set No. 768001 is shown at Long Marston in summer 2022. Some limited commercial operations were undertaken in late 2021, but no firm contrcats have been signed. On the right is the TPMV coach. All window positions, including opening hoppers, have been retained. Bi-parting doors remain operational for vehicle loading. The original toilet compartment has been removed. **Both: CJM**

Class:	769 'Flex' Northern		
Number range:	769424/431/434/442/448/450/456/458	Internal layout:	2+2, 2+3
Former number range:	319424/431/434/442/448/450/456/458	Seating:	Total - 268S
Built by:	As 319 - BREL York		DTS(A)-62, PMSO-75, TSOL-58, DTS(B)-73
	As 769 - Wabtec, Loughborough	Gangway:	Within set, emergency end doors
Years introduced:	As 319 - 1987-1988, As 769 - 2018-2020	Toilets:	PTSOL 1 universal
Max speed:	100 mph (161km/h)	Brake type:	Air (Westcode)
Formation:	DTSO(A)+MSO+PTSOL+DTSO(B)	Bogie type:	DTSO, TSOL - BREL T3-7
Vehicle numbers:	DTSO(A) - 77337-77453 [odd] series		PMSO - BREL P7-4
	PMSO - 62914-62972 series	Power collection:	25kV ac overhead or diesel
	TSOL - 71795-71877 series	Traction motor type:	4 x GEC G315BZ
	DTSO(B) - 77336-77452 [even] series	Horsepower:	1,438hp (1,072kW)
Vehicle length:	DTSO - 65ft ¾in (19.83m)	Diesel engine:	2 x MAN D2876 LUE631 of 523hp (390kW)
	PMSO, TSOL - 65ft 4¼in (19.93m)	Diesel alternator:	ABB
Height:	11ft 9in (3.58m)	Coupling type:	Outer - Tightlock, Inner - Bar
Width:	9ft 3in (2.82m)	Multiple restriction:	Within class and Class 319
Weight:	Total - 155.3 tonnes	Door type:	Bi-parting sliding
	DTSO(A) - 36.5 tonnes, PMSO - 50.6 tonnes,	Construction:	Steel
	TSOL - 31 tonnes, DTSO(B) - 37.2 tonnes	Operator:	Northern Rail
		Owner:	Porterbrook Leasing

Rolling Stock Review : 2023-2024

Class 768 / 769 'Flex'

Above, Left & Lower Left: In December 2016, Porterbrook announced a partnership with Northern Rail to convert Class 319 EMUs into Bi-mode Multiple Units (BMUs), by fitting a MAN D2876 diesel engine below each driving car, powering an ABB alternator. This became the Porterbrook 'Flex' project. Engineering was undertaken by Wabtec (Brush) at Loughborough, with the first eight sets going to Northern. The project was plagued with problems and sets commissioned for diesel only operation by Transport for Wales were withdrawn in early 2023 and a fleet for Grat Western which were converted and under test were abandoned in early 2023. The long term future of the Northern sets is also in doubt. Above set No. 769424 is seen at Southport, with its DTSO(B) nearest the camera. The left illustration shows the Pantograph Motor Standard Open (PMSO). The lowered pantograph is at the far end and the boxes between the bogies house the transformer and traction equipment. Seating in this vehicle is for 77 in the 2+3 layout. Internally the Northern '769s' feature a Class 319 interior, with train seating for 268 standard class. One universal toilet compartment is located in the intermediate Trailer Standard. All: **CJM**

Bottom: Each former Class 319 driving car is fitted with an underslung 'raft' carrying one MAN D2876 diesel engine of 523hp (390kW), powering an ABB alternator. As the original driving cars were trailer vehicles, sufficient space was available to install this extra equipment. On the far left is the fuel tank, with a brown filler port, moving to the right is the battery and control equipment, the silver radiator is towards the middle with the power unit/alternator raft to its right. Power generated by the power pack passes by a train line to the MSO vehicle, which retains all traction equipment. **CJM**

Right: Class 769 front end equipment. 1: Forward facing camera, 2: Emergency front end door, 3: Destination indicator, 4: White marker light, 5: Headlight, 6: Red tail light, 7: Lamp bracket, 8: Warning horns, 9: Tightlock coupling, 10: Emergency uncoupling lever, 11: Emergency air supply, 12: Coupling air/pneumatic connection. **CJM**

Key Facts

- Major project between Porterbrook and Brush to reuse some of the large number of EMUs which become available after new stock was introduced.
- Internal refitting carried-out at Wolverton or Eastleigh Works, traction upgrade carried out at Brush/Wabtec, Loughborough.
- Northern sets operate from 25kV AC or diesel.
- Sets originally operated by Transport for Wales were withdrawn in summer 2023, proposed use by Great Western was abandoned in spring 2023.
- In 2023 TfW and GWR sets stored at Long Marston.

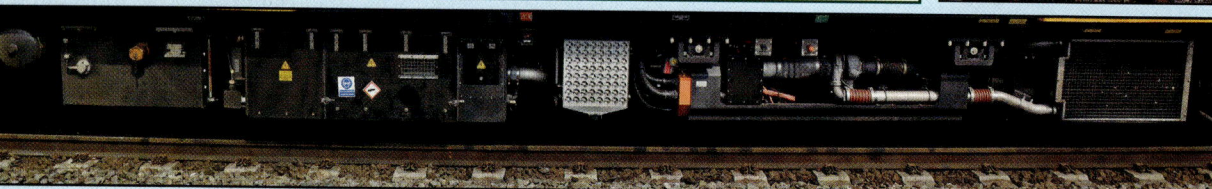

2023 Fleet

768001	ROG
768002	ROG
769365	ROG
769366	ROG
769424	NOR
769431	NOR
769434	NOR
769442	NOR
769448	NOR
769450	NOR
769456	NOR
769458	NOR

For history and pictures of this class, see Modern Locomotives Illustrated Issue 243

'Metro' Class 777

Left: Class 777 front end layout. 1: High level marker light, 2: Destination indicator, 3: Emergency evacuation door, via built-in steps, 4: Emergency door handle, 5: Forward facing camera, 6: Rear marker strip, 7: Front end white marker/head lights, 8: Emergency door release (behind cover), 9: Dellner coupling, 10: Warning horns inside coupling pocket, 12: Dellner electrical connections. **CJM**

Class:	777
Number range:	777/0 - 777001-777053, 777/1 - 777140-777152
Built by:	Stadler, Bussnang, Switzerland. Body shells from Szolnok in Hungary, Siedlce in Poland and Altenrhein, Switzerland
Introduced:	2020-2023
Max speed:	75mph (121km/h) 60mph (97km/h) on battery
Formation:	DMS(A)+MS(A)+MS(B)+DMS(B)
Axle arrangement:	2 (Bo) (Bo) (Bo) 2
Vehicle numbers:	DMS(A) 427001-427053 MS(A) 428001-428053 MS(B) 429001-429053 DMS(B) 430001-430053
Length:	Train 213ft 2in (64.97m) DMS - 62ft 3in (19m) MS - 44ft 7in (13.6m)
Height:	12ft 6in (3.82m)
Width:	9ft 3in (2.82m)
Weight:	99 ton
Internal layout:	2+2
Seating:	Total - 162 (+22TU, 28P) DMS(A) - 53S (+4TU, 5P), MS(A) & MS(B) - 38S (+7TU, 9P) DMS(B) - 53S
Standing space:	302
Gangway:	Within set, emergency end doors
Toilets:	Not fitted
Brake type:	Air (regenerative)
Bogie type:	Jakobs
Power collection:	750V dc third rail, Traction batteries (for shunting)
Traction motor type:	6 x TSA 470hp (350kW)
Horsepower:	Max - 2,820hp (2,100kW) Cont - 2,011hp (1,500kW)
Coupling type:	Outer - Dellner 12, Inner - Jakobs
Multiple restriction:	Within class
Door type:	Twin sliding plug, 1 pair on driving and two pairs on intermediate cars
Construction:	Aluminium
Operator:	MerseyRail
Owner:	Liverpool City Region
Notes:	Can be 25kV ac overhead fitted 777/0 are straight electric sets 777/1 fitted with 160kW/h Lithium Titanate Oxide traction batteries, for non-electrified line use, initally between Kirkby and the new Headbolt Lane station

Below: In December 2016, Mersey Travel ordered 52 (later amended to 53, dual-voltage Class 777s, as fleet replacement for the Mersey Electric system, from Stadler. Trains commenced delivery to the UK in 2020, but were extensively delayed by production slow-downs, the Covid-19 pandemic and trade union issues relating to the method of door operation. Sets entered limited service from the start of 2023 with a progressive deployment on all routes by early 2024. The Class 777 driving cars only have one set of bi-parting doors on each side, while intermediate vehicles have two. The trains incorporate retractable 'edge fillers' bridging the gap between train and platform when doors open. Cab ends include an off-set emergency door to conform with tunnel evacuation regulations. The Liverpool City Region Combined Authority, which owns the trains, has a long-term plan to extend the network to Warrington and even Wrexham and thus the sets have a dual power capability, with an easy addition of 25kV AC overhead if required. Two sub-classes of 777 exist, 777/0 are standard 750V dc electric units and 777/1 which have full dc electric systems plus traction batteries for non-electrified route operation, such as serving the new Headbolt Lane station an extension of the Kirkby line. Standard electric set No. 777008 is seen at Kirkdale with a service to Kirkby. **CJM**

Below: Class 777 low-density interior, gives the appearance of a lot of space, shown in this through train view. All seats have access to a charging plug and USB socket. A good passenger information system is provided. A slight problem for some passengers will be ramps in the floor to bridge the articulated bogies. **CJM**

Key Facts

- Ordered for modernisation of Mersey Electric network to replace Class 507 and 508 fleets.
- Stadler articulated four-vehicle 'Metro' platform trains.
- All standard class, no toilets, fitted with disabled seating area and cycle parking points.
- Fitted with emergency front-end off-set door.
- Seven Class 777/1 sets are bi-mode, with electric and battery power.
- No yellow warning ends.

2023 Fleet

777/0

777001	MER	777028	MER
777002	MER	777029	MER
777003	MER	777030	MER
777004	MER	777031	MER
777005	MER	777032	MER
777006	MER	777033	MER
777007	MER	777034	MER
777008	MER	777035	MER
777009	MER	777036	MER
777010	MER	777037	MER
777011	MER	777038	MER
777012	MER	777039	MER
777013	MER	777041	MER
777014	MER	777043	MER
777015	MER	777045	MER
777016	MER	777047	MER
777017	MER	777049	MER
777018	MER	777051	MER
777019	MER	777053	MER
777020	MER	**777/1**	
777021	MER	777140	MER
777022	MER	777142	MER
777023	MER	777144	MER
777024	MER	777146	MER
777025	MER	777148	MER
777026	MER	777150	MER
777027	MER	777152	MER

Rolling Stock Review : 2023-2024

Class 777   'Metro'

Above: Each 777 has two near identical Motor Standard (MS) articulated coaches formed between the driving cars, these have two paired of sliding plug doors and seat 38, plus seven pop up and nine perch seats. Each vehicle has a wheelchair parking position. **CJM**

Above & Below: Interior and exterior of passenger doors, showing door mounted control buttons. Both: **CJM**

Above Left: In 2022 it was announced that seven Class 777s would be equipped with traction batteries to allow operation on short non-electrified routes and extensions to the Merseyrail system. The first line to benefit from this technology will be the extension of the Kirkby line to Headbolt Lane, opening in 2023. The battery equipped sets are classified 777/1 and carry battery logos on the bodyside. Otherwise these are standard sets. No. 777146 is seen at Hall Road on 20 July 2023 with a test and training run. **CJM**

Left: Class 777 driving cab layout, with the emergency tunnel evacuation door arrangement to the right. The driving position uses a combined one stalk power/brake controller, on the driver's left side. Most other information is shown on the two angled screen displays. CCTV displays showing the train-platform interface are located to the left of the main windscreen. **CJM**

Right: Articulated bogie arrangement using a Jacob bogie. The bodywork on the lower vehicle is slightly different on each coach to allow movemnt. **CJM**

'HydroFlex' — Class 799

Class:	799/2
Number range:	799201
Former number range:	319382
Built by:	BREL York
Year built:	1990
Rebuilt as 799:	2021
Rebuilt by:	Porterbrook, Long Marston
Max speed:	100 mph (161km/h)
Formation:	DTSO(A)+MSO+TSOL+DTSO(B)
Vehicle numbers:	DTSO(A) - 77975 MSO - 63094 TSOL - 71980 DTSO(B) - 77976
Vehicle length:	DTSO(A) - 65ft ¾in (19.83m) DTSO(B) - 65ft ¾in (19.83m) MSO - 65ft 4¼in (19.93m) TSOL - 65ft 4¼in (19.93m)
Height:	11ft 9in (3.58m)
Width:	9ft 3in (2.82m)
Weight:	Total - 153.3 tonnes DTSO(A) - 42 tonnes MSO - 50.6 tonnes TSOL - 31 tonnes DTSO(B) - 29.7 tonnes
Internal layout:	Special
Seating:	2+2, 2+3, Lounge, Board Room
Gangway:	Within set, emergency end doors
Toilets:	1
Brake type:	Air (Westcode)
Bogie type:	DTSO, TSOL - BREL T3-7 MSO - BREL P7-4
Power collection:	25kV ac overhead and Hydrogen fuel cell
Traction motor type:	4 x GEC G315BZ
Horsepower:	1,438hp (1,072kW)
Coupling type:	Outer - Tightlock, Inner - Bar
Multiple restriction:	Not fitted
Door type:	Bi-parting sliding
Construction:	Steel
Owner/Operator:	Porterbrook Leasing

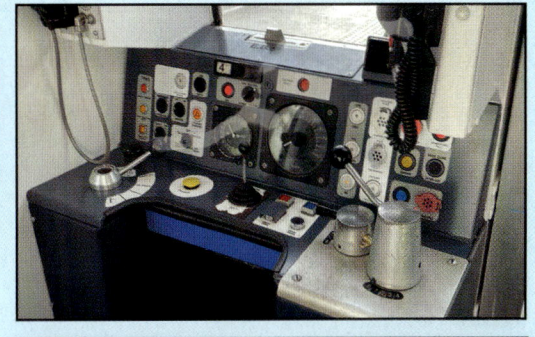

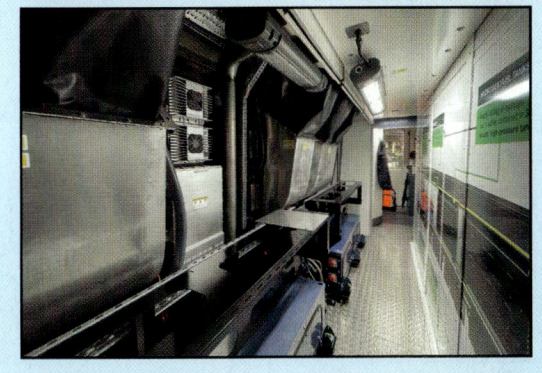

Top, Above, Below & Right: The use of alternative sources of traction power for locos and multiple units is now ramping up, with several projects ongoing using battery, hydrogen fuel cells or a combination of both. In November 2021, the Porterbrook Hydroflex Class 799/2, a rebuild of Class 319 No. 319382 was launched. Fitted with a hydrogen/battery independent power system, working along side the trains main 25kV AC overhead power system. The rebuild was undertaken at the Long Marston facility, with driving coach 77975 converted to a hydrogen vehicle, housing three large hydrogen multi-cylinder tanks at the inner end, adding 16 tonnes to the coaches weight. Top and above, the train is seen at Honeybourne on the Long Marston branch, the hydrogen car at the top shows the revised body and roof detail. The fuel cells, producing the electricity, are located inward of the cab end and are shown in the image at the bottom right of this page. The front end shown below remains much as in its Class 319 days. On the hydrogen coach underframe, a large traction battery is slung, this is charged from the hydrogen fuel cells, with the battery then providing power to the train line for traction, much in the same way as the transformer does under normal electric conditions. In the driving cab (shown top right, an Independent Power (IP) panel has been placed on the non-driving side, indicating the operation/status of the hydrogen/fuel cell system. The set retains its 25kV AC power system. The '319s' original 3rd rail DC power pick up system has been removed. Internally, the opportunity was taken to rebuild and demonstrate various different layout possibilities, most of used reclaimed material from other Porterbrook rebuild projects. The image middle right shows a board room set up for one of the intermediate coaches The view above shows the set from the DTSO(B) vehicle No. 77976. All: **CJM**

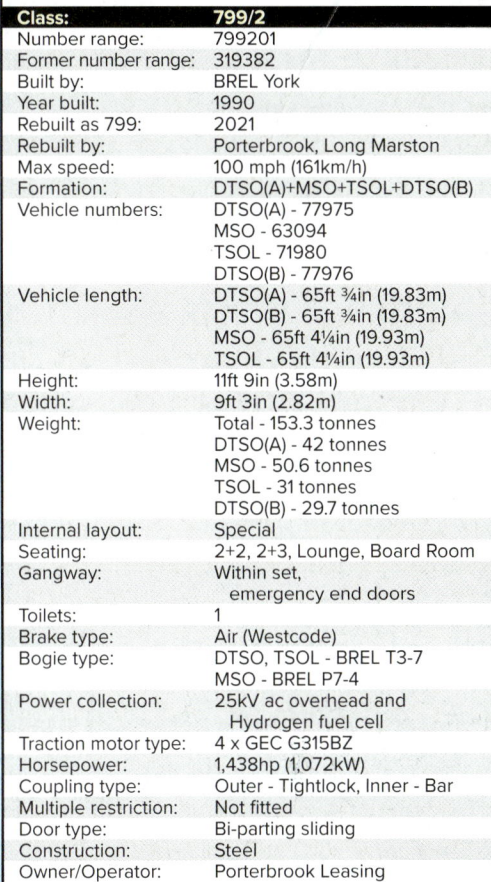

Class 800-810

For history and pictures of this class, see Modern Locomotives Illustrated Issue 247

Class:	800/0	800/1	800/2
Traction:	Bi-mode	Bi-mode	Bi-mode
Number range:	800001-800036	800101-800113	800201-800210
Introduced:	2017-2018	2017-2018	2018-2019
Built by	Hitachi, Kasado and Newton Aycliffe	Hitachi, Kasado and Newton Aycliffe	Hitachi, Kasado and Newton Aycliffe
Type:	-	Azuma	Azuma
Formation:	PDTS+MS+MS+MC+PDTRBF	PDTS+MS+MS+TSRB+MS+ TS+MC+MF+PDTRBF	PDTS+MSRB+MS+MC+PDTRBF
Vehicle numbers:	PDTS - 811001-811036 MS - 812001-812036 MS - 813001-813036 MC - 814001-814036 PDTRBF - 815001-815036	PDTS - 811101-811113 MS - 812101-812113, MS - 813101-813113 TSRB - 814101-814113, MS - 815101-815113 TS - 816101-816113, MC - 817101-817113 MF - 818101-818113, PDTRBF - 819101-819113	PDTS - 811201-811210 MSRB - 812201-812210 MS - 813201-813210 MC - 814201-814210 PDTRBF - 815201-815210
Train length:	528ft 2in (162m)	794ft 0in (242m)	528ft 2in (162m)
Vehicle length:	Driving - 83ft 2in (25.35m) Intermediate - 82ft 0in (25.0m)	Driving - 83ft 2in (25.35m) Intermediate - 82ft 0in (25.0m)	Driving - 83ft 2in (25.35m) Intermediate - 82ft 0in (25.0m)
Width:	8ft 10in (2.7m)	8ft 10in (2.7m)	8ft 10in (2.7m)
Seating:	Total - 36F/290S, PDTS - 56S, MS - 88S, MS - 88S MC - 18F/58S, PDTRBF - 18F	Total - 101F/510S, PDTS - 48S, MS - 88S, MS - 88S TSRB - 72S, MS - 88S, TS - 88S, MC - 30F/38S, MF - 56F, PDTRBF - 15F	Total - 48F/255S, PDTS - 56S, MSRB - 72S, MS - 88S, MC - 30F/38S, PDTRBF - 18F
Internal layout:	2+1F, 2+2S	2+1F, 2+2S	2+1F, 2+2S
Gangway:	Within set	Within set	Within set
Toilets:	6	9	6
Weight:	Total - 250.5t PDTS - 47.8t, MS - 50.1t, MS - 50.3t MC - 50.6t, PDTRBF - 51.7t	Total - 429.5 PDTS - 47.7t, MS - 50.5t, MS - 50.3t, TSRB - 41.0t, MS - 50.3t, TS - 38.3t, MC - 49.1t, MF - 50.6t, PDTRBF - 51.7t	Total - 250.5t PDTS - 47.8t, MSRB - 50.1t, MS - 50.3t MC - 50.6t, PDTRBF - 51.7t
Brake type:	Air (regenerative)	Air (regenerative)	Air (regenerative)
Power:	25kV ac overhead or underslung diesels	25kV ac overhead or underslung diesels	25kV ac overhead or underslung diesels
Diesel type:	3 x MTU 12V 1600 R80L at 700kW	5 x MTU 12V 1600 R80L at 700kW	3 x MTU 12V 1600 R80L at 700kW
Diesel output:	Motor car - 938hp (700kW) per engine Train - 2,814hp (2,100kW)	Motor car - 938hp (700kW) per engine Train - 4,690hp (3,500kW)	Motor car - 938hp (700kW) per engine Train - 2,814hp (2,100kW)
Electric output:	12 x Hitachi of 303hp (226kW) = 3,636hp	20 x Hitachi of 303hp (226kW) = 6,060hp	12 x Hitachi of 303hp (226kW) = 3,636hp
Max speed:	125mph (201km/h)	125mph (201km/h)	125mph (201km/h)
Coupling type:	Outer - Dellner 10, Inner - Semi-auto	Outer - Dellner 10, Inner - Semi-auto	Outer - Dellner 10, Inner - Semi-auto
Multiple restriction:	Class 800-802	Class 800-802	Class 800-802
Door type:	Single leaf sliding	Single leaf sliding	Single leaf sliding
Construction:	Aluminium	Aluminium	Aluminium
Operator:	5-car bi-mode sets for Great Western	9-car bi-mode sets for LNER	5-car bi-mode sets for LNER
Owner:	Agility Trains	Agility Trains	Agility Trains

2023 Fleet

800/1					
800002	GWR	800108	LNE	801103	LNE
800003	GWR	800109	LNE	801104	LNE
800004	GWR	800110	LNE	801105	LNE
800005	GWR	800111	LNE	801106	LNE
800006	GWR	800112	LNE	801107	LNE
800007	GWR	800113	LNE	801108	LNE
800008	GWR			801109	LNE
800009	GWR	**800/2**		801110	LNE
800010	GWR	800201	LNE	801111	LNE
800011	GWR	800202	LNE	801112	LNE
800012	GWR	800203	LNE		
800013	GWR	800204	LNE	**801/2**	
800014	GWR	800205	LNE	801201	LNE
800015	GWR	800206	LNE	801202	LNE
800016	GWR	800207	LNE	801203	LNE
800017	GWR	800208	LNE	801204	LNE
800018	GWR	800209	LNE	801205	LNE
800019	GWR	800210	LNE	801206	LNE
800020	GWR			801207	LNE
800021	GWR	**800/3**		801208	LNE
800022	GWR	800301	GWR	801209	LNE
800023	GWR	800302	GWR	801210	LNE
800024	GWR	800303	GWR	801211	LNE
800025	GWR	800304	GWR	801212	LNE
800026	GWR	800305	GWR	801213	LNE
800027	GWR	800306	GWR	801214	LNE
800028	GWR	800307	GWR	801215	LNE
800029	GWR	800308	GWR	801216	LNE
800030	GWR	800309	GWR	801217	LNE
800031	GWR	800310	GWR	801218	LNE
800032	GWR	800311	GWR	801219	LNE
800033	GWR	800312	GWR	801220	LNE
800034	GWR	800313	GWR	801221	LNE
800035	GWR	800314	GWR	801222	LNE
800036	GWR	800315	GWR	801223	LNE
		800316	GWR	801224	LNE
		800317	GWR	801225	LNE
800/1		800318	GWR	801226	LNE
800101	LNE	800319	GWR	801227	LNE
800102	LNE	800320	GWR	801228	LNE
800103	LNE	800321	GWR	801229	LNE
800104	LNE			801230	LNE
800105	LNE	**801/1**			
800106	LNE	801101	LNE		
800107	LNE	801102	LNE		

Top & Above: *Great Western Railway operates, 36 five-car and 21 nine-car, Agility Trains owned Class 800 bi-mode sets. The five car units have three powered vehicles (the intermediates), while the nine car sets have five powered coaches. In the top view, GW five-car Class 800/0 No. 800003 is seen at Gloucester from its first class end. These '800' sets can be recognised by having white insets to the cab door handles, door frames and bogie/body brackets. In the above image, is an Agility Trains Class 800/3 Motor Standard (MS), these vehicles seat 88 in the 2+2 style in a mix of airline and group layouts. The '800/3s', Nos. 800301-800321 operate alongside the Class 800/0s, providing 71 first and 576 standard class seats. All GW sets usually operate with their first class at the London end. First class is identified by a silver band and white blocks at cant rail height. Both:* **CJM**

Class 800-810

800/3	801/1	801/2
Bi-mode	Electric	Electric
800301-800321	801101-801112	801201-801230
2017-2019	2017-2019	2018-2019
Hitachi, Kasado and Newton Aycliffe	Hitachi, Kasado and Newton Aycliffe	Hitachi, Kasado and Newton Aycliffe
	Azuma	Azuma
PDTS+MS+MS+TS+MS+TS+MS+MF+PDTRBF	PDTS+MSRB+MS+MC+PDTRBF	PDTS+MS+MS+TSRB+MS+TS+MC+MF+PDTRBF
PDTS - 821001-821021	PDTS - 821101-821112	PDTS - 821201-821230
MS - 822001-822021, MS - 823001-823021	MSRB - 822101-822112	MS - 822201-822230, MS - 823201-823230
TS - 824001-824021, MS - 825001-825021	MS - 823101-823112	TSRB - 824201-824230, MS - 825201-825230
TS - 826001-826021, MS - 827001-827021	MC - 824101-824112	TS - 826201-826230, MC - 827201-827230
MF - 828001-828021, PDTRBF - 829001-829021	PDTRBF - 825101-825112	MF - 828201-828230, PDTRBF - 829201-829230
794ft 0in (242m)	528ft 2in (162m)	794ft 0in (242m)
Driving - 83ft 2in (25.35m)	Driving - 83ft 2in (25.35m)	Driving - 83ft 2in (25.35m)
Intermediate - 82ft 0in (25.0m)	Intermediate - 82ft 0in (25.0m)	Intermediate - 82ft 0in (25.0m)
8ft 10in (2.7m)	8ft 10in (2.7m)	8ft 10in (2.7m)
Total - 71F/576S	Total - 48F/255S,	Total - 101F/510S,
PDTS - 48S, MS - 88S, MS - 88S, TS - 88S, MS - 88S	PDTS - 56S, MSRB - 72S, MS - 88S,	PDTS - 48S, MS - 88S, MS - 88S
TS - 88S, MS - 88S, MF - 56F, PDTRBF - 15F	MC - 30F/38S, PDTRBF - 18F	TSRB - 72S, MS - 88S, TS - 88S, MC - 30F/38S
		MF - 56F, PDTRBF - 15F
2+1F, 2+2S	2+1F, 2+2S	2+1F, 2+2S
Within set	Within set	Within set
9	6	9
Total - 429.2t	Total - 238.7t	Total - 404.6t
PDTS - 47.8t, MS - 50.1t, MS - 50.3t, TS - 41.0t,	PDTS - 47.8t, MSRB - 52.1t, MS - 43.5t	PDTS - 47.7t, MS - 50.5t. MS - 43.5t, TSRB - 43.0t
MS - 50.3t, TS - 38.3t, MS - 49.1t, MF - 50.6t,	MC - 44.1t, PDTRBF - 51.2t	MS - 43.5t, TS - 38.3t, MC - 42.6t, MF - 43.8t
PDTRBF - 51.7t		PDTRBF - 51.7t
Air (regenerative)	Air (regenerative)	Air (regenerative)
25kV ac overhead or underslung diesels	25kV ac overhead	25kV ac overhead
5 x MTU 12V 1600 R80L at 700kW	MRSB 1 x MTU 12V 1600 R80L at 700kW	MSO (822xxx) 1 x MTU 12V 1600 R80L at 700kW
Motor car - 938hp (700kW) per engine	(Emergency power)	(Emergency power)
Train - 4,690hp (3,500kW)		
20 x Hitachi of 303hp (226kW) = 6,060hp	12 x Hitachi of 303hp (226kW) = 3,636hp	20 x Hitachi of 303hp (226kW) = 6,060hp
125mph (201km/h)	125mph (201km/h)	125mph (201km/h)
Outer - Dellner 10, Inner - Semi-auto	Outer - Dellner 10, Inner - Semi-auto	Outer - Dellner 10, Inner - Semi-auto
Class 800-802	Class 800-802	Class 800-802
Single leaf sliding	Single leaf sliding	Single leaf sliding
Aluminium	Aluminium	Aluminium
9-car bi-mode sets for Great Western	5-car electric sets for LNER	9-car electric sets for LNER
Agility Trains	Agility Trains	Agility Trains

Above: *The 21 Great Western operated Class 800/3s are based at the Hitachi depot at North Pole, just outside London Paddington. The sets are mainly deployed on London-Bristol, South Wales and Cotswold line duties, but frequently work to the West Country, especially the Paignton line working through services to London, set No. 800314 is seen from its first class end at Cardiff General.* **CJM**

Right: *LNER operates a fleet of 13 nine-car (800/1) and 10 five-car (Class 800/2) bi-mode sets, to the same specification as the GW sets but with LNER fixtures and fittings. The sets sport LNER Azumas white and red livery and are based at Doncaster depot. The fleets principally operate on routes where some non-electrified operation is needed. Five car set No. 800204 is captured south of York in June 2022. Note the plated body windows adjacent to the first class catering area.* **CJM**

Rolling Stock Review : 2023-2024

Class 800-810

Class:	802/0	802/1	802/2
Traction:	Bi-mode	Bi-mode	Bi-mode
Number range:	802001-802022	802101-802114	802201-802219
Introduced:	2017-2018	2017-2018	2018-2019
Built by:	Hitachi, Kasado, Japan and Pistoia, Italy	Hitachi, Kasado, Japan and Pistoia, Italy	Hitachi, Kasado, Japan and Pistoia, Italy
Type:			Nova 1
Formation:	PDTS+MS+MS+MC+PDTRBF	PDTS+MS+MS+TS+MS+TS+ MS+MF+PDTRFB	PDTS+MS+MS+MS+PDTF
Vehicle numbers:	PDTS - 831001-831022 MS - 832001-832022 MS - 833001-833022 MC - 834001-834022 PDTRBF - 835001-835022	PDTS - 831101-831114 MS - 832101-832114, MS - 833101-833114 TS - 834101-834114, MS - 835101-835114 TS - 836101-836114, MS - 837101-837114 MF - 838101-837114, PDTRBF - 839101-839114	PDTS - 831201-831219 MSO - 832201-832219 MSO - 833201-833219 MSO - 834201-834219 PDTF - 835201-835219
Train length:	528ft 2in (162m)	794ft 0in (242m)	528ft 2in (162m)
Vehicle length:	Driving - 83ft 2in (25.35m) Intermediate - 82ft 0in (25.0m)	Driving - 83ft 2in (25.35m) Intermediate - 82ft 0in (25.0m)	Driving - 83ft 2in (25.35m) Intermediate - 82ft 0in (25.0m)
Width:	8ft 10in (2.7m)	8ft 10in (2.7m)	8ft 10in (2.7m)
Seating:	Total - 36F/290S, PDTS - 56S, MS - 88S, MS - 88S MC - 18F/58S, PDTRBF - 18F	Total -101F/526S, PDTS - 48S, MS - 88S, MS - 88S, TS - 88S MS - 88S, TS - 88S, MS - 88S, MF - 56F, PDTRBF - 15F	Total - 24F/320S, PDTS - 56S, MS - 86S, MS - 88S, MS -88S PDTF - 24F
Internal layout:	2+1F, 2+2S	2+1F, 2+2S	2+1F, 2+2S
Gangway:	Within set	Within set	Within set
Toilets:	5	10	5
Weight:	Total: 252.8t PDTS - 48.0t, MS - 50.9t, MS - 51.1t MC - 51.5t, PDTEBF - 51.3t	Total: 430.3t PDTS - 47.7t, MS - 50.1t, MS 50.3t, TS - 41.0t, MS - 50.3t, TS - 38.3t, MS - 50.3t, MF - 50.6t PDTRBF - 51.7t	Total: 251.5t PDTS - 48.0t, MS - 50.9t, MS - 51.1t MS - 51.3t, PDTF - 50.2t
Brake type:	Air (regenerative)	Air (regenerative)	Air (regenerative)
Power:	25kV ac overhead or underslung diesels	25kV ac overhead or underslung diesels	25kV ac overhead or underslung diesels
Diesel type:	3 x MTU 12V 1600 R80L at 700kW	9 x MTU 12V 1600 R80L at 700kW	3 x MTU 12V 1600 R80L at 700kW
Diesel output:	Motor car - 938hp (700kW) per engine Train - 2,814hp (2,100kW)	Motor car - 938hp (700kW) per engine Train - 4,690hp (3,500kW)	Motor car - 938hp (700kW) per engine Train - 2,814hp (2,100kW)
Electric output:	12 x Hitachi of 303hp (226kW) = 3,636hp	20 x Hitachi of 303hp (226kW) = 6,060hp	12 x Hitachi of 303hp (226kW) = 3,636hp
Max speed:	125mph (201km/h)	125mph (201km/h)	125mph (201km/h)
Coupling type:	Outer - Dellner 10, Inner - Semi-auto	Outer - Dellner 10, Inner - Semi-auto	Outer - Dellner 10, Inner - Semi-auto
Multiple restriction:	Class 800-802	Class 800-802	Class 800-802
Door type:	Single leaf sliding	Single leaf sliding	Single leaf sliding
Construction:	Aluminium	Aluminium	Aluminium
Operator:	5-car bi-mode sets for GWR West of England	9-car bi-mode sets for GWR West of England	5-car bi-mode sets for TransPennine Trains
Owner:	Eversholt Leasing	Eversholt Leasing	Angel Trains

Above: *LNER also operates a fleet of 12 five-car and 30 nine-car all-electric Class 801s, these only have one emergency diesel engine mounted below the MSO (822xxx) on nine-car sets and the MRSB on five-car sets, recognisable by the fuel filler on the underframe. The sets are identical to the bi-mode fleet. The five-car sets are based at Doncaster while the 30 nine-car sets are based at Bounds Green. Set No. 801201 is seen south of York.* **CJM**

Left: *After the Agility Trains order for IEP stock was placed, Great Western, in partnership with Eversholt, ordered extra near identical sets. These are classified as 802, with 22 five-car (802/0) and 14 nine-car (802/1) trains in service. The first two five-car and the first nine-car set were built in Japan, the remainder were built at the Hitachi works in Pistoia, Italy. Five-car sets Nos. 802019 and 802014 are seen south of Didcot. All GW sets are painted in GW green with mid-grey passenger doors.* **CJM**

Class 800-810

Key Facts

- IET stock, replaced HST stock on Great Western and HST and Class 91 stock on East Coast routes.
- First vehicles of each sub class built in Kasado, Japan, remainder of 800, 801s, 803s, 805s and 807s built in Newton Aycliffe and 802s in Pistoia, Italy.
- First Class 800s entered service in autumn 2017 with entire fleet of 800, 801 and 802s in traffic on GWR by May 2019 and on LNER by summer 2020.
- Class 802/2 and 802/3 sets for TransPennine and Hull Trains entered service in 2019-2020.
- Class 803 stock for Open Access operator Lumo commenced operation in 2021. Class 805, 807, 810 stock for AWC and EMR under final test in 2023.

802/3
Bi-mode
802301-802305
2019-2020
Hitachi, Kasado, Japan, and Pistoia, Italy
Paragon
PDTS+MS+MS +MC+PDTF

PDTS - 831301-831305
MS - 832301-832305
MS - 833301-833305
MC - 834301-834305
PDTF - 835301-835305
794ft 0in (242m)
Driving - 83ft 2in (25.35m)
Intermediate - 82ft 0in (25.0m)
8ft 10in (2.7m)
Total - 43F/284S,
PDTS - 50S, MS - 88S, MS - 88S
MC - 18F/58S, PDTF - 25F

2+1F, 2+2S
Within set
5
Total: 248t
PDTS - 48.0t, MS - 49.6t, MS - 50.4t,
MC - 50.4t, PDTF - 49.6t

Air (regenerative)
25kV ac overhead or underslung diesels
3 x MTU 12V 1600 R80L at 700kW
Motor car - 938hp (700kW) per engine
Train - 2,814hp (2,100kW)
12 x Hitachi of 303hp (226kW) = 3,636hp
125mph (201km/h)
Outer - Dellner 10, Inner - Semi-auto
Class 800-802
Single leaf sliding
Aluminium
5-car bi-mode sets for Hull Trains
Angel Trains

803
Electric
803001-803005
2021-2022
Hitachi, Kasado, Japan and Newton Aycliffe
Lumo
PDTS+MS+MS+MS+PDTS

PDTS - 841001-841005
MS - 842001-842005
MS - 843001-843005
MS - 844001-844005
PDTS - 845001-845005
528ft 2in (162m)
Driving - 83ft 2in (25.35m)
Intermediate - 82ft 0in (25.0m)
8ft 10in (2.7m)
Total - 394S + 4 Tip-up
PDTS(A) - 52 + 2 TU, MS - 94S, MS - 94S
MS - 94S, PDTS(B) - 60S + 2TU

2+2
Within set
5
Total: 229.7t
PDTS(A) - 47.7t, MS - 45.0t, MS - 44.2t
MS - 45.1t, PDTS(B) - 47.8t

Air (regenerative)
25kV ac overhead
-
-
-
12 x Hitachi of 303hp (226kW) = 3,636hp
125mph (201km/h)
Outer - Dellner 10, Inner - Semi-auto
Class 802, 803
Single leaf sliding
Aluminium
5-car all electric sets for First Group / Lumo
Beacon Rail

2023 Fleet

802/0		802102	GWR	802212	TPT
802001	GWR	802103	GWR	802213	TPT
802002	GWR	802104	GWR	802214	TPT
802003	GWR	802105	GWR	802215	TPT
802004	GWR	802106	GWR	802216	TPT
802005	GWR	802107	GWR	802217	TPT
802006	GWR	802108	GWR	802218	TPT
802007	GWR	802109	GWR	802219	TPT
802008	GWR	802110	GWR		
802009	GWR	802111	GWR	**802/3**	
802010	GWR	802112	GWR	802301	FHT
802011	GWR	802113	GWR	802302	FHT
802012	GWR	802114	GWR	802303	FHT
802013	GWR			802304	FHT
802014	GWR	**802/2**		802305	FHT
802015	GWR	802201	TPT		
802016	GWR	802202	TPT	**803**	
802017	GWR	802203	TPT	803001	LUM
802018	GWR	802204	TPT	803002	LUM
802019	GWR	802205	TPT	803003	LUM
802020	GWR	802206	TPT	803004	LUM
802021	GWR	802207	TPT	803005	LUM
802022	GWR	802208	TPT		
		802209	TPT		
802/1		802210	TPT		
802101	GWR	802211	TPT		

Left: *The 14 nine-car Class 802/1s are usually deployed on the main Paddington to West of England services to Paignton, Plymouth and Penzance, the sets also operate other routes as needed. Each nine-car GW set has seating for 101 first and 526 standard class passengers. Set No. 802108 passes Denchworth on on 21 February 2023.* **CJM**

Left Middle: *The Class 801 (all electric) sets can be identified by their motor cars not having a fuel intake on the underframe, as shown in this view of MS 823210 from set 801210.* **CJM**

Left Lower: *Great Western have maintained their policy of applying names to many of its trains, with a large number of IETs now carrying names, many after people who made significant contributions to helping others during the Covid pandemic, selected in a GWR/BBC 'Make a Difference' campaign. The name Kieron Griffin & Ocha as applied to No. 802010 is illustrated.* **CJM**

Right: *Class 800 roof detail of driving car, showing pantograph well with the pantograph in the lowered position.* **CJM**

Rolling Stock Review : 2023-2024

Class 800-810

Above Left: Standard class interior of Great Western Railway operated Class 800/0, seating is mainly in the airline style, with around four 'group seats' per vehicle. This view shows the general layout. PIS equipment is provided above end doors and electronic 'traffic light' seat reservations above seats. Armrests are in the raised position. **CJM**

Below Left: First class interior of Great Western Railway Class 800 / 802, showing the 2+1 seating layout with a mix of airline and group seating in a Motor First (MF) vehicle seating 56 with a standard design toilet at one end. **Antony Christie**

Below: Two universal toilet compartments are formed within each IET set, both at the inner ends of driving cars, one for first and one for standard class passengers (of identical layout). The toilet in the first class end of 800305 is shown. **CJM**

Above Left: On LNER Class 800 and 801 sets, the seating, while of the same style as found on the Great Western fleet, is covered in a different moquette and gives the impression of a more comfortable seat. The standard class uses bright and mid-tone red, a colour pallette chosen by the previous operator Virgin. Seating is in 2+2 in standard class with a mix of airline and group layout. Each window has a pull down sun blind, but these frequently cause arguments amongst passengers as one blind covers more than one seating bay. **CJM**

Above Right: The LNER first class seating again uses the same style of seat as on the Great Western sets, but with a deep brown/red moquette. Although a high proportion of seats are of the group layout, the airline seats have a good quality fixed table. Full power supply for laptop and telephone charging is provided. **CJM**

Class 800-810

Above: Class 800 and Class 802 Bi-mode driving cab. (Not applicable to 'all-electric' Class 801 sets). 1: Cab lighting control panel. 2: In-cab CCTV monitors, to show exterior or interior, forward facing image or a view of the pantograph. The driver can select which camera he requires. If a passenger alarm is activated the nearest camera in the train is activated. 3: Cab ventilator, 4: Left side door controls, housing a signal button, two door release buttons, a door close and interlock button, indicator, and a train door control button which hands door control to an on board member of staff, 5: Camera button, when depressed records the CCTV image, 6: Track Protection and Warning System (TPWS) controls, 7: Indicators for Line Volts, Vacuum Circuit Breaker, Safety Systems Isolated and Pantograph Auto Dropper Device, 8: Emergency brake plunger, 9: Snow brake button, 10: Direction (Master) switch, (Off, Forward, Neutral, Reverse), 11: Master key socket, 12: Combined power/brake controller, 13: Drivers Reminder Appliance (DRA), 14: Speedometer incorporating Automatic Train Protection, 15: Duplex air gauge, main reservoir (yellow) and brake cylinder (red), 16: Brake force gauge, 17: AWS indicator, 18: European Train Control System (ETCS) screen (Not in use), 19: Coupling control panel for coupling and uncoupling, 20: Wheel slip/slide warning and train wash push-buttons, 21: ETCS acknowledge button (Not in use), 22: AWS reset button, 23: Right side door controls, housing a signal button, two door release buttons, a door close and interlock button, indicator and a train door control button which hands door control to an on board member of staff, 24: Drivers safety device hold over button, 25: Train fault acknowledge button, 26: Horn valve. 27: Train Management System (TMS) screen, 28: Electric mode control panel, with electric select, pantograph up and pantograph down buttons, 29: GSM-R radio, 30: Windscreen wash/wipe panel and demist facility, 31: Diesel mode control panel, with engine start button, diesel set up button, engine stop button and emergency engine stop button, 32: GSM-R radio handset, 33: Cab-train/staff audio communication control panel and handset. **CJM**

Above: The outer couplings, hidden behind side rotary doors, is of the Dellner 10 type, providing physical, pneumatic and control connectivity between sets. The attachment and detachment process is controlled from the driving cab. The coupling, via an adapter plate can be used to rescue a failed train if needed. **CJM**

Right Above & Right Below: Two very different designs of lightweight bogie can be found below the 800 breed stock. The bogies which carry traction equipment below the MSO and MCO vehicles are of a heavier design and carry two traction motors per bogie, illustrated above on vehicle 835112. The trailer bogies on the TSO vehicles on nine car sets are of the inner-frame bolster-less type, as illustrated below. The bogies fitted below the driving trailer coaches are of the heavier motor design but are unpowered.
Both: **CJM**

Rolling Stock Review : 2023-2024

Class 800-810

Above & Inset: *First TransPennine Express was the third operator to introduce Hitachi IET Class 802 stock from 2019, with a fleet of 19 five-car Class 802/2 sets based at Edinburgh Craigentinny, but are maintained at Newcastle Heaton. They are deployed on Scotland to Manchester and Liverpool services via the East Coast route through Leeds and York. The sets are identified by TransPennine as their 'Nova 1' fleet. The design seats 24 first and 320 standard class passengers and display TransPennine grey, blue, turquoise and mauve livery, the sets do not have yellow ends. Above, No. 8022018 is seen south of York from its standard class driving trailer. In the above inset, the DTF from set No. 802219 is shown, note the blanked out saloon windows adjacent to the food preparation area. All TPTE Class 802s were constructed at the Hitachi plant in Pistoia, Italy.* **Both: CJM**

Left: *The TransPennine Express Class 802s, carry their full European Vehicle Numbers (EVNs), these are applied just below the standard coach number. In this view MSO No. 832218 from set No. 802218 is shown. The cables linking between vehicles on the 80x designs have raised concern of the safety authorities, with the thought they could be used as a 'ladder' to climb onto coach roofs. A new design of between vehicle connections is under development. Low down on the bodywork, to the right of the passenger door is the emergency door release valve.* **Antony Christie**

Above Left & Above Right: *Of all the 800 breed interiors the First TransPennine Express sets look the most impressive in terms of colour pallet and presentation. The left image shows the 2+1 first class layout, consisting of 24 seats in the 2+1 style. Seats are mainly in the group layout with provision for a wheelchair at the vestibule/toilet end. All seats have a fixed table. In the above right view, the standard class, set out in the 2+2 layout in a mix of group and airline is shown. Pull down window blinds, a passenger information system and an electronic seat reservation system is fitted.* **Both: Antony Christie**

Class 800-810

Above, Right & Right Lower: *Open access operator Hull Trains, owned by First Group, introduced five, five-car Class 802/3s for use on King's Cross to Hull services at the end of 2019. This sub-class seat 43 first and 284 standard class passengers. First class is located in one driving car (DTF) and a one-third section of the adjoining Motor Composite (MC). Sets carry a mid blue-livery, offset by a lengthwise colourful route associated pictogram (except the final set which carries Hull Trains 21st celebration livery, shown right). The fleet is known as the 'Paragon' fleet by Hull Trains and in 2022-2023 all were named after famous people from Hull. The above image shows set No. 802302 Jean Bishop (The Bee Lady) from its Driving Trailer Standard end. No. 802305 is seen right in 21st anniversary livery, from its Driving Trailer First end, carrying the stick on name The Humber Bridge. The right lower image shows Motor Composite (MC) No. 834303 a vehicle seating 18 first (far end) and 58 standard class seats. All:* **CJM**

Below: *Stick-on nameplate Amy Johnson as applied to No. 802301.* **Antony Christie**

Above Left & Above Right: *The interior of the Hull Trains '802s' is based on earlier 80x fleets, with first class seating (shown above right) using the 2+1 style, with most seats in group layouts with tables. This is a view of Motor Composite Open (MCO) coach, looking towards the doors leading into the standard class area. Above left is the standard class seating in a MSO vehicle, seats are a mix of airline and group style around fixed tables. The trains are fitted with electronic seat reservation equipment and luggage racks at coach ends. Both:* **CJM**

Rolling Stock Review : 2023-2024   183

Class 800-810

Class:	805	807	810
Traction:	Bi-mode	Electric	Bi-mode
Number range:	805001-805013	807001-807010	810001-810033
Introduced:	2023-2024	2023-2024	2023-2024
Built by:	Hitachi, Kasado, Japan and Newton Aycliffe	Hitachi, Kasado, Japan and Newton Aycliffe	Hitachi, Newton Aycliffe
Type:	AT300	AT300	AT300SXR
Formation:	PDTS+MS+MS+MC+PDTRBF	PDTS+MS+MS+TS+MS+MC+PDTF	DPTF+MC+TS+MS+DPTS
Vehicle numbers:	PDTS - 861001-861013 MS - 862001-862013 MS - 863001-863013 MS - 864001-864013 PDTF - 865001-865013	PDTS - 871001-871010 MS - 872001-872010, MS - 873001-873010 TS - 874001-874010, MS - 875001-875010 MC - 876001-876010, PDTF - 877001-877010	DPTF - 851001-851033 MCO - 852001-852033 TSO - 853001-853033 MSO - 854001-854033 DPTS - 855001-855033
Train length:	528ft 2in (162m)	TBA	TBA
Vehicle length:	Driving - 83ft 2in (25.35m) Intermediate - 82ft 0in (25.0m)	Driving - 83ft 2in (25.35m) Intermediate - 82ft 0in (25.0m)	Driving - 78ft 9in (24m) Intermediate - 78ft 9in (24m)
Width:	8ft 10in (2.7m)	8ft 10in (2.7m)	8ft 10in (2.7m)
Seating:	Total - tba, PDTS - tba, MS - tba, MS - tba MS - tba, PDTF - tba	Total -tas, PDTS - tba, MS - tba, MS - tba, TS - tba MS - tba, MC - tba, PDTF - tba	Total - 301 - 47F/254S DPTF - tba, MC - tba, TS - tba, MS - tba DPTS - tba
Internal layout:	2+1F, 2+2S	2+1F, 2+2S	2+1F, 2+2S
Gangway:	Within set	Within set	Within set
Toilets:	5	7	5
Weight:	Total: tba PDTS - tba, MS - tba, MS - tba MS - tba, PDTF - tba	Total: tba PDTS - tba, MS - tba, MS tba, TS - tba, MS - tba, MC - tba, PDTF - tba	Total: tba DPTF - tba, MC - tba, TS - tba MS - tba, DPTS - tba
Brake type:	Air (regenerative)	Air (regenerative)	Air (regenerative)
Power:	25kV ac overhead or underslung diesels	25kV ac overhead	25kV ac overhead or underslung diesels
Diesel type:	3 x MTU 12V 1600 R80L at 700kW	-	4 x MTU 12V 1600 R80L at 735kW (986hp)
Diesel output:	Motor car - 938hp (700kW) per engine Train - 2,814hp (2,100kW)	Train - 4,690hp (3,500kW)	985hp (700kW) below DPTS, MS, MC & DPTF Train - 2,955hp (2,203kW)
Electric output:	12 x Hitachi of 303hp (226kW) = 3,636hp	16 x Hitachi of 303hp (226kW) = 4,848hp	8 x Hitachi of 335hp (250kW) = 2,680hp
Max speed:	125mph (201km/h)	125mph (201km/h)	125mph (201km/h)
Coupling type:	Outer - Dellner 10, Inner - Semi-auto	Outer - Dellner 10, Inner - Semi-auto	Outer - Dellner 10, Inner - Semi-auto
Multiple restriction:	Class 805 & 807	Class 805 & 807	Class 810
Door type:	Single leaf sliding	Single leaf sliding	Single leaf sliding
Construction:	Aluminium	Aluminium	Aluminium
Operator:	5-car bi-mode sets for Avanti West Coast	7-car electric sets for Avanti West Coast	5-car bi-mode sets for East Midlands Railway
Owner:	Rock Rail	Rock Rail	Rock Rail

Above, Below and Inset: *In late 2021, FirstGroup introduced a fleet of five Hitachi AT300 all-electric Class 803s, for use on a new London King's Cross to Edinburgh Waverley 'open access' low price service, known as Lumo. The trains, financed by Beacon Rail are all electric with no diesel back-up, a battery provides hotel power if the overhead power supply is lost. The sets operate between Edinburgh and King's Cross, calling at Stevenage, Newcastle and Morpeth on route, with an end to end journey time of around 4 hours. The sets have all one class seating in the 2+2 style, mainly set out in the airline style. No on-board catering is provided. The sets are air-conditioned, have power sockets by each seat and wi-fi. The sets are painted in mid-blue with no yellow warning ends. Above, set No. 803004 is seen south of York, left is Motor Standard 842005, these vehicles seat 94 and have one standard design toilet compartment. The interior (inset), is very cramped with the highest number of seats per vehicle of any 80x design, this fits in with the low cost high-capacity, no frills business plan of the operator. Airline style seats have fold down tables attached to the seat ahead.*
CJM / Antony Christie (2)

Class 800-810

Above & right: *In spring 2023, the first Class 805 and 807 sets for Avanti West Coast emerged from the Hitachi factory in Newton Aycliffe and commenced testing on the Old Dalby test track and on the West Coast main line. Above, set No. 805001 passes Crewe on 9 February 2023 with a test run, below Motor Standard (MS) No. 864001 is seen from the driving car end. Both:* **Cliff Beeton**

Above: *In July 2023, the first of 33 five car Class 810 IETs were completed at the Hitachi factory at Newton Aycliffe for East Midlands Trains to replace Class 222 stock. These IETs are very different from previous builds with vehicles shorter at 78ft 9in (24m), to enable operation on the Midlands Main Line route. Each set will seat 47 first and 254 standard class passengers and be based at a modernised Derby Etches Park depot. Showing its redesigned front end, set No. 810001 is seen at Newton Aycliffe.* **Stephen Barker**

2023 Fleet

805		810004	EMR
805001	AWC	810005	EMR
805002	AWC	810006	EMR
805003	AWC	810007	EMR
805004	AWC	810008	EMR
805005	AWC	810009	EMR
805006	AWC	810010	EMR
805007	AWC	810011	EMR
805008	AWC	810012	EMR
805009	AWC	810013	EMR
805010	AWC	810014	EMR
805011	AWC	810015	EMR
805012	AWC	810016	EMR
805013	AWC	810017	EMR
		810018	EMR
807		810019	EMR
807001	AWC	810020	EMR
807002	AWC	810021	EMR
807003	AWC	810022	EMR
807004	AWC	810023	EMR
807005	AWC	810024	EMR
807006	AWC	810025	EMR
807007	AWC	810026	EMR
807008	AWC	810027	EMR
807009	AWC	810028	EMR
807010	AWC	810029	EMR
		810030	EMR
810		810031	EMR
810001	EMR	810032	EMR
810002	EMR	810033	EMR
810003	EMR		

Rolling Stock Review : 2023-2024

Coaching Stock

Mk 1

Type:	RF (AJ11)	RKB (AK51)	RBR (AJ41)	RMB (AN21)	FO (AD11)	TSO/SO (AC21)
Number range:	325	1566	1651-1730, 1953-1961	1813-1882	3045-3150	3798-5044
Former range:	2907	-	-	-	-	-
Introduced:	1961	1960	1960-1961	1960-1962	1955-1963	1959-1962
Built by:	BR Swindon	Cravens	Pressed Steel/BRCW BR Swindon	BR Wolverton	BR Doncaster, Swindon BRCW	BR York, Wolverton
Seating:	24F	6F	11-21 chairs*	44S	42S	64S
Speed:	100mph (161km/h)	100mph (161km/h)	100mph (161km/h)	100mph (161km/h)	100mph (161km/h)	100mph (161km/h)
Brake type:	Air	Air	Air	Air	Air/Dual	Air/Dual
Heating:	Electric/2	Electric/1	Electric/2	Electric/3	Electric/3	Electric/4
Bogie type:	B5	B5	B5, COM	COM	B4, COM	COM, B4
Length:	64ft 6in (19.65m)	64ft 6in (19.65m)	64ft 6in (19.65m)	64ft 6in (19.65m)	64ft 6in (19.65m)	64ft 6in (19.65m)
Height:	12ft 9½in (3.90m)	12ft 9½in (3.90m)	12ft 9½in (3.90m)	12ft 9½in (3.90m)	12ft 9½in (3.90m)	12ft 9½in (3.90m)
Width:	9ft 3in (2.82m)	9ft 3in (2.82m)	9ft 3in (2.82m)	9ft 3in (2.82m)	9ft 3in (2.82m)	9ft 3in (2.82m)
Weight:	34 tonnes	41 tonnes	37-39 tonnes	37-38 tonnes	33-36 tonnes	33-37 tonnes
Operator:	WC	WC	RV, LS, ER, WC	RV, WC, LS	RV, SR, WC, LS, ER	NY, WC, SR, RV, LS
Notes:		Bar car	* Depending on style			

Type:	GEN (AX51)	FK (AA11)	BFK (AB11)	SK (AA21)	BCK (AB31)	BSK (AB21)
Number range:	6311-6313	13227-13320	17013-17041	18756	21096-21269	35089-35486
Former range:	From NDA	-	14013-14041	24756	-	-
Introduced:	1956-1958	1959-1962	1961	1962	1961-1964	1959-1963
Built by:	Pressed Steel, Cravens	BR Swindon	BR Swindon	BR Derby	BR Swindon, Derby	BR Wolverton
Seating:	-	42F	24F	48S	12F/18S or 12F/24S	24S
Speed:	100mph (161km/h)	100mph (161km/h)	100mph (161km/h)	100mph (161km/h)	100mph (161km/h)	100mph (161km/h)
Brake type:	Air	Dual	Dual/vacuum	Dual/vacuum	Dual	Dual
Heating:	-	Electric/3	Electric/2	Electric/4	Electric/2	Electric/2
Bogie type:	B4	B4/COM	COM	COM	COM	B4/COM
Length:	57ft 0in (17.37m)	64ft 6in (19.65m)	64ft 6in (19.65m)	64ft 6in (19.65m)	64ft 6in (19.65m)	64ft 6in (19.65m)
Height:	12ft 9in (3.88m)	12ft 9½in (3.90m)	12ft 9½in (3.90m)	12ft 9½in (3.90m)	12ft 9½in (3.90m)	12ft 9½in (3.90m)
Width:	9ft 3in (2.82m)	9ft 3in (2.82m)	9ft 3in (2.82m)	9ft 3in (2.82m)	9ft 3in (2.82m)	9ft 3in (2.82m)
Weight:	37.5 tonnes	33-37 tonnes	36 tonnes	36 tonnes	36-37 tonnes	33-37 tonnes
Operator:	LS, WC, VS	SR, WC, LS	VI	WC	NY, SR, RV, WC	NY, SR, RV, LS
Notes:		13320 now FO	Support Coaches	Harry Potter World		35469 has generator

Type:	Pullman (Various types)
Number range:	213-354
Introduced:	1925-1961
Built by:	MCW, BRCW, Metro-Cam
Seating:	Various
Speed:	75-100mph (121-161km/h)
Brake type:	Air
Heating:	Steam/electric
Bogie type:	Gresley/ Commonwealth / B5
Length:	Various
Height:	Various
Width:	Various
Weight:	Various
Operator:	LS, WC, VS, VT
Notes:	

Right: A number of various types of Mk1 coach are still in main line use with various charter operators. Mk1 First Kitchen No. 325, operating with West Coast is a former Royal Train coach No. 2907. In 2023 it operates in the Northern Belle set. It has a full kitchen and 24 first class seats.
Antony Christie

Above: A number of former Pullman Car Co Mk1 and earlier vehicles are in service, mainly with the private charter operators. In the main these retain their distinctive umber and cream Pullman colours and trademark inward opening doors. Pullman Car 335 is seen, a Pullman Kitchen originally built for the East Coast Main Line in 1960. Today, it is operated by Tyseley-based Vintage Trains and operates as private owner vehicle 99361. It has 30 standard class seats and one toilet compartment (near end). **Antony Christie**

Coaching Stock

GEN (AX51)
6310
975325 (81448)
1958
Pressed Steel

-
125mph (201km/h)
Air
-
B5
57ft 0in (17.37m)
12ft 9in (3.88m)
9ft 3in (2.82m)
42 tonnes
RV
Former HST generator

RK(RBR) (AK51)
80041-80044
16XX
Orig: 1960, RK 1989-2014
Pressed Steel
-
100mph (161km/h)
Dual, Air
Electric/2
COM
64ft 6in (19.65m)
12ft 9½in (3.90m)
9ft 3in (2.82m)
39 tonnes
RV, LS

Right Top: To provide high-quality catering on land cruise and charter trains, a fleet of RBRs are in service numbered in the 16xx and 17xx series No. 1671 is seen in ex works BR blue and grey colours. These cars should have 23 unclassified seats, but many now have different seating arrangements. This coach is owned by Riviera Trains. **Spencer Conquest**

Right Middle Upper: The most common of the loco-hauled Mk1 passenger vehicles is the SO or TSO, seating 64 in the 2+2 style with a centre aisle. No. 4905 is owned by West Coast Railway and carries their standard maroon livery. It is mounted on Commonwealth bogies. **Antony Christie**

Right Middle Lower: A number of corridor brake vehicles have been adapted to act as steam loco support vehicles, with accommodation for support staff and a small workshop in the former brake area. BSK No. 35461 is owned by Loco Services of Crewe and can be found operating with any of the LSL steam fleet. It is painted in carmine and cream colours and mounted on Commonwealth bogies. **CJM**

Right Lower: Another steam loco support vehicle is BSK 35470, this is owned by Tyseley-based Vintage Trains and displays BR branded chocolate and cream colours. This is a vacuum brake fitted dual heat coach. **Antony Christie**

Right Bottom: To provide the high level of catering expected on the upmarket land cruise trains, specialist catering vehicles are in service. Four former RBR vehicles were rebuilt between 1989-2019 as full kitchen cars and numbered in the 80041-80044 series. Two are owned/operated by Riviera Trains and two by Loco Services Ltd, No. 80043 is shown, owned by Loco Services and branded for use in the Statesman train, it also carries the 'name' Kitchen Car. **Antony Christie**

Rolling Stock Review : 2023-2024

Coaching Stock

Mk 2

Type:	PKF (AP1Z)	PFP (AQ1Z)	PFB (AR1Z)	RFB (AJ1F)	FO (AD1D)	FO (AD1E)
Mark:	2	2	2	2f	2d	2e
Number range:	504-506	546-553	586	1200-1256	3174-3188	3229-3275
Former range:	-	-	-	32, 33 & 34xx	-	-
Introduced:	1966	1966	1966	1973-1975	1971-1972	1972-1973
Built by:	BR Derby	BR Derby	BR Derby	BR Derby	BR Derby	BR Derby
Seating:	-	36F	30F	25F or 26F	42F	36-42F*
Speed:	100mph (161km/h)	100mph (161km/h)	100mph (161km/h)	100mph (161km/h)	100mph (161km/h)	100mph (161km/h)
Brake type:	Air	Air	Air	Air	Air	Air
Heating:	Electric/6	Electric/5	Electric/4	Electric/6	Electric/5	Electric/5
Bogie type:	B5	B4	B4	B4	B4	B4
Length:	66ft 0in (20.12m)	66ft 0in (20.12m)	66ft 0in (20.12m)	66ft 0in (20.12m)	66ft 0in (20.12m)	66ft 0in (20.12m)
Height:	12ft 9½in (3.90m)	12ft 9½in (3.90m)	12ft 9½in (3.90m)	12ft 9½in (3.90m)	12ft 9½in (3.90m)	12ft 9½in (3.90m)
Width:	9ft 3in (2.82m)	9ft 3in (2.82m)	9ft 3in (2.82m)	9ft 3in (2.82m)	9ft 3in (2.82m)	9ft 3in (2.82m)
Weight:	40 tonnes	35 tonnes	37-39 tonnes	37-38 tonnes	34 tonnes	32-35 tonnes
Operator:	WC	WC	WC	RV, LS, WC, ER, NR	WC, LS	LS, WC
Notes:						

Type:	TSO (AC2Z)	SO (AD2Z)	TSO (AC2A)	TSO (AC2B)	TSO (AC2E)	TSO (AC2F)
Mark:	2	2	2a	2b	2e	2f
Number range:	5157-5222	5229-5249	5278-5419	5482-5487	5787-5842	5910-6183
Former range:	-	-	-	-	-	-
Introduced:	1965-1967	1966	1967-1968	1969	1972-1973	1973-1975
Built by:	BR Derby	BR Derby	BR Derby	BR Derby	BR Derby	BR Derby
Seating:	64S	48S	62-64S	62S	64S*	58-64S*
Speed:	100mph (161km/h)	100mph (161km/h)	100mph (161km/h)	100mph (161km/h)	100mph (161km/h)	100mph (161km/h)
Brake type:	Vacuum	Air or Vacuum	Air	Air	Air	Air
Heating:	Electric/4	Electric/4	Electric/4	Electric/4	Electric/5	Electric/5
Bogie type:	B4	B4	B4	B4	B4	B4
Length:	66ft 0in (20.12m)	66ft 0in (20.12m)	66ft 0in (20.12m)	66ft 0in (20.12m)	66ft 0in (20.12m)	66ft 0in (20.12m)
Height:	12ft 9½in (3.90m)	12ft 9½in (3.90m)	12ft 9½in (3.90m)	12ft 9½in (3.90m)	12ft 9½in (3.90m)	12ft 9½in (3.90m)
Width:	9ft 3in (2.82m)	9ft 3in (2.82m)	9ft 3in (2.82m)	9ft 3in (2.82m)	9ft 3in (2.82m)	9ft 3in (2.82m)
Weight:	32 tonnes	32 tonnes	32 tonnes	32 tonnes	33 tonnes	33 tonnes
Operator:	WC, VT	WC	RV, WC, LS	WC, ER	ER	RV, LS, DR, WC, NR
Notes:						* Depending on style

Type:	RLO (AN1F)	RFB (AN1D)	BSOT (AH2Z)	BSO (AE2D)	BSO (AE2E)	BSO (AE2F)
Mark:	2f	2d	2	2d	2e	2f
Number range:	6700-6708	6723-6724	9101-9104, 9391-9392	9479-9493	9497-9509	9513-9539
Former range:	32, 33xx	56, 57xx	-	-	-	-
Introduced:	1973-1974	1971	1966	1971	1972	1974
Built by:	BR Derby	BR Derby	BR Derby	BR Derby	BR Derby	BR Derby
Seating:	25F	30F	23S	22-31S*	28-32S*	32S
Speed:	100mph (161km/h)	100mph (161km/h)	100mph (161km/h)	100mph (161km/h)	100mph (161km/h)	100mph (161km/h)
Brake type:	Air	Air	Vacuum	Air	Air	Air
Heating:	Electric/5	Electric/5	Electric/4	Electric/5	Electric/5	Electric/5
Bogie type:	B4	B4	B4	B4	B4	B4
Length:	66ft 0in (20.12m)	66ft 0in (20.12m)	66ft 0in (20.12m)	66ft 0in (20.12m)	66ft 0in (20.12m)	66ft 0in (20.12m)
Height:	12ft 9½in (3.90m)	12ft 9½in (3.90m)	12ft 9½in (3.90m)	12ft 9½in (3.90m)	12ft 9½in (3.90m)	12ft 9½in (3.90m)
Width:	9ft 3in (2.82m)	9ft 3in (2.82m)	9ft 3in (2.82m)	9ft 3in (2.82m)	9ft 3in (2.82m)	9ft 3in (2.82m)
Weight:	33.5 tonnes	32 tonnes	31 tonnes	33 tonnes	33 tonnes	34 tonnes
Operator:	ER, LS	WC	VT, WC	DR, LS, WC, NR	ER, VS, RV, DR	ER, RV, DR, NR
Notes:	6705 observation car				* Depending on style	* Depending on style

Type:	BUO (AE4E)	FK (AA1A)	BFK (AB1A)	GEN (AX5B)	BFK (AB1D)	BPK (AB5C)
Mark:	2e	2a	2a	2b	2d	2c
Number range:	9800-9810	13440	17056-17102	17105	17159-17167	35508-35511
Former range:	57, 58xx	-	14056-14102	14105, 2905	14159-14167	14130/17130
Introduced:	1972-1973	1968	1967-1968	1977	1971-1972	1969
Built by:	BR Derby	BR Derby	BR Derby	BR Derby	BR Derby	BR Derby
Seating:	31S	42F	24F	-	24S	-
Speed:	100mph (161km/h)	100mph (161km/h)	100mph (161km/h)	100mph (161km/h)	100mph (161km/h)	100mph (161km/h)
Brake type:	Air	Vacuum	Air	Air	Air	Air
Heating:	Electric/4	Electric/4	Electric/4	Electric/5	Electric/5	Electric/4
Bogie type:	B4	B4	B4	B5	B4	B4
Length:	66ft 0in (20.12m)	66ft 0in (20.12m)	66ft 0in (20.12m)	66ft 0in (20.12m)	66ft 0in (20.12m)	66ft 0in (20.12m)
Height:	12ft 9½in (3.90m)	12ft 9½in (3.90m)	12ft 9½in (3.90m)	12ft 9½in (3.90m)	12ft 9½in (3.90m)	12ft 9½in (3.90m)
Width:	9ft 3in (2.82m)	9ft 3in (2.82m)	9ft 3in (2.82m)	9ft 3in (2.82m)	9ft 3in (2.82m)	9ft 3in (2.82m)
Weight:	33.5 tonnes	33 tonnes	32 tonnes	46 tonnes	33.5 tonnes	32.5 tonnes
Operator:	ER, LS	WC	WC, LS, VT	RV	LS, WC	LS, EL
Notes:						Brake Power Kitchen

Left: A popular choice for charter trains is the BR Mk2 air conditioned First Open (FO) coach design. These are set out in the 2+1 seat style with all seats in group layout. No. 3325, a Mk2f built in 1973 is operated by Riviera Trains and painted in BR blue/grey livery. It is fitted with central door locking, has 42 seats and two toilets. **Spencer Conquest**

Coaching Stock

FO (AD1F)
2f
3278-3438
-
1973-1975
BR Derby
42F
100mph (161km/h)
Air
Electric/5
B4
66ft 0in (20.12m)
12ft 9½in (3.90m)
9ft 3in (2.82m)
34 tonnes
WC, RV, LS, ER
* Depending on style

TSOT (AG2C)
2c
6528
5592
1969
BR Derby
55S
100mph (161km/h)
Air
Electric/4
B4
66ft 0in (20.12m)
12ft 9½in (3.90m)
9ft 3in (2.82m)
32.5 tonnes
WC

DBSO (AF2F)
2f
9701-9714
95xx
1974
BR Derby
30S
100mph (161km/h)
Air
Electric/5
B4
66ft 0in (20.12m)
12ft 9½in (3.90m)
9ft 3in (2.82m)
34 tonnes
DR, NR, LS, ER

Right Top: In early 2023, Loco Services released to traffic a fully restored Driving Brake Standard Open (DBSO) originally used on the Class 47/7 powered Edinburgh-Glasgow high speed services. The vehicle with fully operational TDM push-pull equipment has been restored to BR ScotRail blue stripe livery. It now sports a high level marker light. It is illustrated at Wigan. **Spencer Conquest**

Right Middle: Eight Mk2f Buffet First vehicles are in use with Riviera, Loco Services, West Coast and Eastern Rail Services. These are numbered in the 12xx series and seat 25 in the 2+1 style with a small catering section at one end. No. 1211 operated by Loco Services is part of the Statesman train and in 2023 carried the name Snaefell. **Antony Christie**

Right Lower: In 2023 seven Mk2e First Open (FO) coaches were in traffic, owned by both West Coast and Loco Services. The four operated by West Coast are used in the Northern Belle formation, No. 3247 is shown. These coaches have modified door windows, to reflect the Pullman nature of the coach. These vehicles seat 42 and have two standard design toilets. **Antony Christie**

Below: An original 1966 built early Mk2 Brake Standard Open 9398 was rebuilt as an Brake Standard Open with a catering trolley position and numbered 9101. It has 23 seats of first class style, with a full guards brake compartment towards the middle. It is owned by Vintage Trains at Tyseley. It currently carries chocolate and cream livery. **Antony Christie**

Coaching Stock

Mk 3

Type:	RFB (AJ1G)	GFW (AJ1F)	SLEP (AU4G)	SLE/SLED (AQ4G)	FO (AD1G)	FO (AD1H)
Mark:	3a	3a	3a	3a	3a	3a
Number range:	10211-10259	10271-10274	10501-10616	10729 &10734	11018-11048	11066-11101 series
Former range:	405xx HST 10xxx, 11xxx	102xx series	-	-	-	-
Introduced:	1975-1979	As GFW - 2012	1981-1983	1980-1985	1975-1976	1985
Built by:	BR Derby	BR Derby	BR Derby	BR Derby	BR Derby	BR Derby
Seating:	18F-24F*	30F	12 Comps	SLE 13 Comps, SLED 11 Comps	47F-48F	37-48F
Speed:	110mph (177km/h)	110mph (177km/h)	110mph (177km/h)	110mph (177km/h)	110mph (177km/h)	110mph (177km/h)
Brake type:	Air	Air	Air	Air	Air	Air
Heating:	Electric/14	Electric/14	Electric/7	Electric/11	Electric/6	Electric/6
Bogie type:	BT10	BT10	BT10	BT10	BT10	BT10
Length:	75ft 0in (22.86m)	75ft 0in (22.86m)	75ft 0in (22.86m)	75ft 0in (22.86m)	75ft 0in (22.86m)	75ft 0in (22.86m)
Height:	12ft 9in (3.89m)	12ft 9in (3.89m)	12ft 9in (3.89m)	12ft 9in (3.89m)	12ft 9in (3.89m)	12ft 9in (3.89m)
Width:	8ft 11in (2.71m)	8ft 11in (2.71m)	8ft 11in (2.71m)	8ft 11in (2.71m)	8ft 11in (2.71m)	8ft 11in (2.71m)
Weight:	39.8 tonnes	40.1 tonnes	41 tonnes	42-43 tonnes	34.3 tonnes	36.5 tonnes
Operator:	DB, ER, GW	CR	ER, LS, DB, GW	WC	CR, DB	Off-lease
Notes:	* Depending on style	Plug door fitted				

Type:	TSO (AC2G)	BSO (AE1H)	DVT (NZ)
Mark:	3b	3b	3b
Number range:	12005-12185 §12602-12627	17173-17175	82101-82152 82301-82309
Former range:	-	-	-
Introduced:	1975-1977	1986	1988
Built by:	BR Derby	BR Derby	BR Derby
Seating:	70-85S*	36S	
Speed:	110mph (177km/h)	110mph (177km/h)	110mph (177km/h)
Brake type:	Air	Air	Air
Heating:	Electric/6	Electric/5	Electric/5
Bogie type:	BT10	BT10	T4
Length:	75ft 0in (22.86m)	75ft 0in (22.86m)	75ft 0in (22.86m)
Height:	12ft 9in (3.89m)	12ft 9in (3.89m)	12ft 9in (3.89m)
Width:	8ft 11in (2.71m)	8ft 11in (2.71m)	8ft 11in (2.71m)
Weight:	34.3 tonnes	36 tonnes	45 tonnes
Operator:	CR, GW	GW	CR, DB
Notes:	* Depending on style § Plug door fitted		

Above: One of the few operators of Mk3 day coaches is Chiltern Railways, with in 2023 two or three diagrams each day on the Marylebone to Birmingham route, formed of Mk3s and Class 68 locos. TSO No. 12615, shows the plug door design as fitted to this fleet. The TSOs seat between 69-72 depending on layout. Coaches are painted in Chiltern main line colours of silver and grey. **CJM**

Left: West Coast Railway operate two Mk3 sleeper cars No. 10729 and 10734 on the Northern Belle train, to provide staff accommodation. The vehicles carry Northern Belle livery and when not in use is usually kept at Carnforth. No. 10729 is seen from its compartment side. The coach carries the name Crewe. **Antony Christie**

Coaching Stock

Right: The only train operator to still use Mk3 sleeping car stock is Great Western, deployed on their Night Riviera service linking London Paddington with Penzance. A fleet of Mk3 sleeper and day saloons are based at Long Rock depot in Penzance. Sleeping car with Pantry SLEP No. 10594 is shown from its corridor side. This design has 12 compartments which can either accommodate one or two passengers. At one end of an attendants compartment, at the other end two toilet compartments are located. **Antony Christie**

Left Upper: The Real Charter Train Co took on a number of ex-Anglia Mk3s, these are available for hire to charter operators. FO No. 11101 with a disabled access toilet (near) is seen. **Antony Christie**

Above & Right: With a large number of high-quality Mk3 vehicles available, many were purchased for further use by various operators. Loco Services Ltd of Crewe purchased a large number and have restored many to a very high standard, painting them in InterCity or ScotRail colours. All vehicles have central door locking and internally have been upgraded with high-quality decor, seating and facilities. In the above view Mk3b First Open (FO) No. 11077 is shown in full InterCity colours. This coach seats 48 and is fitted with retention toilets. The image right shows Mk3 Driving Van Trailer (DVT) No. 82139, one of two owned by Loco Services Ltd and used at the remote end of Mk3 charter trains powered by Class 86, 87 or 90 traction. Both: **Antony Christie**

Right: Chiltern Railways has a fleet of six Mk3 Driving Van Trailers (DVTs) for working at the remote end of Class 68 powered Mk3 sets. These DVTs have been modified from their original West Coast days and now sport a diesel-generator set in the van at the cab end for train auxiliaries. One of the original luggage van doors has been removed and replaced by a grille panel. The vehicles carry branded Chiltern Railways livery. No. 82302 is shown. **CJM**

Rolling Stock Review : 2023-2024

Coaching Stock

Mk 3 HST

Type:	TGFB (-)	TRFB (GK1G)	TF (GH1G)	TS (GH2G)	TGS (GJ2G)	TCK (GH3G)
Number range:	40601-40626	40701-40755	41026-41209¤	42002-42586¤	44012-44100	45001-45005
Former range:	41xxx	403xx	-	-	-	12xxx loco-hauled
Introduced:	Orig: 1976-1982 TFB - 2018-2020	Orig: 1978-1982 TRFB – 1987	1976-1982	1976-1982	1980-1982	Orig: 1975-1977 CK - 2005-2006
Built by:	BR Derby	BR Derby	BR Derby	BR Derby	BR Derby	BR Derby
Seating:	32F	17F	40-48F*	68-84S*	67S	30F/10S
Speed:	125mph (201km/h)	125mph (201km/h)	125mph (201km/h)	125mph (201km/h)	125mph (201km/h)	125mph (201km/h)
Brake type:	Air	Air	Air	Air	Air	Air
Heating:	Electric	Electric	Electric	Electric	Electric	Electric
Bogie type:	BT10	BT10	BT10	BT10	BT10	BT10
Length:	75ft 0in (22.86m)	75ft 0in (22.86m)	75ft 0in (22.86m)	75ft 0in (22.86m)	75ft 0in (22.86m)	75ft 0in (22.86m)
Height:	12ft 7in (3.84m)	12ft 7in (3.84m)	12ft 7in (3.84m)	12ft 7in (3.84m)	12ft 7in (3.84m)	12ft 7in (3.84m)
Width:	9ft 2in (2.79m)	9ft 2in (2.79m)	9ft 2in (2.79m)	9ft 2in (2.79m)	9ft 2in (2.79m)	9ft 2in (2.79m)
Weight:	38 tonnes	38 tonnes	33 tonnes	33 tonnes	34 tonnes	35 tonnes
Operator:	AS	OLS, LSL, DATS	XC, LSL, OLS	AS, XC, LSL, OLS	XC, LSL, OLS	XC
Notes:			* Depending on operator ¤ Some rebuilt from Mk3 hauled stock	* Depending on operator ¤ Some rebuilt from Mk3 hauled stock		Rebuilt from Mk3 hauled stock

Type:	TS	TGS
Number range:	48101-48150	49101-49117
Former range:	42xxx	44xxx
Introduced:	Orig: 1975-1977 TS: 2018-2020	Orig: 1980-1982 TGS: 2018-2020
Built by:	BR Derby	BR Derby
Seating:	62-84S	71S
Speed:	125mph (201km/h)	125mph (201km/h)
Brake type:	Air	Air
Heating:	Electric	Electric
Bogie type:	BT10	BT10
Length:	75ft 0in (22.86m)	75ft 0in (22.86m)
Height:	12ft 7in (3.84m)	12ft 7in (3.84m)
Width:	9ft 2in (2.79m)	9ft 2in (2.79m)
Weight:	35.6 tonnes	34 tonnes
Operator:	GW	GW
Notes:	Sliding door fitted GW 'Castle' stock	Sliding door fitted GW 'Castle' stock

Above, Below & Inset: Great Western operates a reducing number of HST trailer vehicles, modified with sliding doors and formed into four vehicle 'Castle' sets. The fleet was scheduled to be withdrawn by the end of 2023 but around four sets plus a couple of spares will continue to late 2024. Above is a TGS, with the guards van and slam door at the far end. These vehicles seat 71 in the 2+2 airline style. Inset right is the interior layout, fold down tables are provided, but overall space is very limited. Below is Trailer Standard coach No. 48118, these vehicles seat between 62-84 depending on layout, if a universal style toilet and wheelchair spaces are provided only 62 seats are included. All toilets are of the retention type, the tank and emptying system is on the left of the underframe. All: **CJM**

Left: With a number of spare HSTs after the introduction of InterCity Express stock on Great Western, Scottish Railways took over vehicles to form 26 sliding door sets, Nine 2+4 and 17 2+5 sets, to operate on Inter7City (I7C) services linking the five radial cities of Stirling, Perth, Dundee, Inverness and Aberdeen with Edinburgh and Glasgow. All were refurbished by Wabtec Doncaster with new interiors, sliding external passenger doors, retention toilets and a new Inter7City pictogram livery. A batch of 26 Trailer Guards First Buffet (TGFB) vehicles were rebuilt from FOs, with a guards office and buffet at one end and 32 first class seats at the other. In this view TS No. 42343 (a rebuild from a TGS) seating 74 standard passengers. **CJM**

Coaching Stock

Right, Right Below & Below: Arriva CrossCountry operate five HST sets based at Plymouth Laira, all are refurbished and fitted with retention toilets and sliding passenger doors. The XC HST fleet will be withdrawn from service by October 2023 as part of a major cost cutting exercise. The door system is the same as installed on Great Western and Scottish sets. Right upper is TCK No. 45001, shown from its seating end. Right below is Trailer Guards Standard (TGS) No. 44072, these seat 67 in the 2+2 style with a mix of airline and group layouts. Below is detail of the sliding door, these can be opened by passengers by the yellow surround push button. An emergency door release (green) is to the left of the door. **Antony Christie (2) / CJM**

Right: The CrossCountry HST fleet and extra converted loco-hauled Mk3s sport high quality Primarius seating. In first class this is set out in the 2+1 style with each seat having a table, most seats on the four side are in groups, while most of the single seats are of airline style group.
Antony Christie

Below & Inset: In 2021, Loco Services Ltd re-created the 1960s iconic 'Midland Pullman' train, by rebuilding an HST formation with a Pullman style interior and painting the train in 'Blue Pullman' style Nankin blue and white colours. The high quality train has been a huge success, with several trips each week in the holiday season. The main image shows TGS No. 44078 from the guards end, clearly showing the livery style and non-contrasting hinged passenger doors, controlled by central locking. The inset view below shows the amazing transformation of one of the catering vehicles No. 40801, to create a bar style atmosphere, complete with bar stalls. Both: **CJM**

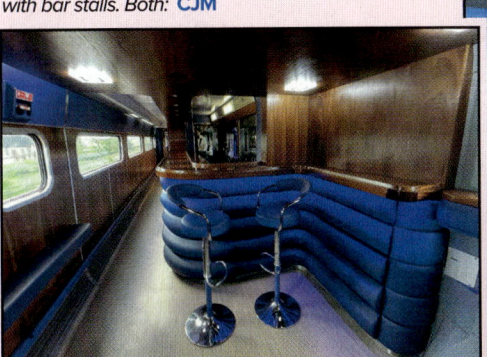

Rolling Stock Review : 2023-2024

193

Coaching Stock

Mk 4

Type:	RSB (AG2J)	FO (AD1J)	FOD (AD1J)	FO (AD1J)	TSOE (AI2J)	TSOD (AL2J)	TSO (AC2J)
Mark:	4	4	4	4	4	4	4
Number range:	10300-10333	11229-11298	11306-11326	11401-11418	12200-12231	12300-12330	12404-12526
Former range:	-	11xxx, 12xxx	11xxx	-	-	-	-
Introduced:	1989-1992	1989-1992	1989-1992	1989-1992	1989-1992	1989-1992	1989-1992
Built by:	Metro-Cammell	Metro-Cammell	Metro-Cammell	Metro-Cammell	Metro-Cammell	Metro-Cammell	Metro-Cammell
Seating:	30S	46F	42F	46F	76S	68S	76S
Speed:	140mph (225km/h)	140mph (225km/h)	140mph (225km/h)	140mph (225km/h)	140mph (225km/h)	140mph (225km/h)	140mph (225km/h)
Brake type:	Air	Air	Air	Air	Air	Air	Air
Heating:	Electric/6	Electric/6	Electric/6	Electric/6	Electric/6	Electric/6	Electric/6
Bogie type:	BT41	BT41	BT41	BT41	BT41	BT41	BT41
Length:	75ft 0in (22.86m)	75ft 0in (22.86m)	75ft 0in (22.86m)	75ft 0in (22.86m)	75ft 0in (22.86m)	75ft 0in (22.86m)	75ft 0in (22.86m)
Height:	12ft 7in (3.84m)	12ft 7in (3.84m)	12ft 7in (3.84m)	12ft 7in (3.84m)	12ft 7in (3.84m)	12ft 7in (3.84m)	12ft 7in (3.84m)
Width:	9ft 2in (2.79m)	9ft 2in (2.79m)	9ft 2in (2.79m)	9ft 2in (2.79m)	9ft 2in (2.79m)	9ft 2in (2.79m)	9ft 2in (2.79m)
Weight:	43 tonnes	42 tonnes	40 tonnes	42 tonnes	39.5 tonnes	39.4 tonnes	40 tonnes
Operator:	LN, TW	LN	LN, TW	LN	LN, TW	LN	LN, TW

Type:	DVT (NZ)
Mark:	4
Number range:	82200-82230
Former range:	-
Introduced:	1988-1989
Built by:	Metro-Cammell
Seating:	-
Speed:	140mph (225km/h)
Brake type:	Air
Heating:	Electric/6
Bogie type:	SIG
Length:	61ft 9in (18.83m)
Height:	12ft 9in (3.88m)
Width:	8ft 11in (2.71m)
Weight:	43.5 tonnes
Operator:	LN, TW

Above, Rght & Below: While the Government run LNER network mainly uses Class 800 and 801 stock, eight Mk4 and Class 91s sets have been retained for the forseeable future. In 2021-2023 the LNER sets have been repainted into an 'Azuma' styled Intercity livery, using oxblood, red and light grey. In the upper image of RSB No. 10300 is shown, these seat 30 in the 2+2 style with a full catering outlet at the opposite (near) end. The image right shows Mk4 Trailer Open Standard End (TSO-E), which is coupled to the loco and does not have a gangway. These coaches seat 76 and have one standard design toilet, vehicle No. 12205 is shown. Below is a Mk4 DVT, No. 82214, these have a Class 91 style cab end, a large luggage van towards the middle with a small guards office at the inner end.
All: **Antony Christie**

Coaching Stock

Right Above, Right Middle & Right Lower: *With a number of spare Mk4 stock available after replacement on the East Coast by IETs, Transport for Wales originally took on three sets to replace Mk3 stock used on the Cardiff-Holyhead and Holyhead-Manchester route. In June 2021, it was announced that a further four five-vehicle sets had been purchased outright by Transport for Wales for use on a new Swansea to Manchester service. These five sets were ones originally planned to be used by open access operator Grand Central on a proposed Blackpool North to Euston service, which was axed before introduction. These sets retain their black livery, while the three original Cardiff-Holyhead sets carry TfW branded Virgin East Coast colours. Extra stock has been added to the TfW fleet in mid-2022 and in 2023 eight sets are on the roster, but only a handful are in service due to teething issues. In the above right image, Trailer Standard Open (TSO) No. 12454 is shown, carrying the previous Virgin/LNER colours with TfW bodyside branding. In the right middle image, Restaurant Buffet Standard (RBS) No. 10325 is seen a vehicle with 30 2+2 style seats and a toilet at the non catering end. Right below is First Open Disabled (FOD) displaying the black livery, with gold passenger doors indicating first class. These refurbished vehicles seat 42 first in the 2+1 layout.* All: **CJM**

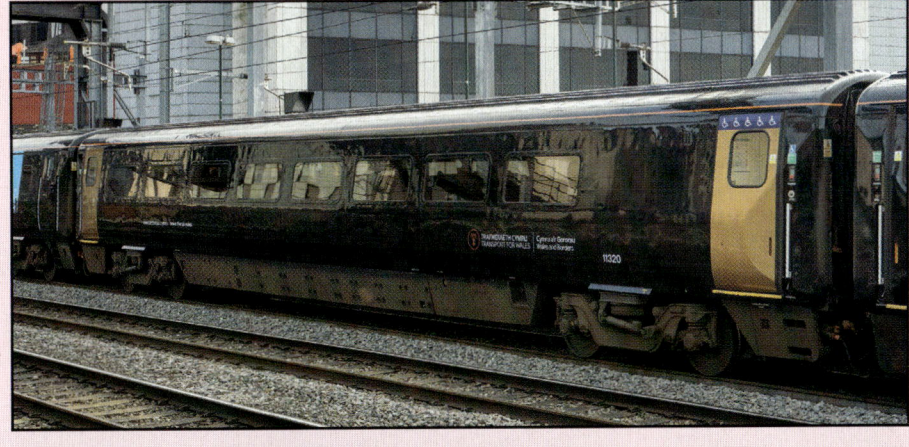

Below Left & Below Right: *In 2023, eight Mk4 Driving Van Trailers were operated by Transport for Wales. Three were the original ones transferred to TfW and sport TfW grey livery with promotional branding for charities. No. 82226 (below left) has Alzheimer's Society Wales branding and is seen at Newport. Below right is DVT 82200, carrying Lest We Forget poppy branded black livery.* Both: **CJM**

Rolling Stock Review : 2023-2024

Coaching Stock

Mk 5

Type:	FO	SO	DTS	SSB	SL	SFA	SLE
Mark:	5a	5a	5a	5	5	5	5
Number range:	11501-11513	12701-12739	12801-12814	15001-15011	15101-15110	15201-15214	15301-15340
Introduced:	2018-2020	2018-2020	2018-2020	2018-2019	2018-2019	2018-2019	2018-2019
Built by:	CAF	CAF	CAF	CAF	CAF	CAF	CAF
Seating:	30F	59-69S	64S	31	28-30	Sleeper 14	Sleeper 20
Speed:	100mph (161km/h)	100mph (161km/h)	100mph (161km/h)	100mph (161km/h)	100mph (161km/h)	100mph (161km/h)	100mph (161km/h)
Brake type:	Air	Air	Air	Air	Air	Air	Air
Heating:	Electric	Electric	Electric	Electric	Electric	Electric	Electric
Bogie type:	CAF	CAF	CAF	CAF	CAF	CAF	CAF
Length:	73ft 5in (22.37m)	73ft 5in (22.37m)	73ft 5in (22.37m)	72ft 10in (22.2m)	72ft 10in (22.2m)	72ft 10in (22.2m)	72ft 10in (22.2m)
Height:	12ft 7in (3.84m)	12ft 7in (3.84m)	12ft 7in (3.84m)	12ft 7in (3.84m)	12ft 7in (3.84m)	12ft 7in (3.84m)	12ft 7in (3.84m)
Width:	9ft (2.75m)	9ft (2.75m)	9ft (2.75m)	9ft 2in (2.79m)	9ft 2in (2.79m)	9ft 2in (2.79m)	9ft 2in (2.79m)
Weight:	32 tonnes	32 tonnes	33 tonnes	32.5 tonnes	35.5 tonnes	35.5 tonnes	35.5 tonnes
Operator:	TransPennine	TransPennine	TransPennine	Caledonian Sleeper	Caledonian Sleeper	Caledonian Sleeper	Caledonian Sleeper
Designation:				Seated Car	Lounge Car	Accessible Car	Sleeper Coach

Above, Above Right, Right & Below: In late-2019, a fleet of CAF-built Mk5a loco-hauled vehicles entered service for TransPennine Express, between Manchester Airport and Scarborough. The five vehicle sets, powered by a hired-in DRS Class 68, are based at Manchester International depot. The fleet consists of 13 trains, with one spare cab car, all carry TPE livery. Right is Driver Trailer Brake Standard No. 12808, these have seating for 64 in the 2+2 style. The cab desk is based on the Class 68. The image above left shows a First Open (FO) coach, these have 30 first class seats in the 2+1 layout with a mix of group and airline style. The coach has a universal access toilet compartment at the inner (near) end and a small catering/staff area at the outer end (far). The first class interior is shown below. Upper right is a Standard Open (SO) No. 12722, this vehicle has 69 seats in the 2+2 style, using a mix of group and airline layouts. One standard design toilet compartment is located at one end. 13 vehicles of the type have cycle stowage space, reducing the seating to 59 plus six tip-up. Plug doors are fitted to all vehicles and destination/route displays are found on the outside at cant rail height. All coaches have wide between vehicle gangways. Low level ventilation grilles are carried just above solebar height on one side of each vehicle. In 2023, the loco hauled sets, which are still not all in daily service are being transitioned onto the Cleethorpes routes.
Antony Christie (2) / CJM (2)

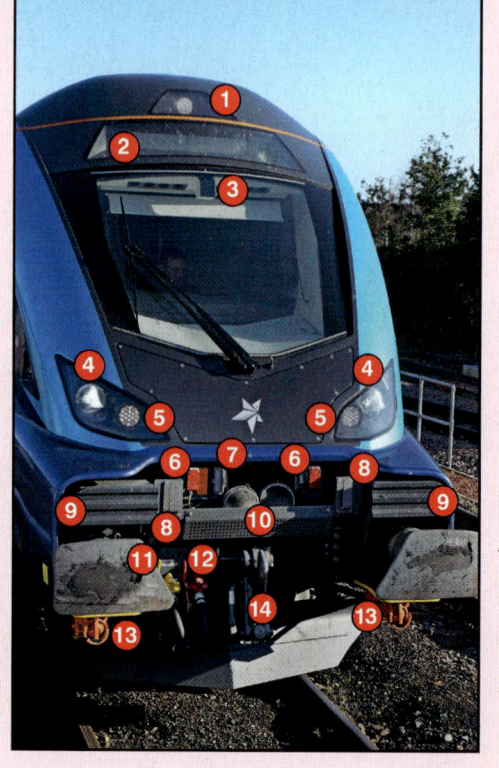

Left: Remote driving facilities are provided by a fleet of Driving Trailer Standard (DTS) coaches. These have a full width driving compartment at the outer end, the desk layout resembles that of a Class 68. Front end equipment positions are 1: High level marker light, 2: Destination indicator, 3: Forward facing camera, 4: Headlight, 5: Combined red tail and white marker light, 6: Control jumper sockets, 7: Class 68 jumper socket, 8: Lamp bracket, 9: Anti-climber plates, 10: Warning horns, 11: Main reservoir pipe (yellow), 12: Air brake pipe (red), 13: Electric train supply jumper sockets, 14: Coupling hook and shackle. **CJM**

Coaching Stock

Above Left & Above Right: In 2019, the overnight Caledonian Sleeper service between London and destinations in Scotland, was taken over by CAF-built Mk5 stock, with sleeping cars, offering either single or double bed facilities, with new levels of comfort for seated passengers. Four types of coach were built. The stock, based at Polmadie, is painted in the Caledonian teal colours, off-set by silver branding. Above Right is Sleeper Seated Brake (SSB) vehicle No. 15001 is shown, these house the train managers office, 31 un-classified seats and one universal toilet compartment. They have one sliding plug door at each end. Above Left, Mk5 sleeper end coach, this has a pair of end doors to form an 'end vehicle'. On the vehicle end jumper cables and sockets are located for inter-vehicle connections, red tail lights are fitted on both sides. The coupling is of the Dellner semi-automatic type, with a 61-pin jumper socket below right for loco-train communication. Vehicle to vehicle and vehicle to loco attachment require manual connection of cables.
Robin Ralston / Barry Edwards

Right Second: To cater for those with disabilities a fleet of 14 fully accessible sleeping cars are operated, these have two accessable berths, two double bed berths with en-suite toilets and two berths with a fold down bed, plus shower compartments and toilets. Coach No. 15214 is shown.
Barry Edwards

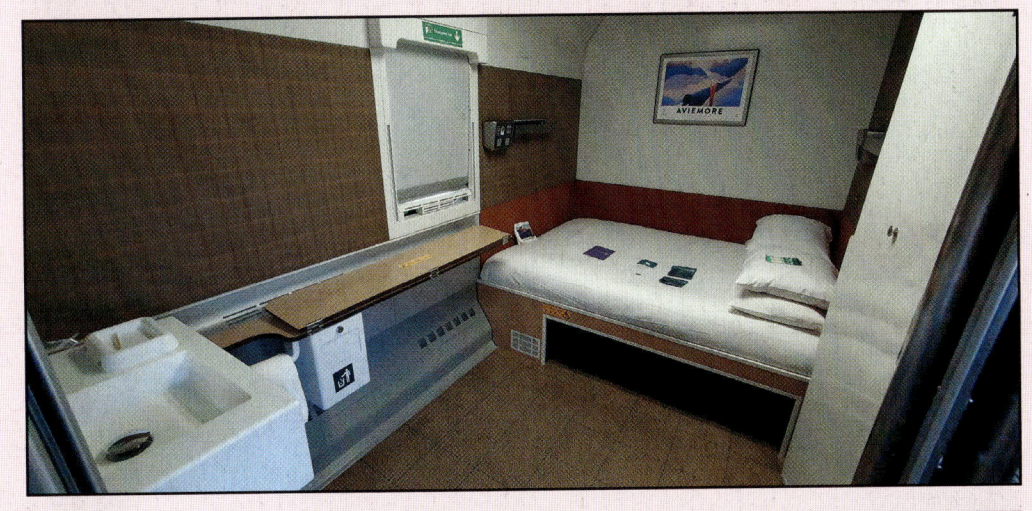

Right Middle: View showing one of the two double bed berths with en-suite toilet in vehicle No. 15206. **Nathan Williamson**

Right Lower: A fleet of 40 ten-berth Sleeping Cars are in service, these house six en-suite and four non en-suite two-berth cabins. The berths are fed by a side corridor, and each cabin has a window. No. 15308 is illustrated from the berth side.
Barry Edwards

Rolling Stock Review : 2023-2024

Royal Train Coaching Stock

Royal Train Fleet List

Number	Type	Design	Lot No.	Introduced (originally built)	Use	Notes
2903 (11001§)	AT5G	Mk3	30886	1977 (1972)	HM The Kings's Saloon	Double vestibule doors one end
2904 (12001§)	AT5G	Mk3	30887	1977 (1972)	Special Saloon	
2915 (10735)	AT5G	Mk3	31002	1985	Royal Household Sleeping Coach	
2916 (40512)●	AT5G	Mk3	31059	1988	HRH The Prince of Wales's Dining Coach	
2917 (40514)●	AT5G	Mk3	31084	1990 (1977)	Kitchen Car and Royal Household Dining Coach	
2918 (40515)●	AT5G	Mk3	31083	1989 (1976)	Royal Household Coach	
2919 (40518)●	AT5G	Mk3	31085	1989 (1977)	Royal Household Coach	
2920 (17109)	AT5B	Mk2b	31044	1986 (1969)	Generator Coach and Household Sleeping Coach	Houses 350kW generator
2921 (17107)	AT5B	Mk2b	31086	1990 (1969)	Brake, Coffin Carrier and Household Accommodation	
2922 (*)	AT5G	Mk3b	31035	1987	HM The King / HRH The Prince of Wales's Sleeping Coach	
2923 (*)	AT5G	Mk3b	31063	1987	HM The King / HRH The Prince of Wales's Saloon Coach	

§ Converted from prototype HST passenger vehicle
● Converted from production HST passenger vehicle
* Purpose built for Royal Train use

Mk2
Vehicle Length: 66ft 0in (20.11m)
Height: 12ft 9½in (3.89m)
Width: 9ft 3in (2.81m)

Mk 3
Vehicle Length: 75ft 0in (22.86m)
Height: 12ft 9in (3.88m)
Width: 8ft 11in (2.71m)

Left: *Converted from BFK No. 17107 (14107) in 1987, Royal Saloon No. 2921 is a brake and staff support coach, usually used to transport the railway officials accompanying the Royal train. Its former brake van now houses a coffin table.* **Nathan Williamson**

Right: *Royal Saloon No. 2904 is a conversion from a prototype HST coach No. 12001, built in 1972, entering Royal Train use in 1977 via a major rebuild at Wolverton Works. For many years it was The Duke of Edinburgh's coach, but today it is known as the Royal Special Saloon. It is seen at Totnes in June 2021.* **Nathan Williamson**

Left: *His Royal Highness the Prince of Wales's dining car, No. 2916. One of a pair of new Royal vehicles built at BREL Derby in 1987 and fitted out at the Wolverton Works Royal Train workshop in 1988.* **Nathan Williamson**

Below: *In 1987-88 two purpose-built Mk3s were assembled at BREL Derby and shipped to Wolverton for fitting out, Nos. 2922 and 2923, introduced for The Prince of Wales's use. The pair are now used by The King and the present Prince and Princess of Wales. No. 2922 is viewed from its ceremonial door vestibule at Newton Abbot in December 2021.* **Nathan Williamson**

Network Rail & Specialist Coaching Stock

Network Rail Fleet

Number	Type	Use
1256 (3296)	AJ1F/RFO	PLPR3
5971	AC2F/TSO	Brake force runner
5981	AC2F/TSO	PLPR2
5995	AC2F/TSO	Brake force runner
6260 (92116)	AX51/GEN	Generator
6261 (92988)	AX51/GEN	Generator
6262 (92928)	AX51/GEN	Generator
6263 (92961)	AX51/GEN	Generator
6264 (92923)	AX51/GEN	Generator
9481	AE2D/BSO	Support vehicle
9516	AE2D/BSO	Brake force runner
9523	AE2D/BSO	Brake force runner
9701 (9528)	AF2F/DBSO	Remote driving car
9702 (9510)	AF2F/DBSO	Remote driving car
9703 (9517)	AF2F/DBSO	Remote driving car
9708 (9530)	AF2F/DBSO	Remote driving car
9714 (9536)	AF2F/DBSO	Remote driving car
9801 (5760)	AN1F/BUO	Support vehicle (Eastern Rail Services)
9803 (5799)	AN1F/BUO	Support vehicle (Eastern Rail Services)
9806 (5840)	AN1F/BUO	Support vehicle (Eastern Rail Services)
9808 (5871)	AN1F/BUO	Support vehicle (Eastern Rail Services)
9810 (5892)	AN1F/BUO	Support vehicle (Eastern Rail Services)
62287	MBS/4CIG	Ultrasonic test car UTU2
62384	MBS/4CIG	Ultrasonic test car UTU1
72612 (6156)	Mk2f/TSO	Brake force runner
72616 (6007)	Mk2f/TSO	Brake force runner
72630 (6094)	Mk2f/TSO	Structure Gauging Train (SGT1)
72631 (6096)	Mk2f/TSO	PLPR1
72639 (6070)	Mk2f/TSO	PLPR4
82111	MK3/DVT	Driving Van Trailer (stored)
82115	MK3/DVT	Driving Van Trailer (stored)
82124	MK3/DVT	Driving Van Trailer (stored)
82129	MK3/DVT	Driving Van Trailer (stored)
82145	MK3/DVT	Driving Van Trailer (stored)
96604 (96156)	Mk1/NVA	Brake force runner
96606 (96213)	Mk1/NVA	Brake force runner
96608 (96216)	Mk1/NVA	Brake force runner
96609 (96217)	Mk1/NVA	Brake force runner
99666 (3250)	Mk2e/FO	Structure Gauging Train (SGT1) (barrier)
971001 (94150)	Mk1/NKA	Tool Van
971002 (94190)	Mk1/NKA	Tool Van
971003 (94191)	Mk1/NKA	Tool Van
971004 (94168)	Mk1/NKA	Tool Van
975025 (60755)	6B Buffet	Inspection Saloon Caroline Operated by Loram
975087 (34289)	MK1/BSK	Recovery train support coach
975091 (34615)	Mk1/BSK	Overhead line test coach Mentor
975464 (35171)	Mk1/BSK	Snowblower coach Ptarmigan
975477 (35108)	Mk1/BSK	Recovery train support coach
975486 (34100)	Mk1/BSK	Snowblower coach Polar Bear
975814 (41000)	HST/TF	NMT Conference coach
975984 (40000)	HST/TRUB	NMT Lecture coach
977868 (5846)	Mk2e/TSO	Radio Survey coach
977869 (5858)	Mk2e/TSO	Radio Survey coach (stored)
977969 (14112)	Mk2/BFK	Staff coach (former Royal Saloon 2906)
977974 (5854)	Mk2e/TSO	Track Inspection coach (TIC 2)
977983 (72503)	Mk2f/FO	Overhead Line Inspection EMV (ex FO 3407)
977984 (40501)	HST/TRFK	NMT Staff coach
977985 (72715)	Mk2f/TSO	Structure Gauging Train (SGT2) (exTSO 6019)
977986 (3189)	Mk2d/FO	Structure Gauging Train (SGT2) (barrier)
977993 (44053)	HST/TGS	NMT Overhead Line Test coach
977994 (44087)	HST/TGS	NMT Recording coach
977995 (40719)	HST/TRFM	NMT Generator coach
977997 (72613)	Mk2f/TSO	Radio Survey Test Vehicle (originally TSO 6126)
999550 -	Mk2 design	Track Recording coach (purpose-built) HSTRC
999602 (62483)	Mk1/4CIG	Ultrasonic Test coach - UTU3
999605 (62482)	Mk1/4REP	Ultrasonic Test coach
999606 (62356)	Mk1/4CIG	Ultrasonic Test coach - UTU4

Note: Mk1 6392 and 6397 also used by Network Rail as brake force runners

■ A large number of ex-revenue earning passenger and van vehicles are in Departmental service, performing a number of tasks to ensure the railway is able to operate safely, many of these are operated by Network Rail, based at Derby. In addition a number of vehicles are used for specialist purposes, such as barrier or escort coaches. These pages show a small selection of the Network Rail vehicles. ■

Above: Former Hastings line buffet car 60755 is currently a Derby-based Network Rail inspection saloon fitted with blue star multiple control equipment. It was previously the Southern Region General Managers saloon. Numbered 975025 it carried Southern green livery and the name Caroline. It is seen at Dawlish Warren. **CJM**

Right: Several former Mk2 DBSO vehicles are operated as part of the Network Rail Infrastructure Monitoring fleet. No. 9701 is illustrated, showing a modified front end with lighting and cameras. **Spencer Conquest**

Rolling Stock Review : 2023-2024

Irish Railways

Class:	071
Number range:	CIE: 071-088
	NIR: 8111-8113
Built by:	General Motors, La Grange, USA
GM model:	JT22CW
Years introduced:	1976-1984
Wheel arrangement:	Co-Co
Gauge:	5ft 3in (1,600mm)
Maximum speed	90mph (145km/h)
Length:	57ft 0in (17.37m)
Height:	13ft 3in (4.04m)
Width:	9ft 5¾in (2.89m)
Weight:	100.6 tonnes
Wheel diameter (new):	3ft 4in (1,016mm)
Min curve negotiable:	164ft (50m)
Engine type:	EMD 12-645E3B or EMD 12-645E3C
Engine output:	2,450hp (1,830kW)
Power at rail:	1,700hp (1,300kW)
Maximum tractive effort:	65,000lb (289kN)
Continuous tractive effort:	43,000lb (192kN)
Cylinder bore:	9⅛ in (230mm)
Cylinder stroke:	10in (250mm)
Transmission type:	Electric
Traction generator:	GM-EMD AR10D3
Auxiliary alternator:	GM-EMD D14
Auxiliary generator:	GM-EMD A-814M1 of 24hp
Traction motor type:	GM-EMD D77B
No. of traction motors:	6
Brake type:	Vacuum and Air
Bogie type:	EMD 'Flexicoil'
Multiple coupling type:	AAR
Fuel tank capacity:	790gal (3,600lit)
Sanding equipment:	Pneumatic
Note:	NIR 111 and 112 modified to provide head end power (later isolated)

Above & Below: *Northern Ireland Railways operate a fleet of three General Motors JT22CW locos, Nos. 111-113. These are identical to the Irish Rail '071' class. The NIR trio are based at Belfast York Road depot and operate engineering trains as required. No freight services operate on the NIR system. Nos. 8111 and 8113 are illustrated at Portadown on 16 March 2023.. The NIR locos are fitted with dual air and vacuum brakes, standard screw shackle couplings, with Nos. 111 and 112 having electric train supply (or head end power) now isolated. Below, we see the Irish Rail version, illustrated by No. 081 at Kildare, these are not fitted with electric train supply. The IR locos are based at Inchicore (Dublin) and can be found system wide working freight and engineering trains. The most common route is on container traffic between Dublin and Ballina and between Ballina and Waterford.* **Gary Adams / CJM**

Left: *Two Irish Rail 071 class locos have been restored to historic liveries. No. 071 (illustrated) carries the original Supertrain orange colours, applied in 2016, while No. 073 (not shown) carries the original Irish Rail orange. No. 071 is seen passing Athlone with a container train from Dublin to Ballina.* **CJM**

2023 Fleet

NIR	
8111	NIR
8112	NIR
8113	NIR

IR	
071	IR
072	IR
073	IR
074	IR
075	IR
076	IR
077	IR
078	IR
079	IR
080	IR
081	IR
082	IR
083	IR
084	IR
085	IR
086	IR
087	IR
088	IR

Irish Railways

Class:	201
Number range:	201-234 (NIR 8208 & 8209)
Built by:	General Motors, Diesel Division, Ontario, Canada
GM model:	JT42HCW
Years introduced:	1994-1995
Wheel arrangement:	Co-Co
Gauge:	5ft 3in (1,600mm)
Maximum speed	100mph (160km/h)
Length:	68ft 9in (20.95m)
Height:	13ft 2in (4.02m)
Width:	8ft 8in (2.64m)
Weight:	109 -112 tonnes
Wheel diameter (new):	3ft 4in (1,016mm)
Min curve negotiable:	262ft (00m)
Engine type:	EMD 12-710G3B
Engine output:	3,200hp (2,400kW)
Power at rail:	2,970hp (2,210kW)
Maximum tractive effort:	43,700lb (194kN)
Cylinder bore:	9⅛ in (230mm)
Cylinder stroke:	11in (280mm)
Transmission type:	Electric
Traction generator:	GM-EMD AR8PHEA/CA6
Auxiliary generator:	GM-EMD5A-8147
Traction motor type:	GM-EMD D43
No. of traction motors:	6
Heating:	Head end power Dayton-Phoenix E7145 of 220-380V three-phase
Brake type:	Vacuum and Air
Brake force:	72 tonnes
Bogie type:	EMD GC
Multiple coupling type:	AAR
Fuel tank capacity:	990gal (4,500lit)
Sanding equipment:	Pneumatic
Note:	Many long term out of service

Above: The sizable fleet of General Motors JT42HCW locos, or 201 class, were introduced in 1994-1995, with Northern Ireland Railways having two locos, with Irish Rail taking 32. The two NIR owned locos Nos. 8208 and 8209 are used on the joint operated 'Enterprise' service linking Belfast and Dublin, the pair sport 'Enterprise' livery. No. 8209 of the NIR fleet shows 'Enterprise' livery passing Laytown. These locos can only operate coupled in one direction when on Enterprise work due to their couplings. The 201 class sport both air and vacuum brakes, standard light clusters of white and red and have US style Association of American Railroad (AAR) jumper sockets enabling multiple operation with all GM builds. **CJM**

Above & Inset: The Irish Rail owned General Motors 201 class (JT42HCW locos) are based at Dublin Inchicore and operate 'Enterprise', Dublin to Cork loco-hauled services and freight. No. 217 shows the Intercity silver and green livery working on a Cork-Dublin service. These locos have screw couplings and dual air vacuum brake connections. The inset image shows a cast nameplate, these are applied above the cab side windows with the names in both English and Gaelic. Both: **CJM**

2023 Fleet

NIR		216	IR
8208	NIR	217	IR
8209	NIR	218	IR
		219	IR
IR		220	IR
201	IR	221	IR
202	IR	222	IR
203	IR	223	IR
204	IR	224	IR
205	IR	225	IR
206	IR	226	IR
207	IR	227	IR
208	IR	228	IR
209	IR	229	IR
210	IR	230	IR
211	IR	231	IR
212	IR	232	IR
213	IR	233	IR
214	IR	234	IR
215	IR		

Right: Many of the Irish Rail-operated 201 class are stored and unlikely to return to service. Some now work freight flows. No. 227 River Laune displays 'Enterprise' livery at Dublin Connolly. Locos on these services usually operate with the loco at the Belfast end of the train, with a remote driving cab car at the Dublin end. **CJM**

Rolling Stock Review : 2023-2024

Irish Railways

2023 Fleet

Class:	3000 (C3K)	4000 (C4K)
Gauge:	5ft 3in (1,600mm)	5ft 3in (1,600mm)
Number range:	3001-3023	4001-4020 (4001-4013 - 3-car sets, 4014-4020 - 6-car sets)
Introduced:	2004-2005	2010-2011, Additional MSs 2021
Built by:	CAF, Zaragoza, Spain	CAF, Zaragoza, Spain
Formation:	DMSO(A)+MSO+DMSO(B)	3-car sets DMSO(A)+MSO+DMSO(B) 6-car sets DMSO(A)+MSO(A)+MSO(B)+MSO(C)+MSO(D)+DMSO(B)
Vehicle numbers:	DMSO(A) - 3301-3323, MSO - 3501-3523 DMSO(B) - 3401-3423	DMSO(A) - 4301-4320, MSO(A) - 4501-4520, MSO(B) - 4614-4620, MSO(C) - 4714-4720, MSO(D) - 4814-4820, DMSO(B) - 4401-4420
Vehicle length:	DMSO - 77ft 8in (23.74m) MSO - 75ft 9in (23.14m)	DMSO - 77ft 8in (23.74m) MSO - 75ft 9in (23.14m)
Height:	11ft 9in (3.62m)	11ft 9in (3.62m)
Width:	9ft 1in (2.76m)	9ft 1in (2.76m)
Seating:	Total - 201 + (15TU) = 216 DMSO(A) - 49 + 9TU MSO - 78 DMSO(B) - 58 + 6TU	3-car - 178 + (34TU) = 212, 6-car - 378 + (64TU) = 442 DMSO - 61 +10TU, MSO(A) - 56 +14TU, MSO(B) - 72 + 8TU, MSO(C) - 72 + 8TU, MSO(D) - 56 + 14 TU, DMSO 61 + 10TU
Internal layout:	Standard - 2+2	Standard - 2+2
Gangway:	Within set	Within set
Toilets:	1U 33xx car, 1S 34xx car	2U, one in MSO(A) and MSO(D)
Weight:	Total - 140.6t DMSO(A) - 45.9t MSO - 45.5t DMSO(B) - 49.2t	Total - 3-car - 142.5t, 6-car - 282.3t DMSO - 48.1t, MSO(A) - 46.5t, MSO(B) - 46.5t, MSO(C) - 46.4t, MSO(D) - 46.9t, DMSO - 47.9t
Brake type:	Air	Air
Bogie type:	CAF	CAF
Engine type (Traction):	One MAN D2876 LUH02 of 453hp (338kW) per car	One MTU 6H1800R84 of 520hp (390kW) per car
Total horsepower (Traction):	1,359hp (1,014kW)	1,560hp (1,170kW)
Engine type (Aux):	One Cummins 6BTS-9 GR1 of 95hp (71kW) per car	Not fitted
Transmission:	Hydraulic	Mechanical ZF 5HP902
Wheel diameter:	34in (864mm)	34in (864mm)
Max speed:	90mph (145km/h)	90mph (145km/h)
Coupling type:	Ends: Scharfenberg auto, Inter: Scharfenberg	Ends: Scharfenberg auto, Inter: Scharfenberg
Door type:	Sliding plug	Sliding plug
Special features:	AWS, TPWS, Sets 3001-3006 fitted with CAWS to operate on cross-border services over Irish Rail Unable to work in multiple with C4K fleet	Unable to work in multiple with C3K fleet

C3K	
3001	NIR
3002	NIR
3003	NIR
3004	NIR
3005	NIR
3006	NIR
3007	NIR
3008	NIR
3009	NIR
3010	NIR
3011	NIR
3012	NIR
3013	NIR
3014	NIR
3015	NIR
3016	NIR
3017	NIR
3018	NIR
3019	NIR
3020	NIR
3021	NIR
3022	NIR
3023	NIR

C4K	
4001	NIR
4002	NIR
4003	NIR
4004	NIR
4005	NIR
4006	NIR
4007	NIR
4008	NIR
4009	NIR
4010	NIR
4011	NIR
4012	NIR
4013	NIR
4014	NIR
4015	NIR
4016	NIR
4017	NIR
4018	NIR
4019	NIR
4020	NIR

Left: In 2004-2005, 23 three-car DMUs built by CAF entered service with Northern Ireland Railways for local services in the Belfast area, based at Belfast York Road depot. The fleet operate all services on the system radiating from Belfast. The C3K sets are unable to operate in multiple with the later built C4K fleet. Units are painted in Translink NI Railways silver and blue livery, with the 30xx sets having a yellow 'rounded' front end. Nos. 3001-3006 are fitted with equipment to allow cross border operations to Irish Rail and can operate to Dublin. Set No. 3022 is seen at Moira. **CJM**

Below: A fleet of 20 three-car CAF DMUs were delivered in 2010-2011, numbered in the 40xx series. Very similar to the 30xx fleet, but have a mechanical transmission, with a slightly revised front end, incorporating a lower square yellow panel, giving easy identification between the two types. The 40xx series can not operate in multiple with the 30xx series. In 2022-23 seven sets (Nos. 4014-4020) were being strengthened to six car formation by the addition of three extra MS coaches to each set. Three-car set No. 4005 is seen arriving at Helens Bay. **CJM**

Irish Railways

Left & Right: C3K and C4K front end layout and equipment. 1: High level marker light, 2: Destination indicator, 3: Combined marker/tail light, 4: Headlight, 5: Coupling electrical connection, 6: Auxiliary power connection, 7: Scharfenberg coupling, 8: Emergency air connection, 9: Forward facing camera. Illustrating the two different styles and livery layout. Both: **CJM**

Right Upper & Right Lower: Intermediate Motor Standard (MS) vehicles. The upper view shows car No. 3522 from set No. 3022, these vehicles seat 78 standard class passengers in the 2+2 style, using a mix of airline and group layouts. The lower image shows a C4K intermediate No. 4505 from set No. 4005. These seat 56 with a universal access toilet compartment at one end, plus two wheelchair spaces and 14 tip-up seats. All vehicles have two pairs of bi-parting plug doors and an electronic destination indicator in located above the middle window. All carry Northern Ireland Translink silver and blue livery. Both: **CJM**

Left: Both driving cars of the C4X fleet have 10 tip-up seats spread through the vehicle, this is the area looking towards the cab, which can be used for cycle parking or for prams and buggies. The door, feeds the driving cab. **CJM**

Right: Interior layout of 40xx series unit, showing the 2+2 style with good quality seats, with an above seat glazed base luggage shelf and a passenger information system by the door positions. **CJM**

Rolling Stock Review : 2023-2024

Irish Railways

Class:	2600
Gauge:	5ft 3in (1,600mm)
Vehicle number range:	2601-2616
Introduced:	1993
Built by:	Tokyu Car, Japan
Formation:	DMSO(A)+DMSO(B) Odd numbers - DMSO(A) Even numbers DMSO(B)
Vehicle length:	DMSO - 20.26m
Vehicle width:	2.90m
Seating:	Total - 129S (58S + 71S)
Internal layout:	Standard - 2+2
Gangway:	Throughout
Toilets:	1U
Weight:	Total - 81.4t DMSO(A) - 41.2t , DMSO(B) - 40.2t
Brake type:	Air
Engine type (Traction):	One Cummins NTA855R1 of 350hp (260kW)
Generator:	Cummins 6B5.9GR
Transmission:	Hydraulic Nilgatta DW14G
Wheel diameter:	33.1in (840mm)
Max speed:	70mph (113km/h)
Coupling type:	Ends: Dellner, Inter: Bar
Door type:	Sliding
Notes:	Operate in pairs

Above: Built by Tokyu Car in Japan, 17 Class 2600 class 'Arrow' vehicles are based at Cork for local services. These two-car sets operate in semi-fixed formations and retain end gangways, they carry standard silver and green livery. No. 2613 is shown at Cork with a Cobh service. **CJM**

Class:	2800
Gauge:	5ft 3in (1,600mm)
Vehicle number range:	2801-2820
Introduced:	2000
Built by:	Tokyu Car, Japan
Formation:	DMSO(A)+DMSO(B) Odd numbers - DMSO(A) Even numbers DMSO(B)
Vehicle length:	DMSO - 20.73m
Vehicle width:	2.90m
Seating:	Total - 85S (39S + 46S)
Internal layout:	Standard - 2+2
Gangway:	Within set
Toilets:	1U
Weight:	Total - 86.5t DMSO(A) - 43.9t , DMSO(B) - 42.6t
Brake type:	Air
Engine type (Traction):	One Cummins NTA855R1 of 350hp (260kW)
Transmission:	Hydraulic Nilgatta DW14G
Wheel diameter:	33.1in (840mm)
Max speed:	75mph (121km/h)
Coupling type:	Ends: Dellner, Inter: Bar
Door type:	Sliding
Notes:	Operate in pairs

Above: Built by Tokyu Car are 10, 2800 class two-car sets based at Limerick for branch line use. These sets have had their original end gangways removed. Nos. 2803 and 2804 are seen approaching Manulla Junction with a Ballina to Manulla Junction service. **CJM**

Class:	22000
Gauge:	5ft 3in (1,600mm)
Formation:	3-car 4-car 5-car
Set numbers:	22001-22010 22011-22030 22031-22040 22046-22063 22041-22045
Vehicle numbers:	222xx, 223xx, 224xx, 226xx, 227xx, 228xx series
Introduced:	2007-2011
Built by:	Hyundai Rotem, South Korea
Formation:	3-car: DMSO+MSO+DMSO 4-car: DMSO+MSO+MSO+DMSO 5-car: DMSO+MSO+MSO+MSO+DRBFO
Vehicle length:	DMSO - 23.5m, MSO - 23m
Vehicle width:	2.84m
Seating:	3-car total - 190S (52S+72S+66S) 4-car total - 262S (52S+72S+72S+66S) 5-car total - 36F/268S (52S+72S+72S+72S+36F)
Internal layout:	First 2+1, Standard - 2+2
Gangway:	Within set
Toilets:	3-car 1U, 1S, 4-car 1U, 2S, 5-car 2U, 3S
Weight:	3-car: 146.1t (49.3+47.1+49.7) 4-car: 193.2t (49.3+47.1+47.1+49.7) 5-car: 240.6 (49.3+47.1+47.3+47.1+49.8)
Brake type:	Air
Engine type (Traction):	One MTU 6H1800 of 480hp (360kW) per car
Transmission:	Hydraulic Voith T211
Wheel diameter:	33.1in (840mm)
Max speed:	100mph (161km/h)
Coupling type:	Ends: Dellner, Inter: Bar
Door type:	Plug

Right Above, Right Below and Inset: Between 2007-2022 a total of 275 Hyundai/Rotem 22000 class DMU vehicles entered service, these are now formed into three, four and five car sets and when the latest vehicles are delivered (2022-2023) some six-car sets will be formed. In 2022, the fleet was formed as three-car (Nos. 001-010/046-063), four-car (Nos. 011-030/041-045), five-car (premium) (Nos. 031-040). The fleet based at Laois depot, can be found throughout the Irish Rail network on main line and local services. In the top view, set No. 17, a four-car formation is seen passing Kildare, bound for Dublin. The above shows five-car 'premium' set No. 33 between Cork and Mallow with a Cork-Dublin service. The inset image shows the interior using a 2+2 seating layout. All: **CJM**

Irish Railways

Class:	29000
Gauge:	5ft 3in (1,600mm)
Set numbers:	29001-29029
Vehicle number range:	291xx, 292xx, 293xx, 294xx
Introduced:	2002-2003
Built by:	CAF, Spain
Formation:	DMSO+MSO+MSO+DMSO
Vehicle length:	DMSO - 20.4m, MSO - 20.2m
Vehicle width:	2.90m
Seating:	Total - 185S (48+40+49+48)
Internal layout:	2+2
Gangway:	Within set
Toilets:	1U, 1S
Weight:	Total - 170.51t (43.56+41.36+42.07+43.52)
Brake type:	Air
Engine type (Traction):	One MAN D2876LUH of 394hp (294kW) per car
Transmission:	Hydraulic
Wheel diameter:	33.1in (840mm)
Max speed:	75mph (121km/h)
Coupling type:	Ends: Dellner, Inter: Bar
Door type:	Sliding

Above: A fleet of 29 four-car CAF-built outer-suburban units of the 29000 class are based at Drogheda and used on Dublin area services. Introduced between 2002-2005, each set has 185 standard class seats in the 2+2 layout, with each coach having two pairs of integral sliding doors. In 2022-2023 sets are currently going through a livery transition, from the old green and blue to the latest two-tone green livery, as displayed on set No. 29409 arriving at Gormanston with a service from Drogheda to Dublin Pearse. For commuter duties, sets either operate on their own or in pairs. **CJM**

Hauled stock

Gauge:	5ft 3in (1,600mm)								
Mk	4	4	4	4	-	-	-	-	3
Type:	Driving Brake Generator	Open Standard	Open First	Buffet	Driving Trailer Brake Open	Open First	Open Standard	Catering	Generator
Number range:	4001-4008	4101-4143	4201-4208	4401-4408	9001-9004	9101-9104	9201-9216	9401-9404	9602-9608
Introduced:	2004-2005	2004-2005	2004-2005	2004-2005	1996	1996	1996	1996	1984-1986
Built by:	CAF	CAF	CAF	CAF	De Dietrich	De Dietrich	De Dietrich	De Dietrich	BREL Derby
Length:	23.81m	23.81m	23.81m	23.81m	23m	23m	23m	23m	23m
Width:	2.85m	2.85m	2.85m	2.85m	2.85m	2.85m	2.85m	2.85m	2.74m
Seating:	0	69S	44F	28U	29F	47F	68-71S	15U	0
Weight:	44.8t	40.6-41.2t	39.6t	42.7t	39.7t	37.7t	37.5t	38.0t	35.6t
Brake type:	Air	Air	Air	Air	Air	Air	Air	Air	Air
Coupling:	Scharfenberg	Scharfenberg*	Scharfenberg	Scharfenberg	Dellner/drop head	Dellner	Dellner	Dellner	Dellner
Max speed:	100mph	100mph	100mph	100mph	100mph	100mph	100mph	100mph	100mph
Door type:	Plug	Plug	Plug	Plug	Plug	Plug	Plug	Plug	Slam
Operation:	InterCity	InterCity	InterCity	InterCity	Enterprise	Enterprise	Enterprise	Enterprise	Enterprise

* End vehicles had drop-headbuck-eye on one end

2 x Cummins 295hp engines for hotel power

Below: Two types of hauled stock exist in Ireland, the De Dietrich-built 'Enterprise' sets working between Belfast and Dublin, or the CAF-built Mk4s used by Irish Rail between Dublin and Cork. The 'Enterprise' fleet consists of four sets, each with a Driving Trailer Brake Open, seating 29 first class passengers in the 2+1 style. No. 9003 is illustrated, showing the standard 'Enterprise' livery. **CJM**

Right Upper, Right Lower & Inset: CAF in Spain built Mk4 loco-hauled vehicles for use on Irish Rail Intercity services, now confined to the Dublin to Cork route. Eight streamlined Driving Brake Generator cars are in service, these have a very sloped cab end. A generator is positioned in the van area to provide hotel power for the train, as the Class 201 locos used to power these services do not have train supply. The image above right shows Open Standard No. 4127, seating 69 with one disabled access toilet compartment and one wheelchair space, at Charleville. Seating is in the 2+2 style, in a mix of aircraft and group styles. To provide catering, a fleet of eight buffet cars are in use, these have catering facilities at one end (near in image) with 28 unclassified seats at the other. The inset shows the interior style, showing the 2+2 layout in standard class. All passenger vehicles have sliding plug doors at coach ends, these are under control of train staff with local passenger activation. All: **CJM**

Rolling Stock Review : 2023-2024

Irish Railways

DART Stock

Gauge:	5ft 3in (1,600mm)			
Class:	8100 class	8500 class	8510 class	8520 class
Formation:	2-car	4-car	4-car	4-car
Numbers driving:	8101-8140	8601-8608	8611-8616	8621-8640
Numbers intermediate:	8301-8340	8501-8508	8511-8516	8521-8540
Introduced:	1983-1984	2000	2001	2003-2004
Built by:	Linke-Hofmann Busch	Tokyu Car	Tokyu Car	Tokyu Car
Formation:	DMSO+DTSO	DTSO+PMSO+ PMSO+DTSO	DTSO+PMSO+ PMSO+DTSO	DTSO+PMSO+ PMSO+DTSO
Vehicle length:	DTMO/DTSO 21m	DTSO 20.73m PMSO 20.50m	DTSO 20.73m PMSO 20.50m	DTSO 20.73m PMSO 20.50m
Vehicle width:	2.90m	2.90m	2.90m	2.90m
Seating:	128S (64S+64S)	160S (40S+40S+ 40S+40S)	160S (40S+40S+ 40S+40S)	160S (40S+40S+ 40S+40S)
Internal layout:	2+2	2+2	2+2	2+2
Gangway:	Within set	Within set	Within set	Within set
Toilets:	No fitted	No fitted	No fitted	No fitted
Weight:	Total: 69.45t	Total: 147.9t	Total: 148.07	Total: 143.3t
Brake type:	Air	Air	Air	Air
Traction:	Electric 1500V dc	Electric 1500V dc	Electric 1500V dc	Electric 1500V dc
Equipment:	4xGEC G314BY of 137kW	4xthree-phase ac induction	4xthree-phase ac induction	4xthree-phase ac induction
Wheel diameter:	33.1in (840mm)	33.1in (840mm)	33.1in (840mm)	33.1in (840mm)
Max speed:	62mph (100km/h)	62mph (100km/h)	62mph (100km/h)	62mph (100km/h)
Coupling type:	Scharfenberg	Scharfenberg	Scharfenberg	Scharfenberg
Door type:	Sliding	Sliding	Sliding	Sliding

Right Above & Right Lower: Commuter services around the Dublin area are provided by the Dublin Area Rapid Transport (DART) system. Based at Fairview depot, the 38 two-car and 16 four-car sets operate in 2, 4, 6 or 8-car formations. The original 1980s two-car sets, built by Linke-Hofmann/Busch are numbered in the 81xx and 83xx series, these have three window cab ends. The Tokyu Car built sets numbered in the 85xx and 86xx series have two window ends. In the above view, No. 8335 arrives at Dublin Connolly, while the lower image shows set 8602 at Dublin Connolly. Both: **CJM**

LUAS Dublin Trams

Gauge:	4ft 8½in (1,435mm)		
Designation:	Citadis TGA301	Citadis TGA401	Citadis TGA502
Set numbers:	3001-3026	4001-4014	5001-5041
Introduced:	2002-2003	2002-2003	2008-2021
Built by:	Alstom	Alstom	Alstom
Formation:	5-section	5-section	9-section
Tram length:	40.8m	40.8m	54.6m
Vehicle width:	2.4m	2.4m	2.4m
Seating:	72	72	88
Weight:	50.9t	50.9t	62t
Brake type:	Air Regen	Air Regen	Air Regen
Traction:	Electric 750V AC	Electric 750V AC	Electric 750V AC
Max speed:	44mph (70km/h)	44mph (70km/h)	44mph (70km/h)
Door type:	Plug	Plug	Plug
Operation:	Red line	Red line	Green line

Right Above, Left & Below: Travel around Dublin City Centre and suburbs is achieved by using the LUAS tram system. This is formed of two lines, the red and green, which interconnect in the middle of the city. All the networks trams have been built by Alstom. Above is a 3000 series five-section tram No. 3002 is seen on the red line in the city centre bound for The Point, these were introduced in 2002-2003. Below left is one of the original 2002 introduced 4000 series five-section vehicles No. 4014 at Museum on the systems red line. Below right is a 5000 series vehicle No. 5024 this is a nine section set operating on the very busy green line. The image left shows the interior of a nine-section 5000 class vehicle, seating is in a mix of styles with single seats and facing pairs. A large amount of standing room is provided and the wide external doors facilitate rapid loading and unloading. 3000 and 4000 series trams operate on the green line and 5000 series vehicles on the red line. All: **CJM**

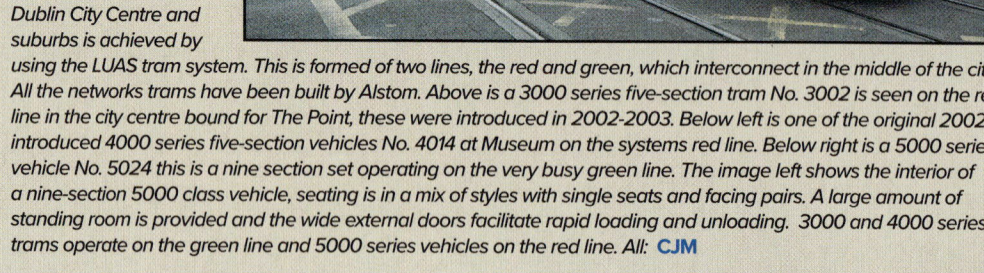

Channel Tunnel Locos

Sub class:	9/0 & 9/1	9/7	9/8
Number range:	9/0 - 9001-9040, 9/1 - 9101-9113	9701-9707 9711-9717*	9801-9838
Previous number range:	-	-	9001-9038
Former class code:	-	-	9/0
Built by:	Brush Traction, Loughborough	Brush Traction, Loughborough	Brush Traction, Loughborough
Years introduced:	1993-1999	2001	1993-1994
Years refurbished:	-	2010-2011	2005-2010
Wheel arrangement:	Bo-Bo-Bo	Bo-Bo-Bo	Bo-Bo-Bo
Weight:	132 tonnes	132 tonnes	132 tonnes
Height (pan down):	13ft 9in (4.20m)	13ft 9in (4.20m)	13ft 9in (4.20m)
Length:	72ft 2in (22.01m)	72ft 2in (22.01m)	72ft 2in (22.01m)
Width:	9ft 9in (2.97m)	9ft 9in (2.97m)	9ft 9in (2.97m)
Wheelbase:	50ft 8in (15.48m)	50ft 8in (15.48m)	50ft 8in (15.48m)
Bogie wheelbase:	9ft 2in (2.8m)	9ft 2in (2.8m)	9ft 2in (2.8m)
Bogie pivot centres:	20ft 9in (6.33m)	20ft 9in (6.33m)	20ft 9in (6.33m)
Wheel diameter:	3ft 9½in (1.15m)	3ft 9½in (1.15m)	3ft 9½in (1.15m)
Power supply:	25kV ac overhead	25kV ac overhead	25kV ac overhead
Traction output (max):	7,725hp (5,760kW)	9,387hp (7,000kW)	9,387hp (7,000kW)
Tractive effort:	69,500lb (310kN)	90,000lb (400kN)	90,000lb (400kN)
Maximum speed:	87mph (140km/h)	87mph (140km/h)	87mph (140km/h)
Brake type:	Air	Air	Air
Brake force:	50 tonnes	50 tonnes	50 tonnes
Train supply:	Electric	Electric	Electric
Multiple coupling type:	Eurotunnel	Eurotunnel	Eurotunnel
Control system:	Asynchronous 3-phase	Asynchronous 3-phase	Asynchronous 3-phase
Traction motor type:	ABB 6PH	ABB 6PH	ABB 6PH
No of traction motors:	6	6	6
Operator:	Eurotunnel	Eurotunnel	Eurotunnel
Sub-class variations:	9/0 - Original as built 9/1 - Freight locos	High-output freight * Rebuilt from 9/1s	Refurbished 9/0 fleet fitted equiped for higher output

2023 Fleet

Class 9/0	9037 EUR	9714 EUR	9802 EUR	9823 EUR
9005 EUR		9715 EUR	9803 EUR	9825 EUR
9007 EUR	Class 9/7	9716 EUR	9804 EUR	9827 EUR
9011 EUR	9701 EUR	9717 EUR	9806 EUR	9828 EUR
9013 EUR	9702 EUR	9718 EUR	9808 EUR	9831 EUR
9015 EUR	9703 EUR	9719 EUR	9809 EUR	9832 EUR
9018 EUR	9704 EUR	9720 EUR	9810 EUR	9834 EUR
9022 EUR	9705 EUR	9721 EUR	9812 EUR	9835 EUR
9024 EUR	9706 EUR	9722 EUR	9814 EUR	9838 EUR
9026 EUR	9707 EUR	9723 EUR	9816 EUR	9840 EUR
9029 EUR	9711 EUR		9819 EUR	
9033 EUR	9712 EUR	Class 9/8	9820 EUR	
9036 EUR	9713 EUR	9801 EUR	9821 EUR	

Right: Class 9 'Shuttle' loco front end equipment. 1: High level marker light, 2: White marker light, 3: Headlight, 4: Red tail light, 5: Grille over warning horns, 6: Control jumper cables, 7: Control jumper receptacles, 8: Air brake pipe, 9: Main reservoir air pipe, 10: Head end power socket, 11: Head end power cable, 12: Coupling hook with screw shackle. Loco No. 9025 illustrated. **CJM**

Below: The fleet of Tri-Bo shuttle locos, together with the rolling stock, is based at Coquelles, Calais, France, and visible from passing trains. Major or extended maintenance is undertaken 'off-site', with movements by road, as Eurotunnel stock is out of gauge on Network Rail tracks. Loco No. 9029 is illustrated at the UK terminal with a passenger shuttle. **Howard Lewsey**

Key Facts

- Class 9s were specifically designed to operate Channel Tunnel 'shuttle' trains.
- Used one either end of passenger or freight 'shuttle' services.
- Class 9/0s have a small cab at slab end for shunting purposes.
- 24 Class 9/0 locos, now refurbished and uprated as Class 9/8.
- Fleet of 10 Mak DE1004 locos used for emergency rescue or special duties, Nos. 0001-0005 fitted with TVM430 cab signalling for use over HS2.

Right: In addition to the passenger loco and stock fleet, Eurotunnel operates a fleet of 10 Krupp/MaK Bo-Bo pilot or rescue locos. The first five, Nos. 0001-0005 were built new for Eurotunnel, with No. 0006-0010 purchased from DB-Cargo, Netherlands. Nos. 0001-0005 are certified for operation as lead locos over HS1 and are fitted with TVM430 cab signalling, the ex-DB locos can operate over, but not lead on UK tracks. Nos. 0002, 0007 and 0002 are seen leading a Class 374 transit move near Hothfield. **Antony Christie**

0001 (21901)	EUR	
0002 (21902)	EUR	
0003 (21903)	EUR	
0004 (21904)	EUR	
0005 (21905)	EUR	
0006 (21906)	[6456] EUR	
0007 (21907)	[6457] EUR	
0008 (21908)	[6450] EUR	
0009 (21909)	[6451] EUR	
0010 (21910)	[6457] EUR	

Built: 0001-0005 1991-92	Horsepower: 0001-0005
0006-0010 1990-91	1,275hp
Length: 16.4m	0006-0010
Height: 3.89m	1,580hp
Width: 2.99m	Transmission: Electric BBC
Engine: MTU 12V396tc	

Main Line Certified Preserved Traction

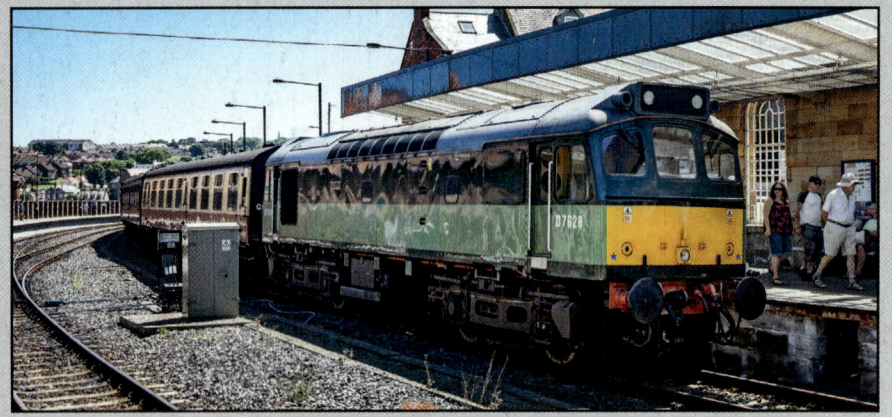

Class 25

Number	Location	Notes
25278 (D7628)	North Yorkshire Moors Rly	Restricted

A small number of preserved diesel locos have been restored to a high standard to allow re-certification to main line standards and allow operation on Network Rail tracks. Class 25 No. 25278, based at the North Yorkshire Moors Railway has limited certification to operate through services between Grosmont and Whitby. The loco is restored to 1960s two-tone green livery with a small yellow end. It is seen at Whitby. The loco is also hauling preserved NYMR Mk1 stock. **Antony Christie**

Sub-class:	25/3	Weight (operational):	76 tonnes	
TOPS:	25278	Height:	12ft 8in (3.86m)	
Original number:	D7628	Width:	9ft 1in (2.76m)	
Former class codes:	D12/1, later 12/1	Length:	50ft 6in (15.39m)	
Built by:	Beyer Peacock	Maximum speed:	90mph (145km/h)	
Introduced:	1965	Brake type:	Dual	
Wheel arrangement:	Bo-Bo	Sanding equipment:	Pneumatic	
		Coupling restriction:	Blue Star	
Engine type:	Sulzer 6LDA 28B			
Engine horsepower:	1,250hp (932kW)			
Power at rail:	949hp (708kW)			
Main generator type:	AEI RTB 15656			
Aux generator type:	AEI RTB 7440			
Number of traction motors:	4			
Traction motor type:	AEI 253AY			

Class 40

Number	Location	Notes
40013 (D213)	Loco Services Ltd	Full certificate
40145 (D345)	Class 40 Pres Soc	Full certificate

Currently two Class 40s are certified for full main line use, D213 with Loco Services (with disc headcode) and No. D345 owned by the Class 40 Preservation Society with a four-character headcode box. Both see frequent trips on main line metals and are popular with enthusiasts and luxury train operators. In 2023 both examples carry 1960s BR green. The pair, led by No. D213 are seen near Cockwood Harbour, Devon powering a charter on 30 October 2021. **CJM**

Class:	40	Weight (operational):	136 tonnes	
TOPS:	40013, 40145	Height:	12ft 10½in (3.92m)	
Original number:	D213, D345	Width:	9ft (2.74m)	
Former class codes:	D20/1, later 20/1	Length:	69ft 6in (21.18m)	
Built by:	EE Vulcan Foundry	Maximum speed:	90mph (145km/h)	
Introduced:	1959, 1961	Brake type:	Dual	
Wheel arrangement:	1Co-Co1	Sanding equipment:	Pneumatic	
		Coupling restriction:	Blue Star	
Engine type:	EE 16SVT MkII			
Engine horsepower:	2,000hp (1,491kW)			
Power at rail:	1,550hp (1,155kW)			
Main generator type:	EE822			
Aux generator type:	EE911-2B			
Number of traction motors:	6			
Traction motor type:	EE526-5D			

Class 45

Number	Location	Notes
45118 (D67)	Loco Services Ltd	Awaiting certificate (due summer 2023)

It has been many years since a member of the 'Peak' family has operated on the main line, but if all goes well, by the end of 2023, LSL-owned No. 45118 The Royal Artilleryman *will return to the main line, being available for charter and private train operation. The loco will be based at Crewe. How it might look, in this view No. 45118 is seen in BR rail blue pulling into Cheltenham Spa on 27 August 1981 with the 08.11 Bristol to Edinburgh.* **NEP/CJM-C**

Sub-class:	45/1	Weight (operational):	135 tonnes	
TOPS:	45118	Height:	12ft 10¼in (3.9m)	
Original number:	D67	Width:	8ft 10½in (2.71m)	
Former class codes:	D25/1, later 25/1	Length:	67ft 11in (20.70m)	
Built by:	BR Crewe Works	Maximum speed:	90mph (145km/h)	
Introduced:	1960	Brake type:	Dual	
Wheel arrangement:	1Co-Co1	Sanding equipment:	Pneumatic	
		Coupling restriction:	Blue Star	
Engine type:	Sulzer 12LDA28-B			
Engine horsepower:	2,500hp (1,864kW)			
Power at rail:	2,000hp (1,491kW)			
Main generator type:	Crompton CG426A1			
Aux generator type:	Crompton CAG252A1			
Number of traction motors:	6			
Traction motor type:	Crompton C172A1			
Heating:	Electric Index 66			

Main Line Certified Preserved Traction

Class 52

Number	Location	Notes
D1015	Diesel Traction Group	Full certificate

While several 'Western' diesel-hydraulic locos are preserved, only one, No. D1015 *Western Champion* owned by the Diesel Traction Group and based on the Severn Valley Railway is main line certified. However, it has not operated on the main line for a few years following an engine failure. It is seen sporting small yellow panel rail blue livery at the Severn Valley Railway in 2022. **Spencer Conquest**

Class:	52	Wheel arrangement:	C-C	Sanding equipment:	Pneumatic
TOPS:	-	Weight (operational):	108 tonnes	Coupling restriction:	Not fitted
Original number:	D1015	Height:	12ft 11¾in (3.96m)	Engine type:	2 x Maybach MD655
Former class codes:	D27/1, later 27/1	Width:	9ft (2.74m)	Total engine horsepower:	2,700hp (2,031kW)
Built by:	BR Swindon Works	Length:	68ft (20.73m)	Power at rail:	2,350hp (1,752kW)
Introduced:	1963	Maximum speed:	90mph (145km/h)	Transmission:	Voith L630rU
		Brake type:	Dual		

Class 55

Number	Location	Notes
D9000 (55022)	Loco Services Ltd	Full certificate
D9002 (55002)	National Collection	Full certificate
D9009 (55009)	Deltic Preservation Soc	Full certificate

Some of the most popular preserved main line locos are the 'Deltic' Type 5s, in 2023 three were main line registered, with No. D9000 (55022) and D9009 (55009) being frequently seen on the main line, another three locos are preserved on private railways. No. 55009 (D9009) *Alycidon* is illustrated in rail blue livery attached to a West Coast Railway Class 57 at Crewe on 30 March 2023. **Spencer Conquest**

Class:	55	Weight (operational):	100 tonnes	Engine type:	2 x Napier D18.25 Deltic
TOPS:	55022, 55002, 55009	Height:	12ft 11in (3.94m)	Total engine horsepower:	3,300hp (2,461kW)
Original number:	D9000, D9002, D9009	Width:	8ft 9½in (2.68m)	Power at rail:	2,460hp (1,834kW)
Former class codes:	D33/1, later 33/2	Length:	69ft 6in (21.18m)	Main generator type:	2 x EE829-1A
Built by:	EE Vulcan Foundry	Maximum speed:	100mph (161km/h)	Aux generator type:	2 x EE913-1A
Introduced:	1961	Brake type:	Dual	Number of traction motors:	6
Wheel arrangement:	Co-Co	Sanding equipment:	Pneumatic	Traction motor type:	EE538A
		Coupling restriction:	Not fitted		

Class 117 / 121

Number / Class	Location	Notes
51356 / 117	Swanage Railway	Restricted certificate
51388 / 117	Swanage Railway	Restricted certificate
59486 / 117	Swanage Railway	Restricted certificate
55028 / 121	Swanage Railway	Restricted certificate

While there are a large number of preserved first generation DMMUs in existence, only four vehicles are currently certified for main line operation, a three-car Class 117 set and a Class 121 'Bubble'. These are based on the Swanage Railway and operate a limited Swanage to Wareham service, as well as operating domestic services on the Swanage Railway. The three-car Class 117 is seen at Corfe Castle from its DMBS, 51356 end. **Mark V. Pike**

Class 142

Number / Class	Location	Notes
142003 - 55544+55594	LSL Crewe	Certified

Loco Services Ltd, based at Crewe have three Class 142 'Pacer' sets on their books, with one, No. 142003 certified for main line operation. In summer 2023 an overhaul and repaint was completed at Arlington Fleet Services at Eastleigh and the set returned to Crewe for use. It is seen passing Winchester on the SW main line. **Mark V. Pike**

Hastings set

Number / Class	Type	Location	Notes
60116 / 6L 1012	DMBS	St Leonards	Certified
60118 / 6L 1013	DMBS	St Leonards	Certified
60119 / 6L 1013	DMBS	St Leonards	Certified
60528 / 6L 1013	TS	St Leonards	Certified
60529 / 6L 1013	TS	St Leonards	Certified
69337 / BIG 2210	RSB	St Leonards	Certified x EMU
70262 / CEP 1524	TS	St Leonards	Certified x EMU

The St Leonards based Hastings Diesels Ltd, own a number of ex-Southern Region DEMU vehicles of which five are main line certified, together with two ex EMU vehicles. These are used for charter services and are painted in BR Southern green. DMBS No. 60118 is seen at St Leonards in mid-2022. **CJM**

modernrailways.com
THE ONLINE HOME OF MODERN railways

Modernrailways.com is the online home of rail industry content, brought to you by *Key Publishing*, publishers of *Modern Railways* magazine.

modernrailways.com

You'll find all the latest industry news, written with authority, at modernrailways.com - plus detailed analysis, in-depth features and website exclusives.

UNLIMITED access to this exciting online content from our dedicated team starts from just £29.99/year for UK customers. And registering couldn't be simpler. Visit **modernrailways.com** for instant access to the latest *Modern Railways* magazine features and industry-leading content.

FREE ACCESS
FOR ALL MODERN railways SUBSCRIBERS*

VISIT
modernrailways.com/subscribe

*Free access available for a limited time only

SIGN UP TODAY!

We value your feedback! Let us know your thoughts on modernrailways.com – drop us a line at subs@keypublishing.com today